Wireless Networks

Series Editor

Xuemin Sherman Shen
University of Waterloo, Waterloo, Ontario, Canada

More information about this series at http://www.springer.com/series/14180

Senchun Chai • Zhaoyang Wang • Baihai Zhang •
Lingguo Cui • Runqi Chai

Wireless Sensor Networks

Springer

Senchun Chai
School of Automation
Beijing Institute of Technology
Beijing, Beijing, China

Zhaoyang Wang
School of Artificial Intelligence
Beijing Technology and Business University
Beijing, Beijing, China

Baihai Zhang
School of Automation
Beijing Institute of Technology
Beijing, Beijing, China

Lingguo Cui
School of Automation
Beijing Institute of Technology
Beijing, Beijing, China

Runqi Chai
Transport & Manufacturing
Cranfield Univ, Sch of Aerospace
Cranfield, Bedfordshire, UK

ISSN 2366-1186 ISSN 2366-1445 (electronic)
Wireless Networks
ISBN 978-981-15-5756-9 ISBN 978-981-15-5757-6 (eBook)
https://doi.org/10.1007/978-981-15-5757-6

This Springer imprint is published by the registered company Springer Nature Singapore Pte Ltd.
The registered company address is: 152 Beach Road, #21-01/04 Gateway East, Singapore 189721, Singapore

Preface

Wireless sensor networks, as capable of collaboratively monitoring, sensing, and collecting information from various environments or monitored objects through various integrated microsensors, can realize wireless transition for users in a self-organized and multi-hop way. The application of wireless sensor networks builds the connection among physical world, computer world and ternary world of human society. Today, the world has witnessed the great development of wireless sensor networks and the related research exhibits a promising prospective, which will bring a new information revolution in the information era.

The purpose of this book is to provide a comprehensive presentation of wireless sensor networks. Not only we introduce the general knowledge of wireless sensor network such as background, application, etc., but also analyze the working principle from the aspects of routing establishment, localization methods and coverage control. This book also contains advanced knowledge on complex network theory and other research on frontier disciplines such as "small-world phenomenon" related to wireless sensor networks.

The book is divided into six chapters: Chapter 1 introduces a set of basic concepts on wireless sensor networks such as background and application. Chapter 2 discusses energy balanced routing protocols and the corresponding algorithms are given. Chapter 3 presents several localization algorithms of wireless sensor networks. Chapter 4 shows algorithms of coverage control in wireless sensor network. Chapter 5 talks about the community detection of complex network. The last chapter is an introduction about small-world phenomenon.

This work was supported by the National Natural Science Foundation of China under Grant No. 61573061. The research contents in this book are multidisciplinary cross-connection with both theory and practice closely integrated. The issues considered in the book constitute a foundation for the fields of science and technology, notably systems research, automatic control and automation, embedded computing technology, communication technology, to name a few. The familiarity of the problems considered, their analyses and solutions should be useful for virtually all graduate students, researchers, and scholars in these areas, as well as many practitioners.

Thanks to the research and work conditions provided by Beijing Institute of Technology and the persistent supports from our laboratory, which helped us accomplish the related research. We would like to acknowledge the generous contribution made by our graduate students, *Qiao Li, Jun Li, Ahmed Abd Elrahim Khamis, Guan Ye, Shi Zhang, Longfei Wen, Zixiao Guan, Peiyi Xu, Jiaming Zhang, Yong Hu, Zhaozhan Song, Bowen Mu, Yuchen Xu, Yunao Li, Zikai Wang, et al.*, for their valuable research achievements. We have carefully reviewed all the contents of the book. All the authors would very much appreciate any comments and criticisms from the readers to improve the quality of this work of future editions.

Beijing, China	Senchun Chai
Beijing, China	Zhaoyang Wang
Beijing, China	Baihai Zhang
Beijing, China	Lingguo Cui
Cranfield, UK	Runqi Chai

Contents

Chapter 1
Introduction

1.1 Background of Wireless Sensor Networks

Wireless sensor networks (WSNs) exhibit promising development prospects in both economic and technical levels because of their ability to provide a variety of services such as search-and-rescue operations, logistics, vehicle routing and intruder detection. The services provided by WSNs are based on collaboration among small energy-constrained sensor nodes. Due to the large-scale application of WSNs, the demand for organization strategies of sensor nodes are increasingly urgent. Research on WSNs have focused on several research topics and emerged as an important new area in wireless technology. This chapter covers the network characteristics, origin and development, architecture and applications of WSNs.

1.1.1 Basic Concepts and Network Characteristics

With the development of micro-electro-mechanical systems and distributed information processing techniques, low-power, low-cost and large-scale WSNs as an emerging technology, have widened the functional application, which improves the ability of cognitive world. A WSN is a multi-hop, self-organizing network. It composes of a large number of tiny and cheap sensor nodes randomly distributed in the application area in a cooperative manner, which can achieve cooperative sensing, collecting and processing information of monitoring objects in the network coverage area. As the collected information transmits to users, WSNs become a new way of information acquisition and processing. They are characterized by wide range distribution, large amount of nodes, low cost of single sensor and strong network dynamics. Therefore, they have been widely used in military, environmental monitoring, medical care, smart home and other fields.

S. Chai et al., *Wireless Sensor Networks*, Wireless Networks,
https://doi.org/10.1007/978-981-15-5757-6_1

WSNs have different characteristics with the traditional wireless networks in terms of application requirements, design goals and technical requirements. The data transmission and communication is realized by each node of WSNs in a multi-hop routing manner. The characteristics of WSNs are summarized as following.

1. Large scale

In order to complete the monitoring task, a large number of sensor nodes should be deployed in the monitoring area. The number of network nodes in per unit of monitoring area is much larger than that of the traditional wireless networks. Because of the network nature, dense monitoring obtains high-precision monitoring information.

2. High dynamics

In WSNs, the network topology changes at any time. Many reasons may lead to the variability of network topology. On the one hand, a single node may be affected by environmental interference, insufficient energy, etc. causing nodes malfunctioning. Meanwhile, wireless communication channel is unstable, which may cause interruption among nodes. On the other hand, the observer and perceived object are mobile, which affects connection among nodes. The variable network topology requires WSN systems adapt to various changes and be reconfigurable with dynamic systems.

3. Limited capabilities of communication, computing and storage

A sensor node is a miniature-embedded device. Considering energy consumption in large-scale WSNs, the design of the sensor node needs to meet the requirements of low price and low power consumption. These requirements stop sensors from assembling processing with strong capabilities. Processors and small-capacity memories cannot perform well in complicated calculations.

4. Centerless and self-organizing

There is no central node in the WSNs. If one sensor node fails, the stability of the entire network isn't affected due to its self-organizing ability. In the applications of WSNs, the locations of nodes and communication information between nodes cannot be set in advance. These nodes may also fail due to their own energy exhaustion or environmental factors. The number of nodes in the network would decrease correspondingly, so the network topology would change. Changes require that the nodes in the network not only have the ability of autonomous management and automatic configuration, but also quickly complete the system layout and construction through distributed algorithms.

5. Data center

The WSNs are a task-oriented network centered by data, which can inquire and transmit clue. When users query events by WSNs, they are concerned about the location and information of the event, rather than transmitted node and path. There is no necessary relationship between the number of nodes in the network and the location of the nodes. Therefore, WSNs are data-centric networks.

6. Limited energy

Nodes are small and only powered by batteries. Because the distributed environment is complicated, it is impossible to supplement the energy by replacing the batteries in practice. Once the battery of some node in the network is insufficient, the node would be invalid and abandoned. Therefore, energy saving has become a key issue for WSNs. It is vital to make full use of the sensor energy and provide services without affecting the function of nodes.

1.1.2 Research and Development

To understand the tradeoffs in today's WSNs, it is helpful to briefly examine their history. Like many advanced technologies, the origin of WSNs can be seen in military and heavy industrial applications, far removed from the light industrial and consumer WSN applications that are prevalent today. The first wireless network that bore any real resemblance to a modern WSN is the Sound Surveillance System (SOSUS), developed by the United States Navy in the 1950s to detect and track Soviet submarines. This network used submerged acoustic sensors – hydrophones – distributed in the Atlantic and Pacific oceans. This sensing technology is still in service today, albeit serving more peaceful functions of monitoring undersea wildlife and volcanic activity.

Because of high scientific value and promising application prospects, WSNs have received widespread attention all over the world especially in regions like the United States, Japan, China, Europe and Russia.

The development of WSNs has now gone through four generations. The first generation of WSNs was born in the late 1970s. The sensor nodes formed a sensor network with a star topology in a point-to-point transmission way. With the development of related disciplines, sensor nodes have enhanced the ability to process multiple signals. As serial or parallel ports connected the nodes, the comprehensive processing capability of sensor network information improves and forms the second generation of WSNs. The third generation of WSNs appeared in the late 1990s. With the fieldbus technology introduced in sensor networks, nodes that obtain multiple signals through intelligent methods form intelligent sensor networks. Nowadays, the fourth-generation WSN is developing rapidly. The fourth-generation nodes have multi-function and multi-signal acquisition capabilities, and the network can quickly connect through wireless self-organization.

In the competition of WSN technology, the earliest research on WSNs was carried out in the United States. The US Department of Defense firstly applied sensor networks in military field. In the early 1950s, the US military system SOSOUS (Sound Surveillance System), a wireless sensor-based system, monitored the movement of submarines in the former Soviet Union. In the 1980s, the

U.S. Navy used ground base stations and radar systems to monitor and track air targets, so the related technologies strongly promoted the development of WSNs. In 1999, the famous American Business Week listed WSNs as one of the most influential technologies in the twenty-first century. In the same year, the concept of "sensors going to the wireless age" was proposed, at the International Conference on Mobile Computing and Networks in the United States, representing a process in which WSNs are gradually facing the public. In 2000, the United States Air Force listed 15 key technologies that would help improve Air Force capabilities in the twenty-first century and sensor technology ranked second. In 2001, the United States Army formulated the "Smart Sensor Network Communication" project, which aimed to analyze the battlefield situation through a large amount of battlefield information collected by sensors, thereby helping Combatants to formulate a combat action plan. After that, the US Army established the "Unattended Ground Sensor Group" and "Battlefield Environment Reconnaissance and Surveillance System" projects. The main purpose was to flexibly deploy sensor nodes in the network and obtain accurate ground information through nodes. In the industrial field, companies such as Dust Networks and Crossbow in the United States have developed "smart dust" projects to commercialize sensor networks. The WSNs developed by Desert Mountain Company regulates the tap water irrigation of the golf course. In 2002, Intel Corporation of the United States released the "New Computing Development Plan Based on Micro Sensor Networks", whose main purpose was to apply sensors in the civilian field. In 2003, the US Science Foundation Board invested a lot of energy and financial resources into the development of WSNs. The fields involved biological sensing and chemical poisoning. Many universities and research institutes have conducted in-depth research on the theoretical basis and key technologies of WSNs. Some of well-known research projects include DARPA project jointly developed by the University of California Los Angeles and the Rockville Automation Center, and the University of California Berkeley, BWRC project, WEBS project, ESP project developed by Purdue University, NMS project and AMPS project developed by MIT and Exscal project developed by Ohio State University.

Countries such as Japan, South Korea, and Russia also quickly entered the research of WSNs. The sixth EU's framework plan includes "information society technology" as one of its priority areas of development, many of which involve research on WSNs. In March 2004, the Ministry of Internal Affairs and Communications of Japan established the "Ubiquitous Sensor Network" survey and research society. The Ministry of Information and Communication of Korea has formulated the "839" strategy for information technology, in which "3" refers to the three major infrastructures of the IT industry, namely, broadband converged networks, ubiquitous sensor networks and next-generation Internet protocols. In the business world, companies such as Philips, Siemens, Ericsson, ZMD, France Telecom, Chipcon and other companies in Europe, NEC, OKI, Sky2 leynetworks, Shikang, Omron and other companies in Japan have carried out WSNs research and achieved good results.

Although it started slightly later than other countries, China also attaches great importance to the development of WSNs. It has deployed relevant research work on the Internet of Things in technology programs such as the National Natural Science Foundation of China, the National 863 Program, the 973 Program, and the National Science and Technology Major Project. Meanwhile, China has also carried out technical research and achieved preliminary results in the fields of chips, communication protocols, network management, collaborative processing, and intelligent computing. At the same time, more and more enterprises and research institutions are paying attention to the development of sensor network technology and launching solutions for the Zigbee protocol of WSNs. Partial breakthroughs have been achieved in the areas of IEEE short-range wireless communications, 3GPP mobile network optimization, and ISO / IEC IoT architecture standard research. Sensor network related technologies in transportation, smart home, smart grid and other fields have been widely used.

In addition, by applying WSNs as the infrastructure for sensing and information collection, a new WSN-based system architecture is constructed, focusing on sensing and collecting environmental information and storing and processing complex data. Applications, industrial production, scientific research, and commercial transactions provide a powerful operating platform that integrates data awareness, mass storage, and intensive processing.

At present, with the rise of the Internet of Things, WSNs have received widespread attention as the most important sensing layer. Universities and research institutes in various countries are conducting in-depth research on multi-disciplinary technology fields including sensor technology, distributed information processing technology, embedded computing technology, and short-range communication technology. The development of the Internet of Things has entered a substantial stage of advancement. The concept of the Internet of Things and related technology products have widely penetrated into various fields of social economy and people's livelihood, and have played a key role in more and more industry innovations. From the perspective of industrial scale, China's Internet of Things has maintained a high growth rate in recent years. The WSN is located in the sensing layer of the Internet of Things, which is the foundation and core of the development of the entire Internet of Things industry. Its industrial scale has also been rapidly increased with the development of the Internet of Things. In addition, a new type of networked intelligent production characterized by IoT convergence innovation is shaping the core competitiveness of the manufacturing industry in the future. It promotes the development of the new industrial revolution, which could bring new challenges to enterprises in the industry.

1.1.3 Network Structure

A WSN consists of sensor nodes, sink nodes and task management nodes. The network architecture is shown in Fig. 1.1. WSN nodes can be randomly deployed in the monitoring area through manual deployment or aircraft deployment according to practical requirement. Each node randomly distributed constitutes a WSN in the

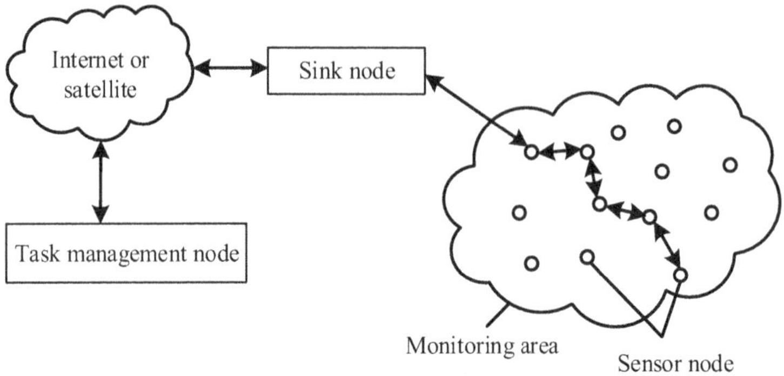

Fig. 1.1 WSN architecture

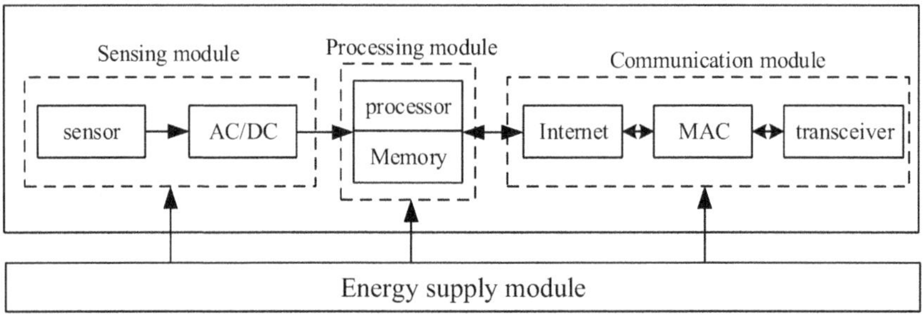

Fig. 1.2 Node structure of WSNs

form of self-organization. The node transmits the received data information to the sink node hop by hop through other sensor nodes in the network. The data information is then transmitted to the task management node via the Internet or satellite. Users can analyze and process the obtained information.

In the WSNs, the sensor node is the basic unit of the networks. The structure of sensor node is shown in Fig. 1.2. A sensor node is mainly composed by a sensor module, a processing module, a communication module, and an energy supply module.

The sensing module is composed of a sensor and an analog-to-digital converter (AC/DC). The task of the sensor is to collect information and convert data of sensing object in the collection monitoring area. The task of the A/D converter is to convert the analog signal into a digital signal, and then transmit the signal to the processing module. The sensor module of the sensor node has two implementation modes. One mode is to integrate various sensors on the node. For example, sensors such as temperature, pressure and humidity are integrated on the node. The advantage of this mode is high integration and small size, which is suitable for sensors with simple circuits, but it has poor scalability and flexibility. The other mode is to connect various sensors with nodes in the form of plug-ins. This mode has the advantage of good scalability and can be flexibly applied to sensors with complex circuits. The processing module is the core module of the sensor node. It consists of a processor and a memory. The main task is to coordinate the operation of the entire node. It is responsible for processing and storing data collected by the node and data sent by other nodes. The task of the communication module is to perform wireless communication with other sensor nodes, exchange control information, receive and send data information collected by the nodes. The energy supply module is especially important for sensor nodes. Its task is to use batteries with limited energy to provide all the energy required for the node to work.

Another important concept in WSNs is the network protocol stack. A network protocol stack includes a physical layer, a data link layer, a network layer, a transport layer and an application layer. The task of the physical layer is to generate carrier frequencies, modulate and demodulate signals. The tasks of the data link layer are media access and error checking. The task of the network layer is to discover and maintain routes so that sensor nodes can communicate with each other. The task of the transport layer is to transmit control data streams to ensure the quality of communication. The task of the application layer is to schedule and distribute data according to different requirements.

The protocol stack of WSNs adopts a cross-layer design method including energy management platform, mobile management platform, and task management platform. The task of the energy management platform is how to save the energy of each protocol layer and extend the survival time of the network. The task of the mobile management platform is to detect and record node movements, and maintain the route from the sensor node to the sink node. The task of the task management platform is to coordinate the tasks of various nodes according to different requirements. These management platforms allow nodes to work together in a more efficient way with less energy consumption, support multitasking and resource sharing.

1.2 Applications of Wireless Sensor Networks

As a new type of network, WSN has huge application value and wide application prospects, which has a profound impact on various fields of humanity. The following shows the important application fields of WSNs.

1.2.1 Military Field

Because WSNs have the characteristics of large node amount, strong fault tolerance, high redundancy and fast self-organization, even if the enemy destroys some of sensor nodes, other nodes can still complete the monitoring task well. The developed military countries attach great importance to the research of WSNs and regard it as an indispensable part of military systems. For example, in the information warfare, the U.S. military randomly distributed a large number of sensor nodes into the enemy's combat area through aircraft and other methods. The nodes can effectively replace manpower to covertly collect enemy's combat information, such as strength, materials, equipment, terrain and other key battlefield information, so as to get a higher chance to win the war.

1.2.2 Environmental Monitoring

With the development of industry, the ecological environment is deteriorating. In order to deal with environmental changes in time and protect the ecological environment, the field of environmental monitoring has introduced the WSNs into the actual monitoring. Because of its small size and large deployment range, WSNs can be used to collect and process data from natural ecosystems, which can effectively replace traditional manual methods and save manpower and resources. Besides, the sensor node can integrate various sensor modules to track the migration of organisms and animals, monitor indicators such as soil, humidity and temperature. Therefore, it can provide an effective basis for research and prediction. For example, Atlantic College and University of California Berkeley have jointly deployed a small WSN on Duck Island. The network consists of 32 sensor nodes. Its main task is to monitor the living habits of the petrel in the island. Similarly, the sensor node also plays an active role in mining mineral resources, monitoring underwater temperature and harmful components underwater.

1.2.3 Medical Care

In the field of medical care, sensors such as heart rate monitoring equipment are installed on patients to track their actions to monitor various physiological data of patients. Based on the data provided by the sensors, doctors can monitor the patient's condition at any time and respond to sudden conditions treatment quickly and accurately. For example, the smart medical room developed by the University of Rochester uses "smart dust" to monitor the resident's sleeping posture, breathing, pulse, activity and so on. Intel Corporation has developed home care technology, which embeds semiconductor sensors into the devices in the home to facilitate the home life of the elderly and the disabled while reducing the burden on medical staff.

1.2.4 Smart Home

Smart homes regard houses as platform and embed sensor nodes in various daily furniture and appliances. These nodes form a network and connect each other through WSNs and Internet, so that people can remotely control home equipment in real time outside the house. For example, by embedding sensors such as light, humidity and temperature in residential equipment, we can obtain data from houses by wireless sensors, so that air conditioning, doors, windows and other home appliances can be automatically controlled.

1.2.5 Industrial Monitoring

In the industrial field, WSNs are also widely used due to their low cost, small size, and strong real-time performance. For different industrial applications, these characteristics play various roles in manufacture and industry.

1. Low cost: The WSNs for machine health monitoring has been widely used in Condition-Based Maintenance (CBM) of machinery, which not only saves labor costs, but also greatly reduces material costs and extends the life of the machine. In addition, wireless sensors can also be placed in a location, which is difficult to reach through a wired system.
2. Small size: Due to high density of server racks in data centers, more and more racks are equipped with wireless temperature sensors to monitor the inlet and outlet temperature of the racks. The American Society of Heating, Refrigerating and Air-Conditioning Engineers (ASHRAE) requires that six temperature sensors are installed in each rack. In this situation, wireless temperature monitoring technology has obvious advantages over traditional wired sensors.
3. Real-time performance: The WSNs equip with data recording to collect various monitoring information data. The implementation process can be as simple as monitoring the temperature in the refrigerator or the water level of overflow tank in nuclear power plant. The statistical information of these data reveals the system's working process, which is more real-time than the traditional recorder system.

Extensive aforementioned application scenarios bring a wider market to WSNs, with the continuous maturation of WSN technology and the gradually increase in market demand. Market research reports show that the role of WSN products in replacing traditional sensors has become more and more apparent.

According to the report from Grand View Research, the global WSNs market size is expected to reach $ 8.67 billion in 2025, with a compound annual growth rate of 14.5%, and this trend is expected to continue. In China, WSN products accounted for about 4.3% of the sensor market in 2014, with a scale of 620 million yuan. By 2019, the proportion of WSNs products in the sensor market reached 10.0%, and the market size is expected to reach 2.42 billion RMB with a compound annual growth rate of 27.1%, so the market prospect is broad.

With the "*Made in China 2025*" in-depth promotion of enterprise production line transformation and communication equipment upgrades, China's demands for WSNs products continue to grow at a high speed, especially in remote areas that require strong network connections. The demand for network infrastructure is expected to drive the growth of the WSNs market. At the same time, in the context of Industry 4.0, intelligent industrial production has become an important means of industrial transformation in the future. The WSN technology, which is one of the important industrial intelligent technologies, wound play an increasingly important role. The market of industrial WSNs would also win broader development opportunities, since the industry's growth rate is expected to grow steadily.

1.3 Hot Topics in Wireless Sensor Networks

The current research hotspots in WSNs focus on five key technologies, i.e. network communication protocols, network localization technology, network coverage control, community detection under complex networks, and small worlds and information dissemination. The research trends of these topics are as follows.

1.3.1 Protocol for Wireless Sensor Networks

The routing protocol is a process to select suitable path for the data to travel from source to destination. The process encounters several difficulties while selecting the route, which depends upon types of network, channel characteristics and the performance metrics.

The data sensed by the sensor nodes in a WSN is typically forwarded to the base station that connects the sensor network with the other networks (may be internet) where the data is collected, analyzed and some action is taken accordingly. Communication types can be divided into mutual communication, single-hop and multi-hop communication. In small-scale networks, the distance between the base station and the nodes is close enough for nodes to communicate with each other. However, it is inapplicable for large-scale WSNs, where the coverage is huge with thousands of nodes deployed in the network. In large-scale WSNs, most sensor nodes are far away from the sink node. They cannot directly communicate with the base station, so multi-hop communication is required. In this way, typical path from the node to the destination node is composed of multi-hop and the intermediate node on the path serves as the forwarding node. Therefore, nodes in a wireless multi-hop network have two functions. First, the nodes act as an end node to generate or receive data. Second, the nodes act as an intermediate node to forward data packets from other nodes. The process of determining the network path from the source node to the destination node is called routing, which is the main responsibility of the network layer.

Routing protocols define how nodes communicate and how information is propagated across the network. The basic classification of routing protocols is shown in Fig. 1.3.

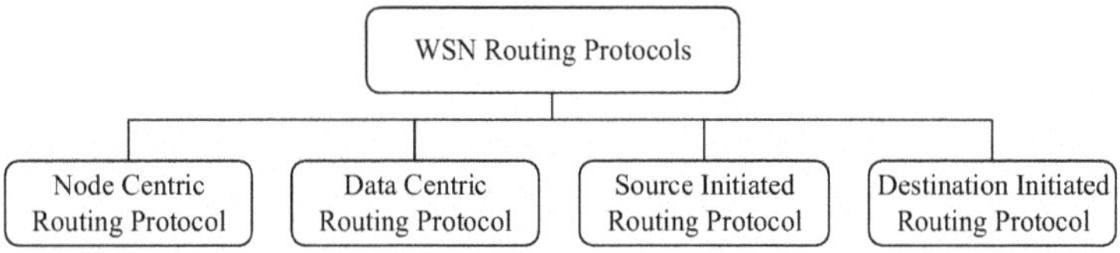

Fig. 1.3 Basic classification of routing protocols

As shown in the figure above, the routing protocol is mainly divided into Node-centric, Data-centric, Destination-initiated (Dst-initiated), Source-initiated (Src-initiated).

In node centric protocols, the destination node is specified with some numeric identifiers. Low energy adaptive clustering hierarchy (LEACH) is a typical node centric protocol. The basic idea of LEACH is to randomly select cluster head nodes in a cyclic manner, to distribute the energy load of the entire network, and to reduce the energy consumption of the network and improve the overall survival time of the network. Compared with general planar multi-hop routing protocols and static layered algorithms, the LEACH clustering protocol can extend the network life cycle by 15%.

Data-centric routing technology is on specified information transmission. The data monitored by WSNs is more valuable than the nodes themselves. Nodes query specific areas to collect data with certain specific characteristics.

Destination-initiated protocol refers to the path setup generation originating from the destination node. It is an evolution of data-centric routing technology where data-centric technology is utilized to collect and disseminate information.

In the source-initiated protocol, the source node advertises when it has data to share and the route is generated from the source side to the destination.

In the design of routing protocols, due to the influence of various characteristics such as network type and channel characteristics, the routing protocols of WSN are significantly different from traditional routing protocols. Specifically, they are mainly reflected in the following aspects:

1. Because there are a large number of sensor nodes in a WSN, it is impossible to assign a universal identifier scheme to these nodes, so classical IP-based protocols cannot be applied in WSNs.
2. Unlike the typical communication network, the data flow from a source node to a specific base station in a WSN is mandatory.
3. In most cases, many sensor nodes would produce the same data, so the created data traffic has significant redundancy, which would waste bandwidth and energy.
4. In addition, wireless nodes are severely limited in bandwidth, capacity and storage

1.3.2 Localization Technology for Wireless Sensor Networks

Localization, i.e. obtaining the position of the sensor, is one of the basic technologies of WSNs. With the popularity of WSNs, Location-Based Service (LBS) has rapidly entered people's daily lives. As a typical wireless self-organizing network, it has shown great application prospects in the fields of environmental monitoring, industrial sensing and diagnostics, battlefield monitoring, and environment-aware computing. Autonomous localization of sensor nodes is critical because localization makes sensed data meaningful. Many applications and services of WSNs rely directly or indirectly on location information. One way to get location information is manual configuration, but this is usually not feasible for large-scale deployments or mobile networks. The world's largest and best-known positioning system is the Global Positioning System (GPS), but this system is not suitable for indoor or underground environments. Although some local positioning systems (LPs) overcome the shortcomings of GPS, they still rely on high-density deployment of base stations that leads to high hardware cost.

Self-localization is an alternate solution of the localization problem, in which sensor nodes can estimate their position by various localization discovery protocols. These protocols share a common characteristic and most of them use a few special nodes, called anchor nodes (also referred as beacon nodes, seeds, references or landmarks), which are assumed to know their own locations (through manual configuration or GPS receivers). These anchor nodes provide position information, in the form of beacon messages, for the benefit of unknown nodes (also referred to as non-beacon nodes, dumb nodes or target). Unknown nodes can utilize the position information of multiple nearby anchor nodes to estimate their own positions.

Most existing positioning schemes consist of two phases: (1) distance/angle estimation. The distance / angle estimation phase refers to obtaining information based on network deployment. This information includes the relative position, angle and distance of the nodes. In distance/angle estimation, the measurement technical indicators mainly include TDOA (Time Difference of Arrival), TOA (Time of Arrival), RSSI (Received Signal Strength Indicator), AOA (Angle of Arrival) and Hops (through common distance measurements technology). (2) Position calculation. In the position calculation, the position of unknown node is estimated based on the available information of the distance or angle and the position of the reference nodes. Common techniques include padding, triangulation, bounding boxes, probabilistic methods and fingerprint recognition.

The existing localization schemes can be divided into two categories, the range-based approaches and the range-free approaches. Among all schemes, each scheme has a corresponding application scenario, but none of them is optimal. In general, range-based methods are expensive and sensitive to environmental noise, which indicates that the network ability to resist interference is poor. On the contrary, the performance of range-free method is not accurate enough and it is easily affected by the network density.

1.3.3 Coverage Control in Wireless Sensor Networks

For self-organized WSNs, the coverage control of network topology affects network's performance. It can be considered as the method of placing the network sensor node and choosing route under the situation where sensor nodes' energy, communication bandwidth, computing resources are limited. The coverage control aims to make all kinds of WSN's resources obtain optimal allocation and improve the various kinds of service quality, such as perception, monitoring, sensing, communications, etc.

The application of optimal coverage control strategy can help network control energy, improve the service quality of perception and prolong the survival time effectively. On the other hand, it also increases the cost of network transmission, management, storage and computing. Therefore, the performance of wireless sensor network's coverage control algorithm is vital for usability and effectivity.

The most applied optimal coverage control strategy can improve perceived service quality and extend network life. But at the same time it also increases the cost of network transmission, management, storage and computing. Therefore, the performance of the coverage control algorithm is critical for WSNs.

At present, the coverage control of WSNs is mainly divided into two categories: static WSNs and dynamic WSNs. Static WSNs are composed of movable sensor nodes. The node movement process only occurs during the network redeployment or self-repair while other times are static. Dynamic WSNs for continuous mobile patrols expand the monitoring range and efficiency of the sensor for a certain period. The nodes of a dynamic mobile sensor network need to ensure energy consumption during the continuous movement of the nodes, so their practical applications are few. In comparison, the energy requirements of static WSNs are much more reasonable. Static WSNs are mainly targeted at randomly deployed or damaged sensor networks. In a randomly deployed sensor network, the partitions of nodes may appear highly redundant and highly dispersed. WSNs can achieve more uniform deployment and coverage through the movement of nodes.

There are three main types of static coverage control strategies: mobile coverage control based on virtual force (potential field), mobile coverage control based on computational geometric segmentation and centralized mobile coverage control.

In the virtual potential field of coverage control problem, the virtual potential field is a virtual force that causes each mobile sensor node to be affected by other nodes and obstacles, moving from high potential energy to low potential, just like the movement of charged particles in an electrostatic field. The nodes are scattered and eventually reach full coverage.

The main idea of the coverage control strategy based on computational geometry is to describe the network structure as graph theory structures such as Voronoi diagrams and Delaunay angles, and try to move the nodes to make the network structure into a regular calculation geometry structure. The Voronoi diagram can divide the planar area into the characteristics of a simple geometric polygon set, so it is widely used in the coverage problem of WSNs.

The main idea of the centralized control algorithm is to move the sensor nodes to the vertices or centers of the regular grid polygons of the expected deployment to achieve full coverage, which not only improves the coverage rate, but also eliminates coverage holes.

Force-based mobile coverage control strategy is easy to understand and can implement distributed algorithms. It is the most commonly used mobile coverage control strategy. However, most of the mobile coverage control strategies based on virtual forces do not consider connectivity, and the phenomenon of node aggregation movement would occur in high connectivity networks. The mobile coverage control algorithm based on computational geometric segmentation needs to know the information of the deployment area in advance, and when the number of nodes is not enough to cover the entire monitoring area, the nodes would be in an oscillating state constantly moving to cover each Voronoi polygon. The centralized mobile coverage control algorithm can achieve nearly perfect coverage. However, it requires global target area information, and uses either a global evolutionary computing strategy, which results in a large amount of calculation and communication overhead, or scanning incremental movement, which moves only one node per movement cycle, resulting in long redeployment. These centralized movement control algorithms are not suitable for large-scale deployment of WSNs. At present, the existing coverage control algorithms of mobile sensor network are mostly control strategies based on force or computational geometry.

1.3.4 Community Detection Algorithms in Complex Networks

In 2002, Newman firstly proposed the concept of community structure. He believes that the connection between nodes in the community is more closer than the connection between nodes in the different communities. Nowadays, community identification of complex networks has become one of the hottest issues in the field of big data research. It has important theoretical and behavioral significance for topology analysis, function analysis and behavior prediction of complex networks. As a result, community structures have attracted widespread attention.

Community detection is the process of accurately identifying the communities belonging to each node in the network, and it is the basis for studying the structural characteristics of the entire community. These community detection algorithms are divided into overlapping community detection algorithms and non-overlapping community detection algorithms. The non-overlapping community detection algorithm can detect several independent communities, and each node uniquely belongs to a community. The overlapping community detection algorithm has better practical significance than the non-overlapping community detection algorithm, and it is more common in large-scale networks. Overlapping nodes are key nodes in the network, so communities are connected to each other. It can also better reflect the real network structure in the real world.

It is a research hotspot in the field of complex networks to accurately and effectively locate community distribution using a given network topology. Specifically, it mainly includes the following aspects:

1. Community detection algorithm based on local aggregate differential evolution

Among many intelligent algorithms, evolutionary algorithms inspired by biological evolution are simple and intuitive global optimization methods. The method constrains the optimization process due to local aggregation operators and topology connections as a priori knowledge. The vector consisting of the community number of each node is regarded as an evolutionary individual, which is simple and intuitive but effective.

2. Community detection algorithm based on SVD

The distance definition based on the diffusion kernel function is used to improve and optimize the community discovery method based on support vector machines. Support vector machines are learning algorithms that are good at solving nonlinear pattern recognition problems in the field of machine learning. The distance definition based on the diffusion kernel function to turns the community classification problem of the original space into a quadratic Programming problem and a community number allocation problem in the mapping space. It uses the neighborhood graph and stable equilibrium point to simplify the community number allocation process.

3. Community detection method based on deep autoencoder and unsupervised over-limit learning machine

This method uses an independent autoencoder in deep learning to output low-dimensional feature embedding representations from the hidden layer through a training process that targets data reconstruction. In the process of preprocessing the input data, it makes full use of the sparse features in the complex network that the community structure generally has to guide the optimization goal, while avoiding large-scale eigenvalue decomposition operations. The unsupervised over-limit learning machine as a second-level embedded processor adds the inherent connection properties in the network as constraints to the dimensionality reduction and feature embedding processes in the form of manifold regularization, prompting the generated low-dimensional embedded representation to include more reasonable classification information.

4. Community structure based WSNs routing protocol and community structure based recommendation algorithm.

Take wireless sensor routing protocol and recommendation system as examples. By establishing a layered routing protocol based on the community structure, member nodes of a WSN only need to communicate through the cluster head node and the sink node of the community, thereby avoiding message broadcast to the entire network during the route establishment process. It effectively reduces system energy consumption. In the recommendation system, using the community discovery algorithm to analyze the user similarity matrix or the item similarity matrix can measure the correlation between users or items from another angle.

With the continuous development of the Internet of Things, big data, cloud computing, and artificial intelligence, more and more complex networks with community structures appear in production and life. Discovering and utilizing the structural characteristics of complex network communities continue to be a hot topic of continuous research in academia and industry.

1.3.5 Small Worlds and Information Dissemination

The small-world phenomenon, the principle that we are all linked by short chains of acquaintances or "six degrees of separation", is a fundamental issue in social networks. It is a basic statement about the abundance of short paths in a graph whose nodes are people, with links joining pairs who know one another. It is also a topic on which the feedback between social, mathematical and computational issues have been particularly fluid.

Nowadays, with the rapid development of computer data processing and computing capabilities, Internet is spreading into every corner of human production and life. For example, the Internet that promotes the era of information explosion, the power network that guarantees the production and life of electricity, and the social network that forms the basis of human society. Scientists have found that a large number of real network structures are very complex, neither regular networks nor random networks. In these actual networks, "small world phenomena" are common. By introducing logical links, Helmy A et al. formed a wireless network based on the "small world phenomenon" and analyzed the small world phenomenon in the wireless network. They also made people enter a new stage in the research of small and medium world phenomena in wireless networks. The network characteristics exhibited by "Small World Network" shorten the route establishment time, reduce the communication overhead, reduce the average energy consumption, improve the efficiency of data processing capabilities, and extend the network lifetime. Therefore, the research of "small world phenomenon" and the popularity of "small world phenomenon" in constructing WSNs routing, data fusion algorithms, navigation algorithms, etc. have attracted great attention from researchers. The research and application of small world phenomenon in WSNs has become extremely important and challenges frontier research directions. At present, research on small-world problems in WSNs and other networks focus on the following areas:

1. Small-world phenomena and characteristics in networks.

In tree-shaped WSNs, small-world phenomena provides theoretical support for preventing the adverse effects and benefit for designing new routing protocols, navigation algorithms and network structures based on small world theory.

2. Spread of viruses in the small world network method (information).

In an actual tree network, many factors may affect the threshold of node infection probability such as node locations, node distribution, communication radius, actual network topology. The infection probability threshold has a certain effect. In the process of analyzing the virus immune strategy, the virus immune strategy is linked with the security mechanism of each layer of the WSNs. However, for actual WSNs, in view of the characteristics of computer virus transmission in the network, it is necessary to propose a more appropriate virus immune strategy to minimize the damage degree of WSNs,

3. Application of small world theory in WSNs

Small world theory is often used to analyze the structural characteristics of real networks, or as a theoretical basis for designing special network structures, so that the constructed network has certain special properties or achieves special purposes. According to the characteristics of the small-world phenomenon in the WSNs, it's better to establish a new structure model that is more in line with the characteristics of the real network structure, and design the new MAC layer, routing layer and other layer protocols based on the model.

In recent years, various countries have increased their research efforts on key technologies of WSNs and further expanded the application range of WSNs. Researching the small-world phenomenon, analyzing the dynamic characteristics of WSNs based on small-world theory, designing hybrid network structures, designing routing and navigation algorithms have attracted more and more researchers' attention and become cutting-edge research directions.

1.4 Outline of the Book

This book is dedicated to show the related concepts and hot issues of WSNs. It mainly introduces network protocols, node localization, coverage control, community detection and small-world characteristics in WSNs.

The book contains six chapters that are summarized as follows.

Chapter 1, (the present chapter) provides a set of basic concepts on WSNs including the background, application and hot topics in WSNs.

Chapter 2, discusses the network routing protocols. This chapter presents methods of energy balanced routing protocols. The corresponding algorithms and routing protocols include, On Demand Clustering Improved LEACH Protocol, Connectivity-Guaranteed and Energy Efficient Routing Protocol, Density-Aware Clustering Protocol based on LEACH-C and AODV based Routing Protocol.

Chapter 3, outlines several localization algorithms of WSNs which include Hop-count-based Expectation of Distance Algorithm, Voronoi-based Localization Algorithm, Heuristic Algorithm for Node Localization, Delaunay Triangulation based Localization Scheme, Neighbor Constraint Assisted Distributed Localization, and Multidimensional Scaling Localization.

Chapter 4, presents approaches or algorithms of coverage control in wireless sensor network. It mainly includes Autonomous Triangular Deployment, Extended Virtual Force Based Approach, and Energy Balanced Redeployment Algorithm, etc.

Chapter 5, This chapter mainly introduces community detection of complex networks. Among them, the community structure is an important research hotspot in the characteristics of complex networks. This chapter mainly introduces community detection by Proximate Support Vector Clustering and Deep Auto-encoded Extreme Learning Machine, Local aggregated differential evolution algorithm, and robustness and fragility of tree-based WSNs.

Chapter 6, The "small world phenomenon" is common in real networks, such as the power grid and the social network that forms the basis of human society. The characteristics of small worlds have also been researched and applied in WSNs. This chapter mainly introduces small world characteristics including tree topologies, epidemic analyses and immunizations in epidemical models.

References

1. Shabbir N, Hassan S R. Routing protocols for wireless sensor networks (WSNs)[J]. Wireless Sensor Networks-Insights and Innovations, 2017.
2. Mesmoudi A, Feham M, Labraoui N. Wireless sensor networks localization algorithms: a comprehensive survey[J]. International Journal of Computer Networks & Communications, 2013, 5(1):748-755.

Chapter 2
Energy Balanced Routing Protocols for Wireless Sensor Networks

The routing protocol builds a mechanism for information transmission from the source node to the destination node. It guides transmitting path information of data packets. In this way, a fast and efficient routing protocol has a significant impact on the performance of WSNs. This chapter mainly discusses the fundamental concepts, fundamental structure and several typical routing protocols. On the basis, we propose several improved protocols, On Demand Clustering Improved LEACH Protocol (ODCL), Connectivity-guaranteed and Energy-efficient Clustering Scheme (CECS), Density-Aware Clustering Protocol based on LEACH-C (DACPL) and Hierarchical AODV Protocol for WSN (HAODV). The protocols all belong to clustering algorithms that effectively improve the energy efficiency of traditional routing algorithm.

2.1 Basic Theory of Routing Protocols

This section focuses on fundamental concepts, major issues in managing the function of routing protocol. To provide theoretical support for following improved routing algorithms, we also discuss some typical routing methods and present LEACH protocol in detail as well.

2.1.1 Fundamentals

Related fundamentals in the design of routing protocols include sensing range, transmission and neighbors set. Here, we give the concepts of these definitions and present a commonly used energy model in most protocols.

S. Chai et al., *Wireless Sensor Networks*, Wireless Networks,
https://doi.org/10.1007/978-981-15-5757-6_2

The sensing range of a sensor node s_i is a circle or a disk of radius r_i including its boundary, centered at ξ_i and defined by the point set

$$D(\xi_i, r_i) = \{\xi : |\xi_i - \xi| \leq r_i\} \tag{2.1}$$

where $|\xi_i - \xi|$ is the Euclidean distance between the locations ξ_i (the location of sensor s_i) and ξ.

The transmission range of a sensor node s_i is a circle or a disk of radius R_i including its boundary, centered at ξ_i, and defined by the point set

$$D(\xi_i, R_i) = \{\xi : |\xi_i - \xi| \leq R_i\} \tag{2.2}$$

The neighbor set of a sensor node s_i is given by

$$N(s_i) = \{s_j : |\xi_i - \xi_j| \leq R_i\} \tag{2.3}$$

where R_i is the radius of the transmission range of sensor s_i. The sensing and communication ranges in randomly distributed nodes are determined by the maximum distance between any two adjacent sensor nodes in the given area.

Let A be an area of the field, a point p belonging to A ($p \in A$) is covered (or sensed) if and only if it belongs to the sensing range of at least one sensor. The area A is covered if and only if every point $p \in A$ is covered.

The dissipation model of radio energy is analyzed in Fig. 2.1. Assuming that the energy consumption issues in sensors are the transmission and reception of data, the energy consumed in transmitting a message of l bits over a distance of d known as the transmission distance is given by

$$E_{TX} = \left(\varepsilon_{amp} d^{\lambda} + E_{elec}\right) l \tag{2.4}$$

where E_{elec} represents the energy consumed by the electronic circuitry of the transceiver, $\varepsilon_{amp} \in \{\varepsilon_{fs}, \varepsilon_{mp}\}$ is the transmitter amplifier in the free space ε_{fs} or the multipath ε_{mp} model, and λ is the path-loss exponent, $\lambda = 2$ if $d \leq d_0$, $\lambda = 4$ if $d > d_0$ where d_0 is known as the crossover distance and it is given by $d_0 = \sqrt{\frac{\varepsilon_{fs}}{\varepsilon_{mp}}}$. So the Eq. (2.4) can be written as

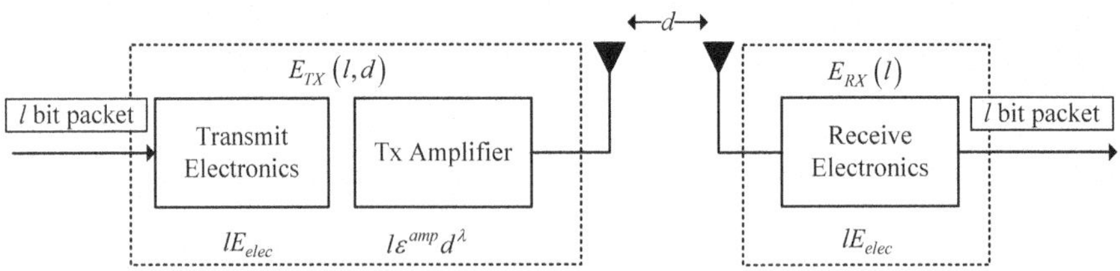

Fig. 2.1 Radio energy dissipation model

$$\begin{cases} E_{TX}(l,d) = kE_{elec} + l\varepsilon_{fs}d^2, \text{if } d <= d_0 \\ E_{TX}(l,d) = kE_{elec} + l\varepsilon_{mp}d^4, \text{if } d > d_0 \end{cases} \tag{2.5}$$

The reception energy is given by

$$E_{RX} = lE_{elec} \tag{2.6}$$

Hence, the total energy consumed by any sensor to receive a message and forward it with distance d is given by

$$\begin{aligned} E_{total} &= lE_{elec} + lE_{elec} + l\varepsilon_{amp}d^{\lambda} \\ &= lE_{elec}\left(2 + \varepsilon_{amp}d^{\lambda}\right) \end{aligned} \tag{2.7}$$

The distance between node i and node j is represented by $D(i,j)$, which is defined as

$$D(i,j) = \sqrt{(x(i) - x(j))^2 + (y(i) - y(j))^2} \tag{2.8}$$

where $i, j = 1, 2, 3 \ldots n$, $i \neq j$ and $(x(i), y(i))$,$(x(j), y(j))$ is the coordinate of node i and j.

2.1.2 Protocol Stack

Since protocol design of WSNs considers node constraints and application requirements, the protocol stack for WSNs consists of five protocol layers as shown in Fig. 2.2. They are physical layer, data link layer, network layer, transport layer and

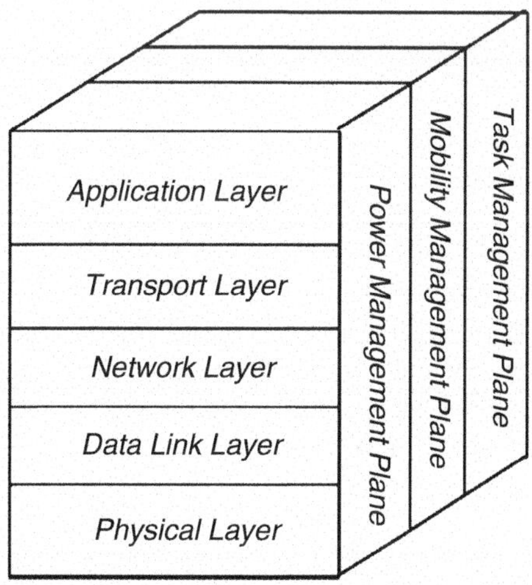

Fig. 2.2 Protocol stack for WSNs

application layer. The application layer contains a variety of application protocols to generate different sensor network applications, and the transport layer function is the delivery of the data required by the application layer reliably. The network layer routes the data from the transport layer, and the data link layer is responsible for data stream multiplexing, data frame transmission and reception, medium access and error control. The physical layer transmits and receives signal over physical communication medium. The protocol stack can be divided across each layer into a group of management planes, including power, task and mobility management planes. The plane function of power management is to manage the power level of the node for sensing, processing and transmission and reception by employing efficient power management mechanisms throughout different protocol layers. The mobility management plane manages the configuration and reconfiguration of nodes to establish and maintain connectivity of the network in case of topology change due to node addition, failure or movement. The task management plane distributes the task among nodes in the sensing field to improve the energy efficiency and prolong the network lifetime.

2.1.2.1 Application Layer

The application layer includes many application layer protocols that perform various sensor network applications, like query-dissemination, sensor localization, synchronization and network security.

2.1.2.2 Transport Layer

The transport layer is responsible for reliable end-to-end data delivery between the nodes and the sink. Due to nodes constraints and different application requirements, the conventional TCP/IP cannot be used for WSNs. That means, sensor networks are application specific so that different applications may have different application requirements. Moreover, in sensor networks data is delivered bidirectionally upstream and downstream. In the upstream delivery, the sensor nodes forward the sensed data to the sink, while in downstream delivery, the data originated from the sink are sent from the sink to the source nodes. The data flows in the two directions may have different reliability requirements, since the data flows in the upstream are loss tolerant, because the sensed data are redundant to some extent. However, in downstream delivery, the data flows are queries, commands and programming binaries sent to sensor nodes, which usually require 100% reliable delivery. Therefore, the unique characteristics of sensor networks and the specific requirements of different applications present many new challenges in the design of transport layer protocols for WSNs.

2.1.2.3 Network Layer

The network layer is responsible for routing sensed data from source sensor nodes to the sink. Sensor nodes can transmit data to sink either directly via single-hop long-range wireless communication or via multi-hop short-range wireless communication. Long-range wireless communication is costly in terms of energy consumption and implementation complexity while short-range wireless communication can significantly reduce energy consumption and signal propagation and channel fading effects inherent in long-range wireless communication. The routing protocols for the traditional wireless networks are not suitable for sensor networks because they do not consider energy efficiency. In addition, data from sensing field toward the sink in sensor networks exhibit a unique many-to-one traffic pattern. The combination of multi-hop and many–to-one communication results in a significant increase in transmit traffic intensity and thus packet congestion, collision, loss, delay and energy consumption as data moves closer toward the sink. This causes the bottleneck problem, thus largely reducing the operational lifetime of the entire network. Therefore, it is very important to consider the energy constraints of the sensor nodes as well as the unique traffic pattern in the design of the network layer routing protocols for WSNs.

2.1.2.4 Data Link Layer

To provide reliable point-to-point and point-to-multipoint transmissions, the data link layer is responsible for data stream multiplexing, data frame creation, medium access and error control. Most important function of the data link layer is medium access control (MAC). MAC aims to share the common communication resources or medium fairly and efficiently among sensor nodes to achieve good network performance in terms of energy consumption, network throughput and delivery latency. MAC protocols for traditional wireless networks cannot be applied to sensor networks because they do not consider the unique characteristics of sensor networks, particularly, the energy constraint. In a cellular system, the main concern is to provide quality of service (QoS) to users, because the base stations have no power limit and users can replace the batteries in their handsets. In MANETs, the mobile nodes are equipped with portable devices powered by replaceable battery. Therefore, the primary concern in sensor networks is energy conservation in order to prolong the network lifetime.

MAC is one of the critical topics in the design of WSNs. WSNs employ MAC protocol to arbitrate access to the shared medium in order to avoid data collision from different nodes and at the same time for the bandwidth resources fairly and efficiently to share among sensor nodes. MAC must consider some factors to improve the performance and provide good network services for different applications. These factors include energy efficiency, scalability, adaptability, channel utilization, latency, throughput and fairness. Energy consumption is the primary factor that affects the operational life time of individual nodes and the entire network. The overall performance of a sensor network highly depends on the energy efficiency of the network. To design an energy efficient MAC protocol, it is important to identify the major sources of energy waste in sensor networks from the MAC perspective. From MAC view of point the energy waste comes from four main sources: collision, overhearing, control overhead, and idle listening. The main fundamental MAC protocols are contention-based MAC, contention-free MAC and hybrid MAC. Carrier sense multiple access (CSMA) represents the most common contention-based MAC protocol, while time division multiple access (TDMA) represents the most common contention-free MAC protocol that is used for sensor networks. TDMA is characterized with its collision-free nature, which significantly improves network energy efficiency under high traffic load. However, it has higher delay and lower throughput under low traffic load due to idle time slots. In addition, TDMA requires accurate synchronization between sensor nodes, and it has limited scalability and adaptability to network changes. In contrast, CSMA is of lower energy efficiency and higher delay under high traffic load, but it can reduce delay and has higher throughput under low traffic load. Depending on specific applications, the MAC protocol can incorporate TDMA or CSMA with other techniques to meet different performance requirements.

2.1.2.5 Physical layer

This layer converts the bit streams from data link layer to signals that are suitable for transmission over the communication medium. Therefore, it must deal with issues such as transmission medium and frequency selection, carrier frequency generation, signal modulation and detection, and data encryption. In addition, it must deal with the design of the underlying hardware, and various electrical and mechanical interfaces. The design of network architecture has great impact on the energy consumption and the operational lifetime of WSNs. Due to energy constraint in sensor nodes and the unique many-to-one traffic pattern, multi-hop short distance communication is preferred. In multi-hop networks, the hierarchical network architecture based on clustering reduces the communication energy consumption, balances traffic load and improves scalability when the network size grows. To perform various network control and management functions, the network protocols must consider the resources constraints and the application-specific nature of sensor networks.

2.1.3 Typical Topology Control Protocols

Based on the network structure of WSNs, we can award routing protocols as flat routing protocols and hierarchical routing protocols.

Flat routing protocols are the routing protocols in networks with a flat structure. The flat structure means that the nodes in the network have the same status in the routing function, and no hierarchical management mechanism is used. The advantage of flat structure routing is that there are no special nodes in the network, network traffic is evenly distributed in the network, and the routing algorithm is easy to implement. The disadvantage is that the scalability is small, which limits the size of the network to a certain extent.

Hierarchical routing protocols are the routing protocols in hierarchical structure. Hierarchical routing protocols use the concept of clusters to divide sensor nodes hierarchically. Several adjacent nodes form a cluster and each cluster has a cluster head. Clusters can communicate with each other through a gateway. The gateway can be a cluster head or another cluster member. The connections between the gateways constitute the upper-layer backbone network, and all inter-cluster communications are forwarded through the backbone network.

Depending on the characteristics and problems of data transmission in WSNs, many improved routing protocols have been proposed. We briefly introduce the existing routing protocols of WSNs and compare them for each category. Among the typical protocols, Flooding protocol and Sensor protocol for Information via Negotiation (SPIN) are flat routing protocols. LEACH, LEACH-C and Power-Efficient Gathering in Sensor Information Systems (PEGASIS) are the hierarchical routing protocols.

2.1.3.1 Flooding Protocol

Flooding protocol is a classic flood-type routing. In this protocol, after a node collects data or receives data from other nodes, it immediately broadcasts to all its neighbors until the transmitted data reaches the destination node or the data expires. The advantage of this routing protocol is that it is simple to implement. Each node only needs to broadcast the received data packets. It does not need to look up the routing table to select the next hop node, and it does not need to use related algorithms to ensure network topology updates and discover new ones routing. But at the same time, there are more serious defects, information explosion and information overlap. Because the protocol transmits data in the form of broadcast, nodes may receive the same data transmitted from multiple neighbor nodes, which is an information explosion. The areas monitored by multiple sensor nodes may intersect. Therefore, the same node may receive data information sent by multiple neighbor nodes for the same monitoring area, which is information overlap.

2.1.3.2 Sensor Protocol for Information via Negotiation (SPIN)

The SPIN protocol is the earliest data-centric routing protocol. It uses a data description-based negotiation mechanism and an energy adaptive mechanism. Under the role of this protocol, there are three types of data passed by the nodes in the WSN: data description (ADV), data request (REQ), and data itself (DATA). After a node generates or receives data, in order to avoid unnecessary overhead caused by blindly broadcasting data, it first sends metadata to all neighboring nodes. Metadata is data that describe the attributes of the data collected by the sensor node. After receiving the data description message, the node that needs data sends a data request message to the source node. Finally, the source node sends the data to the requesting node.

2.1.3.3 Low Energy Adaptive Clustering Hierarchical (LEACH)

The LEACH protocol is an energy adaptive clustering algorithm designed and developed by MIT researchers. This protocol fully combines the characteristics of WSNs and uses clustering mechanisms to reduce redundant data transmission. In addition, the cluster head can perform fusion processing on the data of the nodes in the cluster, and then transmit the data to the base station in a one-hop manner.

The operation of LEACH is divided into many rounds. Each round is divided into two phases, the setup phase and the steady state phase. The setup phase is the cluster formation phase. At this stage, sensor nodes form several clusters according to the LEACH algorithm. The steady state phase is divided into several repeated frames. In each frame, each member node transmits the collected data to its cluster head. The cluster head then transmits the aggregated data to the base station node. In general, in order to reduce the proportion of cluster overhead, the time interval in the stable phase should be much larger than the time used in the setup phase.

2.1.3.4 Low Energy Adaptive Clustering Hierarchical-Centralized (LEACH-C)

The LEACH-C protocol is an improved version of the LEACH protocol. The LEACH-C protocol takes a central control mechanism to optimize network performance. LEACH-C also divides each round into the setup phase and the steady state phase. At the beginning of each round, nodes in the network send status information such as position and energy to the base station and the base station performs clustering uniformly. After receiving the information sent by the node, the base station first calculates the average energy value in the network. Only the nodes with energy state greater than this value can participate in the cluster head selection. The base station then uses the simulated annealing algorithm to perform clustering, and selects cluster head nodes for each cluster.

Finally, the base station broadcasts information including clustering and the cluster head to each node, and the node confirms its identity based on the information. When the received ID is same as its own ID, the node is the cluster head, otherwise it is an ordinary node. Data transmission in the LEACH-C protocol ready phase is the same as in the LEACH protocol. In the LEACH-C protocol, the selection of a cluster is no longer completely random, and clustering is uniformly scheduled by the base station, which is more conducive to balance the energy allocation in the network and extend the network survival time.

2.1.3.5 Power-Efficient Gathering in Sensor Information Systems (PEGASIS)

PEGASIS routing protocol is an improvement on the LEACH routing protocol. PEGASIS uses only one cluster head, and treats the entire network as a cluster. The cluster head can directly communicate with the sink node. Ordinary nodes use the token control chain to transmit data to the cluster head node in a multi-hop manner. The election of cluster head is simple. If there are n nodes in the entire network, each node is numbered. The cluster head in the r-th round is the i-th node, where $i = r \bmod n$. When the node knows that it has become the cluster head of this round through calculation, it broadcasts an ADV message to notify the entire network. When the ordinary nodes send data to the cluster head node, they must first know their geographic coordinates, and then still use the nearest principle to send data to the node closest to them. Multi-hop transmission further reduces the network energy consumption. Because the PEGASIS routing protocol organizes the entire network of nodes into a chain, the cluster head node divides the network into two parts and the detection data is transmitted to the cluster head from two directions. After the cluster head node receives the data, it performs processing fusion and the hop is sent to the sink node, then the next round of election is performed.

The PEGASIS routing protocol avoids frequent cluster head elections and hence greatly reduces the number of data communications. Nodes send data in a short-distance multi-hop mode, which effectively saves energy consumption. Compared with the LEACH protocol, the PEGASIS routing protocol greatly extends the network survival time. However, there are also some problems. Since there is only one cluster head in the entire network, when the cluster head fails, the entire network data cannot be sent to the sink node. Also, high delay is another drawback.

2.1.4 LEACH Routing Protocol

As most improved algorithms in this chapter are referred to LEACH, we firstly discuss the theory of LEACH. LEACH provides a concept of rounds and each round consists of setup phase and steady state phase as shown in Fig. 2.3. In the setup phase, clusters are formed in self-adaptive mode. In the steady state phase, data are transferred to base station. The time of the steady state phase is longer than the time of setup phase to avoid overhead. LEACH randomly selects cluster heads at each round in the setup phase. The steady state phase consists of several frames. For each frame, every node has data to send during its slot time to the cluster head, which in turn sends data to base station after fusion.

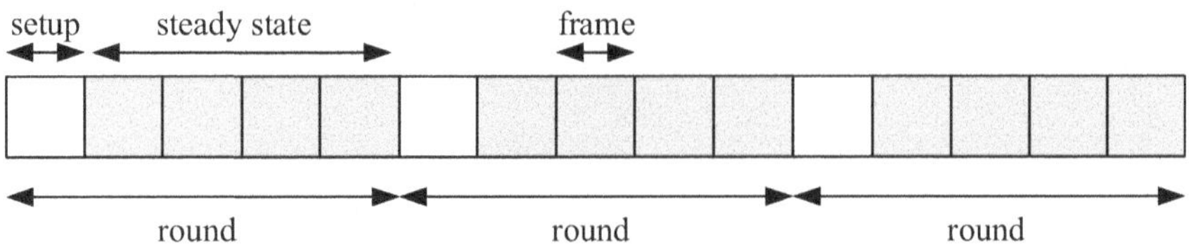

Fig. 2.3 Operation time line of LEACH

2.1.4.1 Setup Phase

The cluster formation algorithm is random, that is, each sensor node in the WSN uses a random autonomous method to elect the cluster head node. The functions of each node in the election algorithm are equal, and the LEACH algorithm is independent, autonomous. Whether a node becomes a cluster head is determined by itself.

According to the percentage of cluster heads in the WSN, one or more cluster heads would be generated based on threshold $T(n)$ as in Eq. (2.9).

$$T(n) = \begin{cases} \dfrac{p}{1 - p^{*}\left(r\text{mod}\left(\dfrac{1}{p}\right)\right)}, & \text{if} \quad n \in G \\ 0, \text{if } n \notin G \end{cases} \tag{2.9}$$

where p is the percentage of cluster heads in the WSN, r is the current round and G represents the group of nodes that had not been selected as a cluster head in the last $\frac{1}{p}$ rounds.

At the beginning, the node generates a random number x, the value of this random number $x \in (0, 1)$, the node issues a message announcing that it has been selected as the cluster head. After the surrounding nodes receive the announcement message, they run the corresponding algorithm to decide whether to join the cluster. If the node has served as the cluster head, we set $T(n)$ as zero at the beginning of this round of election to prevent this node from being selected as the cluster head again.

In fact, $T(n)$ is the average probability that the node which has not become the cluster head becomes the class head in the i-th round. The mathematical analysis is as follows. Suppose that the WSN has N nodes. If k clusters are guaranteed per round, that is, k cluster heads, we can know that $p = k/N$. The conversion mechanism of the cluster head of the LEACH protocol assumes that the initial energy of each node is the same. In order to ensure that each node becomes the cluster head only once in the $\frac{1}{p}$ round, corresponding to the N/k round, it is assumed that the node is at r round and at time t. The probability of becoming a cluster head in the time t and $r + 1$ round is $T_i(t)$.

$$E[\#CH] = \sum_{i=1}^{N} T_i(t) * 1 = k \tag{2.10}$$

When the node is selected as the cluster head by the election algorithm, it broadcasts a message frame at the MAC layer. The LEACH protocol uses the Carrier Sensing Multiple Access (CSMA) protocol at the MAC layer. When the non-cluster head node receives the frame, it chooses whether to join the cluster based on the received signal strength and communication cost.

After joining a cluster, a non-cluster head node sends a join-request frame to register with the cluster-head node, and indicates that the node belongs to the cluster through a hand-shaking protocol.

After finishing cluster head selection, the base station broadcasts notification message including clusters and cluster head IDs of each cluster. The cluster heads send TDMA schedules to their members and the network goes to steady state phase. The whole process of the cluster formation is shown in Fig. 2.4.

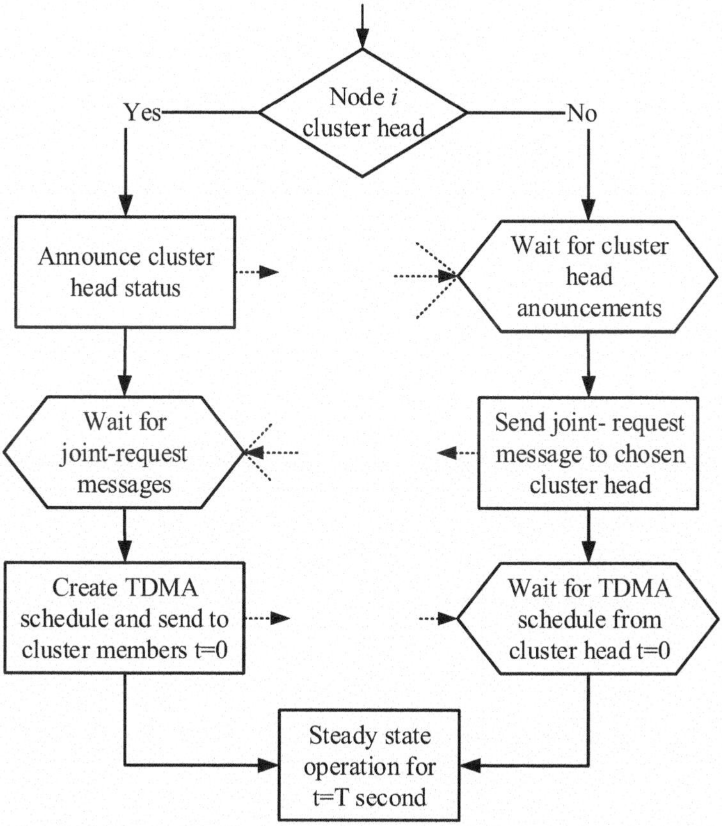

Fig. 2.4 Flowchart of the cluster formation process of LEACH. (Redrawn from Heinzelman, W.B. et al., IEEE Trans. Wireless Commun. 1, 660, 2002)

2.1.4.2 Steady State Phase

The steady state phase is a relatively long phase and the flowchart of the steady state is shown in Fig. 2.5. Nodes in the cluster collect the surrounding environment information, and then process the collected data and forward them to the cluster head. After receiving the data transmitted by the nodes in the cluster, the cluster head processes the data, and then forward the fused data to the base station.

After the cluster head is selected, the cluster head informs the network itself through broadcast. Finally, the cluster head allocates a transmission slot to each node by establishing a TDMA mode schedule. When the time slots of all nodes are used up, the network enters the next round.

The energy model in LEACH uses the radio energy dissipation model. The model is divided into two parts, energy consumption of the transceiver circuitry and the power amplification circuit consumption. The transceiver circuitry depends on the message length. The consumption of power amplification circuit depends on the distance between the transmitting and receiving nodes. The condition of free space mode is $d <= d_0$ and multi-path fading mode needs $d > d_0$, where $d > d_0$ is constant (cross over distance). The cost for the normal node, that is not cluster head, can be expressed as

$$E_{non-CH} = E_{RX}(l) + E_{TX}(l, d) \tag{2.11}$$

where $E_{RX}(l)$ is the reception energy cost, calculated by

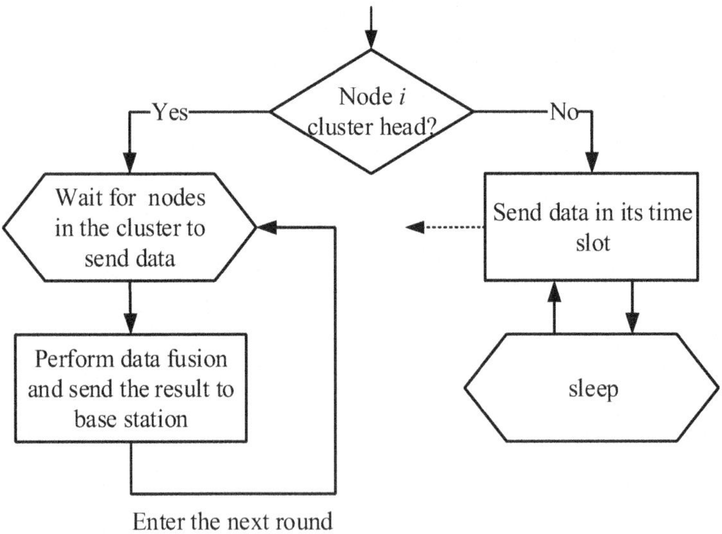

Fig. 2.5 Flowchart of the steady state data transmission phase process of LEACH. (Redrawn from Heinzelman, W.B. et al., IEEE Trans. Wireless Commun. 1, 660, 2002)

$$E_{RX}(l) = E_{elec}l \tag{2.12}$$

$E_{TX}(l, d)$ is the transmission energy cost, stated as

$$\begin{cases} E_{TX}(l,d) = lE_{elec} + l\varepsilon_{fs}d^2, \text{if } d <= d_0 \\ E_{TX}(l,d) = lE_{elec} + l\varepsilon_{mp}d^4, \text{if } d > d_0 \end{cases} \tag{2.13}$$

$$E_{TX}(l,d) = E_{elec}l + \varepsilon_{amp}ld^{\lambda} \tag{2.14}$$

where l is the length of the message in bits, E_{elec} is energy used by the transmitter or receiver circuitry per bit, ε_{amp} is the energy dissipation of the transmitter amplifier, d is the distance between transmitting and receiving nodes and λ is the pass-loss exponent. The energy needed by the cluster head normally consists of three components, which are reception energy, transmission energy and aggregation energy. The total energy consumed by the cluster head can be expressed as

$$E_{CH} = nE_{RX}(l) + nlE_{da} + E_{TX}(c.n.l, d) \tag{2.15}$$

where n represents the number of nodes per cluster, c denotes the aggregation coefficient and E_{da} is energy dissipated for data aggregation. For simulation purposes, the radio model values used in this work are as follow:

Energy dissipated by the transmitter or receiver circuitry is given by

$$E_{elec} = 50\text{nJ/bit} \tag{2.16}$$

Energy dissipation of the transmitter amplifier is given as

$$\varepsilon_{mp} = 0.0013\text{pJ/bit/m}^4 \tag{2.17}$$

$$\varepsilon_{fs} = 10\text{pJ/bit/m}^2 \tag{2.18}$$

Energy dissipated for data aggregation E_{da} is given by

$$E_{da} = 5\text{nJ/bit/signal} \tag{2.19}$$

LEACH has comparative advantages but it ignores some important issues such as:

(a) The residual energy of concerned node is not considered in cluster formation and cluster head selection processes, because in the case that the less energy node is selected as cluster head it could reduce network lifetime.
(b) Distance to base station is not considered, although direct communication of cluster head with base station consumes more energy when the distance is farther away.
(c) Cluster heads are selected randomly and clustering is done in each round.

The LEACH algorithm is the first hierarchical network topology organization algorithm proposed for WSNs. Compared with the way that nodes send data to the sink node or directly send data to base station, the LEACH algorithm has greatly improved the performance, but the algorithm still has certain limitations in some aspects.

The LEACH protocol is organized with clusters, so that the cluster head has to undertake a large number of communication tasks and data fusion. Its energy consumption is faster than that of member nodes in the cluster, hence it is more prone to failure, leading to more frequent clustering.

The clustering process and cluster head election process are an additional overhead for the energy consumption of the entire WSN. If the clustering process and the cluster head election process is frequent, the extra energy consumption increases greatly. Therefore, it must be strictly controlled. Otherwise, the energy consumption of the entire WSN may be wasted in the clustering process and cluster head election process, and the energy utilization of the nodes will decrease.

In addition, due to the limitations of the LEACH algorithm itself, the LEACH protocol is more suitable for smaller-scale WSNs, but is inefficient for large-scale WSNs. Therefore, in the rest of this chapter, we have made corresponding improvements to the energy consumption and connectivity issues in LEACH. In addition, we also propose improvements to the ADOV protocol to make it more suitable for WSNs.

2.2 On Demand Clustering Improved LEACH Protocol

In this section, on demand clustering improved LEACH protocol (ODCLP) [1] is proposed. This protocol balances the energy load among nodes and hence minimizes the energy consumption in the entire network. After receiving location information from nodes, the base station divides the nodes into clusters via the iterative algorithm *K*-means, and then broadcasts the clustering information and centroids to all nodes. Upon receiving this information, every node can decide whether to become a cluster head or not. The protocol considers the communication distance and the residual energy to avoid data loss due to sudden death of cluster head or member nodes. Moreover, the cluster head reselection is relevant to the current cluster head residual energy in order to reduce overheads. The algorithm improves the performance of existing LEACH and demonstrates the advantage of balanced energy consumption, reduction of clustering overhead and improving network reliability.

2.2.1 On Demand Clustering Protocol

In contrast to LEACH, the clusters are formed at first in centralized method, and then the cluster heads are selected in distributed method in each cluster. Therefore, the algorithm can be classified as hybrid scheme. In this protocol, both the cluster formation and the selection methods of cluster head have been modified. For more energy conservation, a multi-hop routing was adopted to forward data to the base station. In this protocol, the energy model is the same as LEACH. Following assumptions are made about the sensors and the network model.

1. The sink node is located inside the sensing field and it has unlimited resources.
2. Sensors are location aware and energy constrained.
3. Each node has a unique identifier and it is initially able to transmit data directly to the base station.
4. A node belongs to only one cluster, but may change its belonging cluster during each round.
5. Sensors are energy-aware and energy-adaptive.
6. Nodes and sink both are stationary after deployment.
7. The cluster is a mixture of single and multi-hop structure.

The same with LEACH, the operation of the algorithm is divided into rounds. Each round consists of setup phase, data transmission phase and cluster head reselection and cluster reformation. The cluster formation, cluster heads selection and multi-hop paths establishment are done successively in the setup phase. In the data transmission phase, the nodes sense and transmit data to the cluster head and then the cluster head forwards data to base station after aggregation.

2.2.1.1 Setup Phase

Cluster Formation: To form proper clusters, the protocol uses *K*-means clustering algorithm. The clustering is highly affected by network topology, nodes distribution and criteria of grouping (geometrical distance in this case).

After initial nodes of K clusters are assigned randomly, the centers of clusters are computed and the nodes are allocated to the clusters with the closest centers. The process is repeated until the cluster centers do not significantly change. Once the cluster assignment is fixed, the mean distance of a node to cluster centers is used as the score. *K*-means clustering produces clusters with minimum sum of distance squared from each node to its cluster centre (cluster energy). The algorithm aims to minimize the squared error function

$$J = \sum_{j=1}^{k} \sum_{i=1}^{n} \left\| x_i^j - c_j \right\|^2 \tag{2.20}$$

where $\left\| x_i^j - c_j \right\|$ represents chosen distance measure between a node x_i^j (node i in cluster j) and cluster centre c_j (centre of cluster j), J is an indicator of the distance of the N nodes from their respective cluster centers.

The algorithm is composed of the following steps:

Step 1. Place K points into the space represented by the nodes that are being clustered. These points represent initial cluster centroids.
Step 2. Assign each node to the cluster that has the closest centroid.
Step 3. When all nodes have been assigned, recalculate the positions of the K centroids.
Step 4. Repeat steps 2 and 3 until the centroids no longer move, i.e. converge to optimal position. This produces a separation of the nodes into clusters, from which the metric to be minimized is calculated. *K*-means is based on finding data clusters in a set of data, by keeping minimized cost function of dissimilarity measure. This dissimilarity measure is chosen as the Euclidean distance as shown in (2.20).

Figure 2.6 shows the flowchart of *K*-means algorithm. The clustering process is carried out by the base station in centralized algorithm after receiving the nodes location data. The base station computes the sum of distances to the neighbor nodes (nodes with the same centroid) for each node and, then sends (broadcasts) this sum of distances and clustering information to all nodes in the network. Therefore, each node can determine its cluster centroid, its intra distance and its neighbor nodes in the cluster (nodes with the same centroid). Every node uses this data in addition to distance from base station to determine its possibility to become a cluster head in its cluster.

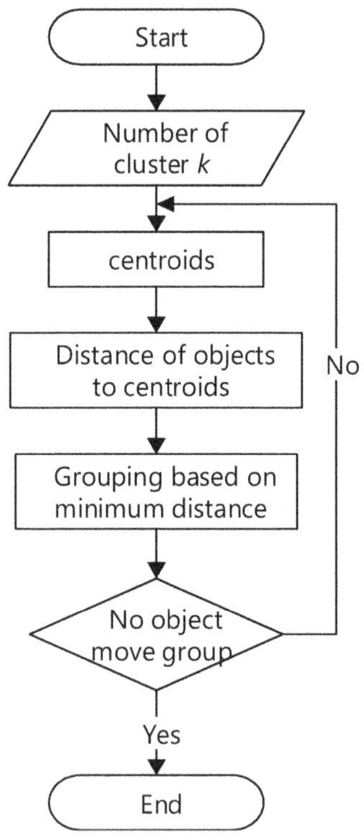

Fig. 2.6 *K*-means algorithm flowchart

Cluster head selection: To overcome the issues mentioned in LEACH sections, the cluster head selection formula in Eq. (2.9) had been extended by inserting the node residual energy, the distance to base station and the sum of distances to neighbors in the same cluster. So, a node with more residual energy, less distance to base station and less sum of distances to its neighbors in the same cluster (intra distance) would be eligible to become the cluster head in that cluster, irrespective of its role in previous rounds. Threshold value in Eq. (2.9) can be rewritten as

$$T(\overline{n}) = T(n) * \frac{E_{resi}}{E_{initial}} * \frac{1}{d_{BS}^2 + d_{Intra}^2} \tag{2.21}$$

where E_{resi} and $E_{initial}$ are the residual and initial energy for each node, while d_{BS}^2 and d_{Intra}^2 are the distance to base station and the sum of distances to neighbor nodes, respectively. The two terms d_{BS}^2 and d_{Intra}^2 can be added together to form the communication cost distance d_{cost}^2 which represents the max communication cost distance for each node when it is selected as a cluster head then the formula can be rewritten as

$$T(\overline{n}) = T(n) * \frac{E_{resi}}{E_{initial}} * \frac{1}{d_{\text{cost}}^2} \tag{2.22}$$

The probability of cluster head increases when a node has high residual energy E_{resi} and less communication cost distance d_{cost}^2 in its cluster. This ensures the optimization of the communication energy for the concerned node n, which in turn increases the probability of the chosen random number r to be less than the modified threshold $T(\overline{n})$.

The cluster heads advertise themselves and their centroids using CSMA MAC protocol, and then all normal nodes belong to their cluster heads according to cluster centroids. After determining to which cluster head they should belong, CSMA MAC protocol is used to send a confirmation message to cluster heads. Each cluster head records its total number of members, determines its low level of residual energy, creates a TDMA schedule based on the number of member nodes and sends the time slots schedule that is employed by members in steady state phase. At this point, the clusters formation is finished.

Establishing multi hop-paths: In this process, the communication radius is defined by energy according to the radio energy dissipation model. Firstly, the hop count of base station is set to 0, then base station broadcasts a message to the cluster heads; the message contains base station's ID and hop count with fixed energy. The hop count within a certain circle area will be set to 1. The hop count is set to 0 as a parent node with hop count is set to 1. Then cluster heads repeat the above process with cluster head, which is one hop count bigger than them. If multiple messages are received, the cluster head of the strong signal is regarded as the parent node. The process stops when operative cluster heads do not receive feedback message from cluster heads, which are one hop count greater than them after broadcasting. Once the cluster heads are selected and the TDMA schedule and multi-hop paths are fixed, data transmission can begin. All member nodes must keep their receivers on during the setup phase in order to hear the advertisements sent by the cluster heads as well as the cluster heads to hear the joint decision messages sent by member nodes.

2.2.1.2 Data Transmission Phase

In this phase, data of cluster is received and fused by the cluster head, and then the cluster head transmits received data to the parent cluster head. The parent cluster head fuses the data and retransmits to next level, the process is repeated till all data reaches the base station.

2.2.1.3 Cluster Head Reselection and Cluster Reformation

Every node checks its energy before sending data and saves its residual energy after sending data to cluster head or to the base station. The selected cluster head continues to work as cluster head until its energy level decreases to the predefined value known as the low level. For any cluster head, if the energy before sending data is below the predefined low-level value, the cluster head informs the base station via a specific bit in the current TDMA frame. The base station informs the entire network nodes about the starting of selection process of cluster head at the next round. Then after receiving the last data of the current frame, the data transmission phase is terminated and the entire system moves to the cluster head selection step in the setup phase. Since there are no dead nodes, the system continues these rounds for K times ($K = N \times P$), because cluster reformation process during this stage may generate the same clusters as the current ones.

After K rounds it is expected that some nodes may die, in fact the value of K in this stage can be varied according to the application and it is not necessary to be equal to the number of clusters. If the base station receives message from any cluster head, it is indicating that the residual energy is below the low level. The base station informs the entire network nodes about the starting of cluster reformation at the next round. Then after receiving the last data of the current frame, the entire system moves to the cluster formation step in the setup phase. When some nodes die, the total number of nodes decreases. The network topology and the nodes distribution are also changed. The system continues these rounds until every node's energy has been depleted. Figure 2.7 shows the flowchart of the algorithm.

2.2.2 *Calculation of Low Level Energy*

We need to know the predefined value of energy before the cluster head and node sending data in the cluster head reselection and cluster reformation phase. Here, we

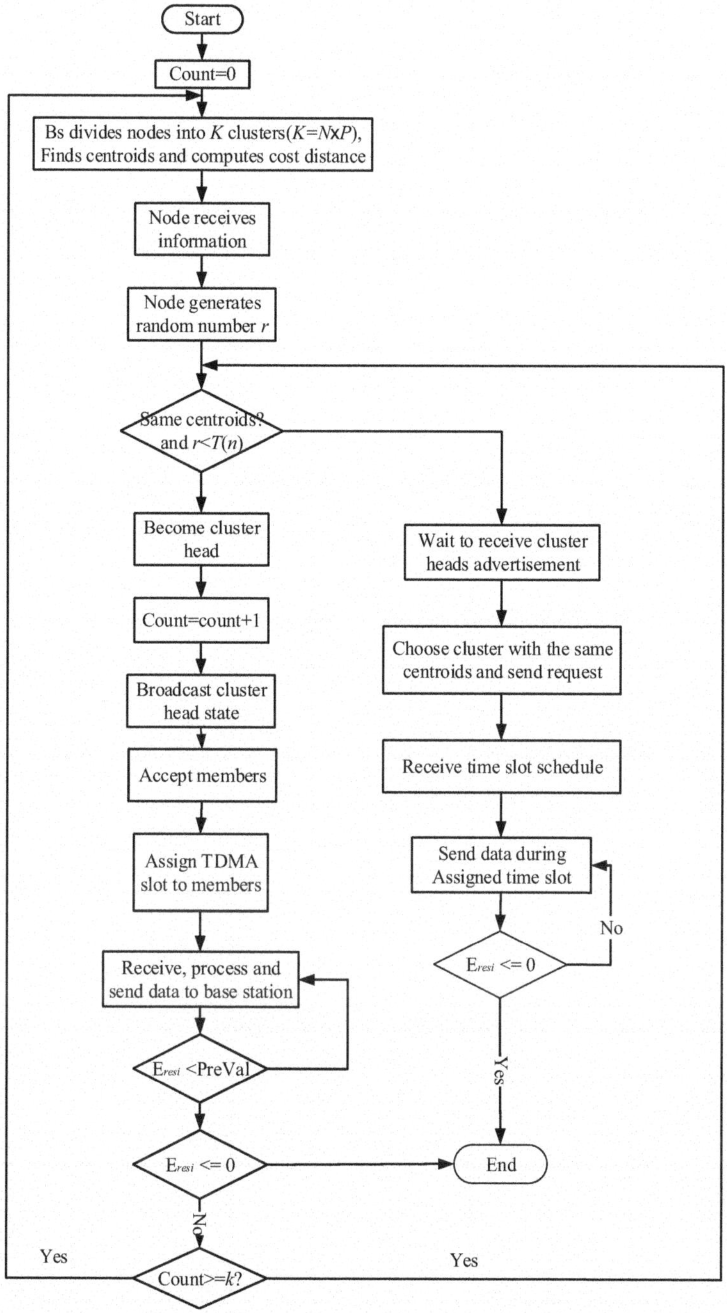

Fig. 2.7 Flowchart of the proposed algorithm

give the calculation method of the low-level energy. They are mainly based on the LEACH energy model.

2.2.2.1 Calculation of Low Level Energy for Cluster Head

The low level energy for cluster head is calculated after every cluster head selection. It is proportional to the number of members and the distance of the cluster head to the base station. Assuming that the cluster head has n members and the distance to the base station is d_{CH} then

$$E_{resi} = E_{\max} - E_{CH} \tag{2.23}$$

where E_{resi} represents the residual energy of the cluster head, $E_{\max}$ is the initial energy or the energy of the cluser head after the previous sending and E_{CH} is the energy consumed by the cluster head to send data to the base station. Same with LEACH, the energy consumption of cluster head is given by

$$E_{CH} = E_{RX} + E_{DA} + E_{TX} \tag{2.24}$$

And it can be rewritten as

$$\begin{cases} E_{CH} = nlE_{elec} + (n+1)\left[lE_{DA} + lE_{elec} + l\varepsilon_{mp}d_{CH}^4\right], \text{if } d > d_0 \\ E_{CH} = nlE_{elec} + (n+1)\left[lE_{DA} + lE_{elec} + l\varepsilon_{fs}d_{CH}^4\right], \text{if } d > d_0 \end{cases} \tag{2.25}$$

Since E_{elec}, E_{RX}, l and ε_{mp} are constants andapplication specific, then

$$E_{CH} \propto n + (n+1) + (n+1)d_{CH}^4 \tag{2.26}$$

$$E_{CH} = c_1 n + c_2(n+1) + c_3(n+1)d_{CH}^4 \tag{2.27}$$

where $c_1 = lE_{elec}$, $c_2 = lE_{elec}$ + lE_{DA} and $c_3 = l\varepsilon_{mp}$, then

$$E_{CH} = n + (n+1)\left[1 + \frac{E_{DA}}{E_{elec}} + \frac{\varepsilon_{mp}d_{CH}^4}{E_{elec}}\right] \tag{2.28}$$

Assuming that $\frac{E_{DA}}{E_{elec}} = \alpha$ and $\frac{\varepsilon_{mp}}{E_{elec}} = \beta_4$

$$E_{CH} = n + (n+1)\left[1 + \alpha + \beta_4 d_{CH}^4\right] \tag{2.29}$$

To ensure that the cluster head has enough energy to continue sending data to the base station, the minimum energy required is

$$E_{\max} \geq 2\left(n + (n+1)\left[1 + \alpha + \beta_4 d_{CH}^4\right]\right) \tag{2.30}$$

Hence, whenever the low level energy of any cluster head becomes less than $2\left(n + (n+1)\left[1 + \alpha + \beta_4 d_{CH}^4\right]\right)$, the cluster head informs the base station. The low level energy of any λ_{CH} can be defined as

$$\lambda_{CH} < 2\left(n + (n+1)\left[1 + \alpha + \beta_2 d_{CH}^2\right]\right), \text{if } d \leq d_0 \tag{2.31}$$

$$\lambda_{CH} < 2\left(n + (n+1)\left[1 + \alpha + \beta_2 d_{CH}^4\right]\right), \text{if } d > d_0 \tag{2.32}$$

whereα, β_4 and β_2 are application specific constants such that $\sigma = \frac{E_{DA}}{E_{elec}}$, $\beta_4 = \frac{\varepsilon_{mp}}{E_{elec}}$, $\beta_4 = \frac{\varepsilon_{fs}}{E_{elec}}$. d_{CH} is the distance from the cluster head to the base station and n is the number of member nodes of the cluster head. Then each cluster head determines its low level energy after each cluster head selection process according to the number of members n and the distance to base station d_{CH}.

Calculation of low level energy for normal node: The low level energy for every node is adjusted after cluster head selection process. In fact, the value is proportional to the distance of the node from the cluster head. If d_n is the distance of the member node from the cluster head, then

$$E_{nresi} = E_{n\max} - E_n \tag{2.33}$$

where E_{nresi} represents the residual energy of the node, $E_{n\max}$ is the initial energy or the energy of the m ember node after the previous sending process and E_n is the energy consumed by the node during sending data to the cluster head. The node energy consumption is given by

$$\begin{cases} E_n = lE_{elec} + l\varepsilon_{mp} d^4, \text{if } d > d_0 \\ E_n = lE_{elec} + l\varepsilon_{fs} d^4, \text{if } d \leq d_0 \end{cases} \tag{2.34}$$

$$E_n = 1 + \frac{\varepsilon_{fs}}{E_{elec}} d_n^2 \tag{2.35}$$

where $\frac{\varepsilon_{fs}}{E_{elec}} = \beta_2$ and d_n is the node distance to the cluster head. To ensure that the node has enough energy to continue sending data to the cluster head, the minimum energy required is

$$E_{n\max} >= 2E_n \tag{2.36}$$

$$E_{n\max} >= 2\left(1 + \frac{\varepsilon_{fs}}{E_{elec}} d_n^2\right) \Rightarrow 2\left(1 + \beta_2 d_n^2\right) \tag{2.37}$$

So, when the energy of node (before sending data to cluster head) becomes less than $2\left(1 + \beta_2 d_n^2\right)$, the node sets specific bit in the current slot with sensed data which is ready to be sent to the cluster head in the current slot. Accordingly, the cluster head informs the base station in order to form new clusters. The low-level energy value for normal node can be defined as

$$\lambda_n < 2\left(1 + \beta_2 d_n^2\right), \text{if } d \le d_0 \tag{2.38}$$

$$\lambda_n < 2\left(1 + \beta_2 d_n^2\right), \text{if } d > d_0 \tag{2.39}$$

where β_2, β_4 are as defined before and d_n is the node distance to the cluster head.

During the first $K(N \times P)$ rounds, the protocol performs only cluster head selection from the existing clusters, and starts cluster reformation in the subsequent rounds. Hence, the protocol avoids the time and energy consumed for clustering. Furthermore, clustering depends on the current cluster heads state. It works on demand, not in each round as in LEACH. It in turn, decreases the overhead caused by dynamic clustering in each round. Finally, most important and critical issue controlled by the low-level energy, is the length of the steady-state period, because it is very critical how to set the time length of each round to prolong the life time and increase the throughput. In this approach, it does not need to define the data transmission period in advance.

2.2.3 *Simulation and Result*

The simulation environment is 100 × 100 meters sensing field containing 100 nodes deployed randomly over the area with the base station located in the centre of the sensing field. The network performance and life time are evaluated according to number of rounds till all nodes die, total data sent to base station during simulation time and time till first node dies out. Figure 2.8 shows the centroids of clusters obtained by *K*-means clustering method at the beginning of the simulation.

Table 2.1 shows the dead nodes, stability period, life time and the data sent to base station for LEACH. This algorithm has two cases: energy balanced periodic clustering and energy balanced on demand clustering. In the energy balanced periodic clustering, *K*-means clustering and cluster head selection are done every round according to the length of data transfer period or until cluster head die.

From Fig. 2.9 and Table 2.1, the network lifetime and the stability period for the algorithm are both twice more than LEACH. There is a significant improvement in the network lifetime and performance due to the clustering technique, minimum communication cost cluster head selection technique and on demand cluster head reselection. The minimum energy clustering technique optimizes the distance between cluster heads and members reducing energy consumption and changes of signal strength. The selection technique of cluster head selects the optimized node as head node, which has the minimal cost in terms of energy consumption while communicating with others.

Fig. 2.8 Sensing field including base station and centroids

Table 2.1 Simulation results

Algorithm	Dead nodes	Stability period	Rounds	Data sent to base station
LEACH	100	101	292	1543
Fixed clustering	100	130	560	2093
Proposed algorithm	100	221	700	2265

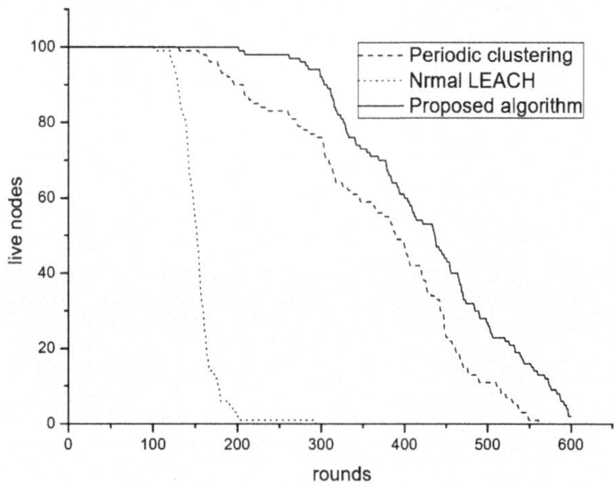

Fig. 2.9 Live nodes vs. rounds

Moreover, on demand reselection of cluster head controls the time length of the steady-state period and permits previous cluster head nodes to work as normal nodes after their energy decreased to the low-level value. Figure 2.10 shows the increase in the total data sent to the base station. It increases approximately by 47% than LEACH as shown in Table 2.1. The reason is that lifetime of the network is extended in the algorithm such that it survives for longer time than in LEACH. Figures 2.11 and 2.12 show the variation of the network energy. In the algorithm, the nodes do not run out of energy suddenly and fast. That is because the former cluster head will function as normal node when a new cluster head is selected. By not performing the clustering process in each round, the number of cluster head selections in algorithm is much less than that in LEACH.

During the network lifetime of the algorithm, by increasing the number of rounds the number of cluster head selection process increases progressively. The reason is that by decreasing energy of nodes, the number of clustering that should be held increases. Furthermore, the low-level energy technique overcomes the problem of data transmission duration, since the length of the steady-state period is critical to

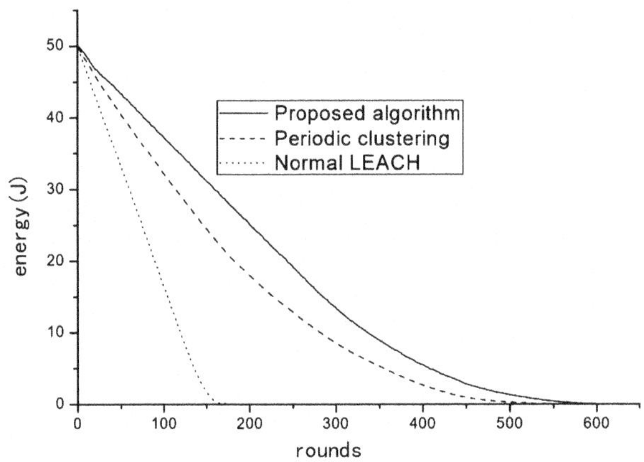

Fig. 2.10 Energy vs. rounds

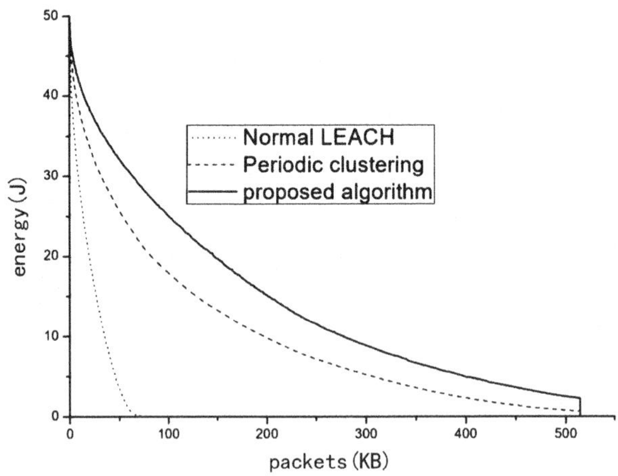

Fig. 2.11 Energy vs. packets

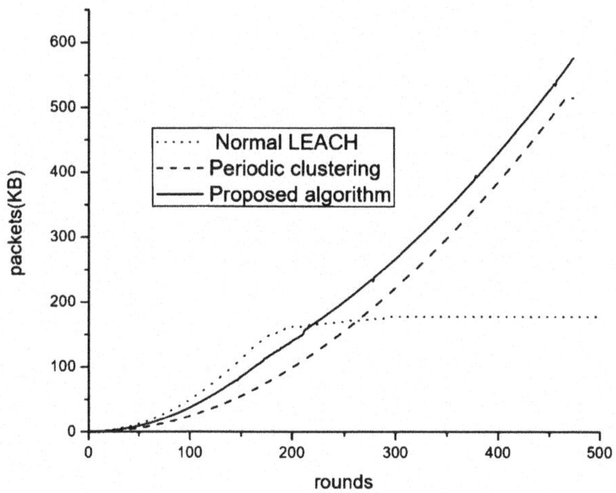

Fig. 2.12 Packets vs. rounds

achieve the energy reduction necessary to offset the overhead caused by the selection process of cluster head. Short steady-state period increases the protocol's overhead, where as long period may lead to cluster energy depletion. From the above results, the algorithm prolongs the lifetime and the stability period of the network, hence, improves the energy and performance efficiency of the wireless network.

The usage of these techniques leads to formation of minimum energy clusters and avoidance of network's holes. The approach does not require clustering every round, but it depends on current cluster head residual energy, which reduces the overhead cost. Moreover, it overcomes the problem of predefinition of the steady-state duration, noting that, the lifetime and throughput of the network are highly affected by the time length of each round.

2.3 Connectivity-Guaranteed and Energy Efficient Protocol

In this section, Connectivity-guaranteed and Energy-efficient Clustering Scheme (CECS) [2] is proposed. The core of CECS is to guarantee the connectivity of the network and save energy cost. It considers the communication range of each sensor node and guarantees the connectivity of the network. Each node only communicates with a close neighbor directly, which can save energy cost.

2.3.1 Connectivity of LEACH

We assume that there is no isolated node in the network. The election of the cluster-head cannot guarantee the connectivity of the network because some nodes maybe cannot communicate with any cluster-head node. Based on the assumptions, we analyze the connectivity of LEACH using mathematical method and propose Theorem 2.1 and Lemma 2.1.

Theorem 2.1 *Let the deployment density of the nodes be* λ*, the area of the sensing field be* S*, and the communication radius of each node be* r*, then the probability denoted by* $P(\lambda, r)$ *that a node cannot communicate with all other nodes in the network is*

$$P(\lambda, r) = \left(1 - \frac{\pi r^2}{S}\right)^{\lambda S - 1} \tag{2.40}$$

When $\pi r^2 \ll S$

$$P(\lambda, r) = \mathrm{e}^{-\lambda \pi r^2} \tag{2.41}$$

Proof As we know that, a sensor node can only communicate with the nodes located in its communication range. Thus, for node i, the probability that node j can communicate with Node i is

$$p_{ij} = \frac{\pi r^2}{S} \tag{2.42}$$

So, the probability that Node j cannot communicate with node i is

$$\overrightarrow{p_{ij}} = 1 - \frac{\pi r^2}{S} \tag{2.43}$$

Since these events that node i cannot communicate with all the remaining nodes are independent, the probability that node i cannot communicate with all the remaining nodes is

$$P_i = \prod_{j=1}^{n-1} \overrightarrow{p_{ij}} = \left(1 - \frac{\pi r^2}{S}\right)^{n-1} \tag{2.44}$$

where n represents the number of the sensor nodes, and $n = \lambda S$, so we have Eq. (2.40).

When $\pi r^2 \ll S$,

$$P(\lambda, r) = \lim_{S\to\infty}\left(1 - \frac{\pi r^2}{S}\right)^{\lambda S-1} = e^{-\lambda\pi r^2} \tag{2.45}$$

So the Theorem 2.1 holds.

Lemma 2.1 *Lemma 2.1 is under the same conditions of Theorem 2.1. In LEACH, let the proportion of cluster-head node be p, the probability that a node cannot communicate with any cluster-head node is*

$$P_{LEACH-uc}(\lambda, r, \rho) = \left(1 - \frac{\pi r^2}{S}\right)^{\rho\lambda S} \tag{2.46}$$

When $\pi r^2 \ll S$

$$P_{LEACH-uc}(\lambda, r, \rho) = \mathrm{e}^{-\rho\lambda\pi r^2} \tag{2.47}$$

Proof We can easily get the above result from the proof of Theorem 2.1.

From Lemma 2.1, if r or λ is too small in LEACH, the probability that the network is not connected could be large. Here, CECS will be an improved routing algorithm to solve this problem.

2.3.2 Connectivity-Guaranteed Protocol

CECS is a clustering-based routing protocol that can guarantee the connectivity of the network. In order to save energy and prolong the network lifetime, we consider the energy information and distance information to select some special sensor nodes. This protocol is used for static network, and there are some assumptions of the sensor network environment.

1. Sensor nodes are randomly deployed.
2. The communication range of each node is a standard disc. A sensor node can only communicate with those nodes in its communication range.
3. Each node periodically senses its nearby environment and would like to send data to the sink.
4. Sensor nodes are energy constrained.
5. There is no isolated node in the network which is the basis of the connectivity of the network.
6. There is no mobility in the network.

CECS is based on the above assumptions. In CECS, besides the cluster-head node, we define two kinds of nodes called entrance nodes and exit nodes, which are used to connect one cluster to another. Being different from LEACH, the connectivity of the network is guaranteed with the special election strategy of cluster-head nodes, entrance nodes and exit nodes. When selecting these sensor nodes, we also consider a number of related factors such as remaining energy and distance information in order to make it more energy-efficient.

As the same with LEACH, the whole process of CECS is divided into several rounds. There are three phases in each round, which are cluster set-up phase, network set-up phase and data transmission phase.

2.3.2.1 Cluster Setup Phase

Initially, each node has a weight of being cluster head node and $cw(i)$ denotes the weight of node i. During the cluster setup phase, whether one node becomes a cluster-head node is decided by the following two situations:

1. If the weight of node i is the biggest of its neighbors, then node i is elected to be a cluster-head node, for example, in Fig. 2.13a.
2. We assume node j is one of the neighbors of node i, if the weight of node i is the biggest of all the neighbors of node j, then node i is elected to be a cluster-head node, for example, in Fig. 2.13b.

The numbers in Fig. 2.13 represents the weights of the nodes. The solid nodes are elected to be cluster-head nodes.

Because one of the functions of cluster-head node is fusing data, those nodes that have fewer neighbors are selected. Thus, we define $cw(i)$ as

$$cw(i) = \frac{1}{\text{num}(i)} \tag{2.48}$$

where num(i) represents the number of neighbors of node i.

After the cluster-head nodes are all selected out, the non-cluster-head nodes keep their receivers on to hear the broadcast information of all the cluster-head nodes, and

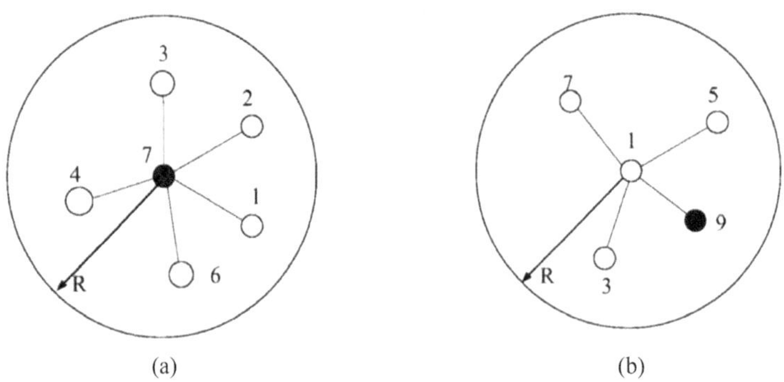

Fig. 2.13 The two situations for the sensor node of being elected to be a cluster-head node (**a**) node i is the biggest one for its neighbours (**b**) node i is the biggest one for all neighbours of node j

choose which cluster they belong to in this round. This decision is based on the received signal strength of the broadcast information. Any non-cluster-head node will belong to the nearest cluster-head node. Then the clusters are set up.

2.3.2.2 Network Setup Phase

In the network set-up phase, only one node is selected as gateway, which is used to send data to the sink. It is similar to the function of the leader in PEGASIS. After the clusters are set up, we need to select two kinds of nodes called entrance nodes and exit nodes to connect the clusters. The two steps of network set-up phase are as follows:

Step 1. Considering the remaining energy information and position information of each node, we select one node as gateway, which has more energy and is nearer to the sink. The priority that node i could become a gateway is denoted by $gw(i)$, that is

$$gw(i) = \text{energy}(i) - \alpha * d(i)^2 \tag{2.49}$$

where energy(i) represents the remaining energy of node i. $d(i)$ represents the distance from the sink. a is a parameter.

Step 2. At the beginning, we regard the cluster, which consists of the gateway as a target cluster. All the remaining clusters connect to the target cluster if they can connect to it. After that, the target cluster sleeps, and the clusters that just connect to the target cluster become the new target clusters. The remaining clusters continue to try to connect to the new target clusters until all the clusters are connected. Considering the clusters being connected, the key point is how to select the exit nodes and entrance nodes to connect these clusters. The two types of nodes are in pairs and they are in the different clusters as shown in Fig. 2.14. For any cluster, there is only one exit node, but there may be several entrance nodes since it may be connected to several other clusters. The two types of nodes involve in data fusion and the exit nodes involve in data transmission, so the

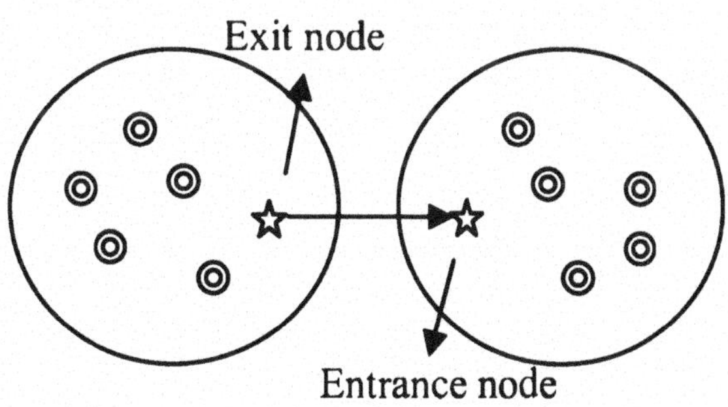

Fig. 2.14 The exit node and the entrance node in different clusters to link two clusters

energy information and the distance information need to be considered and the priority of node *i* becoming an exit node is denoted by *ew*(*i*), that is

$$ew(i) = \text{energy}(i) - \alpha * d(i,j)^2 \tag{2.50}$$

where *d*(*i*, *j*) represents the distance between node *i* and node *j*, which are a pair of exit node and entrance node. α is a parameter.

Since we consider the energy information in the election of gateway, exit nodes and entrance nodes, this phase is dynamic in each round.

2.3.2.3 Data Transmission Phase

For every cluster, the ordinary nodes transmit data to cluster-head node. The cluster-head node receives the data, fuses them with its own, and transmits the result to the exit node. Then the exit node fuses the received data with its own, transmits them to the corresponding entrance node. This process would continue until the sensed data are transmitted to the gateway. The process is shown in Fig. 2.15.

2.3.3 *Connectivity and Energy Analysis*

In CECS, the election strategy of cluster-head node, entrance nodes and exit nodes guarantees the connectivity of the network. In the cluster setup phase, according to two rules of the election of cluster head node, we can see that for any node, either it

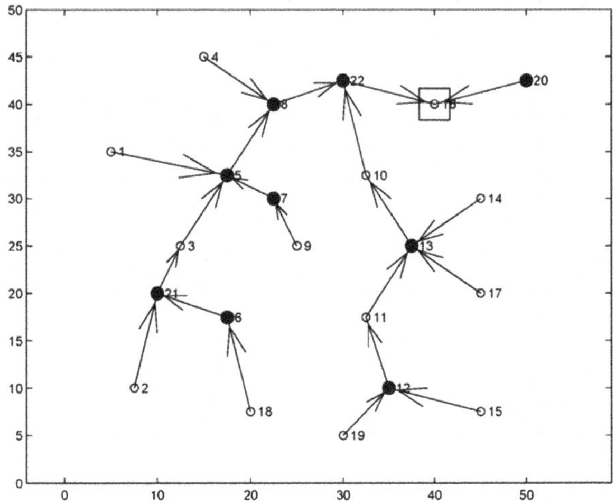

Fig. 2.15 The process of data transmission in CECS

can be elected to be a cluster head, or there is at least a cluster head in all of its neighbors. It means that all the non-cluster head nodes can communicate with at least one cluster head, that is, all the nodes can belong to a cluster.

In the network setup phase, we start from a target cluster that consists of the gateway, which is regarded as an initial network. All the remaining clusters connect to it if they can. Then the clusters, which have just connected, to the initial network become the new target clusters. After that, the clusters, which have not connected to the network, try to connect to the new target clusters. This strategy can guarantee that all of the clusters can connect to the network.

In a word, all the nodes can belong to a cluster and all the clusters can connect to the network. As a result, CECS guarantees the connectivity of the network.

Based on the simple dissipation model of radio energy, we can analyze the energy consumption in CECS, which can be describe as.

Transmitting

$$E_{TX}(l,d) = E_{elec} * l + \varepsilon_{amp} * l * d^2 \tag{2.51}$$

Receiving

$$E_{RX}(l,n) = E_{elec} * l \tag{2.52}$$

Fusion cost

$$E_{DA}(l,n) = E_{da} * l * n \tag{2.53}$$

In the cluster set up phase, $cw(i)$ is used to decide if a node can be elected to be a cluster head node. In Eq. (2.48), it only considers the information of the neighbors. If there are no dead nodes, the clusters need not to be set up frequently. Thus the energy cost of the creation of clusters is reduced. In the network set-up phase, we use the similar method with PEGASIS to connect clusters to each other. When the entrance nodes and exit nodes are selected, we consider the distance information as well as the energy information. The energy cost is saved compared with LEACH. In Eqs. (2.49) and (2.50), there is a parameter a. According to the radio energy dissipation model, if we make linearization of the relationship of energy cost and distance, we can see that for the distance of d, it costs about $\varepsilon_{amp} * k * d^2$. So, we can calculate a more appropriate value of parameter $a = \varepsilon_{amp} * k$.

In PEGASIS, as we all know, it may produce the long-chain, which makes the energy cost large and not even. In CECS, it has been mentioned that the network set-up phase is dynamic, which can balance the energy consumption of the sensor nodes. The network set-up algorithm in CECS can also solve the problem of long-chain in PEGASIS. The following simulation shows that CECS has high performance in energy consumption.

2.3.4 Simulation and Results

To evaluate the performance of CECS, we simulate CECS, PEAGSIS and LEACH with MATLAB. And the simulation of connectivity and energy cost is as follows.

2.3.4.1 Simulation of Connectivity

In this work, we compare the connectivity of CECS with LEACH with the statistical method. At first, we ensure there is no isolated node in the network, and then we carry out the simulation under different parameters.

Table 2.2 shows that the time of connected network under different communication radius r and cluster-head node density $d_c = p * \lambda$ for a field of size 50m × 50m in LEACH and CECS. In addition, we calculate the theoretical value of LEACH using the equations we have analyzed in Sect. 2.3.3.

From Table 2.2, we can see that LEACH could not guarantee the connectivity of the network but CECS can ensure that the network is connected. If node density or the communication radius increases, the connectivity of LEACH will be better.

2.3.4.2 Simulation of Energy Cost

In order to evaluate the energy cost performance of CECS, we simulate other two protocols: LEACH and PEGASIS. The aim is to measure the total remaining energy of the network and the proportion of the dead nodes under various conditions and parameters. Then we obtain some useful results.

In the simulation model, at the beginning, we assume the size of the sensing field is 50m × 50m, the number of the nodes $n = 22$, and the communication radius of each node is $r = 15m$.

Figure 2.16 shows the percentage of dead nodes and the total remaining energy of the network with increasing number of rounds. We can obviously find that CECS has a high performance of reducing energy cost. It can reduce energy cost about 20% compared to PEGASIS.

Then, the parameters are modified to see whether CECS still has a high performance of reducing energy cost. Figure 2.17 shows that the total remaining energy of

Table 2.2 Comparison of the successful time of connected network in LEACH and CECS where the total time is set as 100

Parameters	$d_c = 2\%$ $r = 15$m	$d_c = 4\%$ $r = 15$m	$d_c = 2\%$ $r = 20$m	$d_c = 4\%$ $r = 20$m
LEACH (theoretical value)	38	62	59	83
LEACH	43	67	64	80
CECS	100	100	100	100

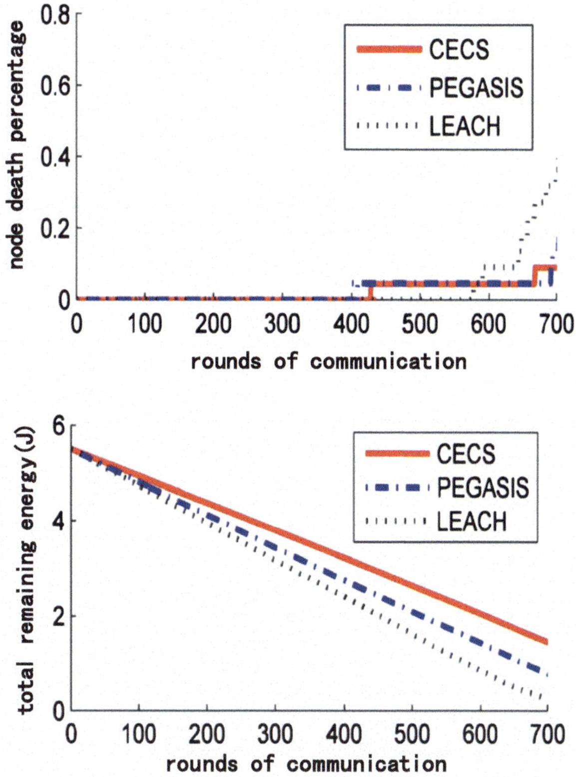

Fig. 2.16 Node death percentage and total remaining energy of the network with increasing number of rounds

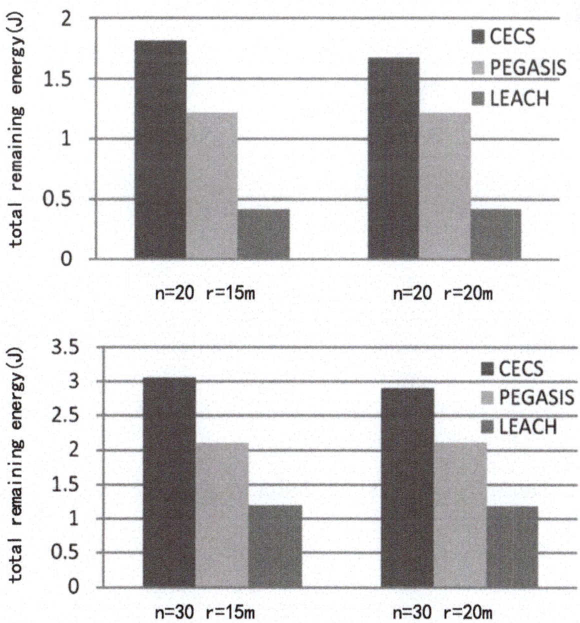

Fig. 2.17 The comparison of total remaining energy of network under different parameters. Hawaii International Conference, Jan (2000)

network after 600 rounds where the numbers of the nodes are 20 and 30 respectively, and the communication radius of the nodes are 15 m and 20 m respectively.

In the algorithm, three types of the nodes are selected. When selecting cluster head nodes, the algorithm can ensure every non-cluster-head node at least communicates with one cluster-head node, which is the basis of connectivity. Simulation results show that CECS can guarantee the connectivity of the network.

While ensuring the connectivity, in order to save energy, we consider some useful factors to select the gateway, entrance nodes and exit nodes. Simulation results show that if the communication radius is appropriate, CECS can reduce energy cost about 20% compared to PEGASIS protocol.

2.4 Density-Aware Clustering Protocol Based on LEACH-C

The transmission distance has notable effects on energy consumption. Most existing clustering algorithms are proposed without considering the density and center of region. Thus, density-aware clustering technique based on region density is proposed to form clusters, in addition to energy-aware method to select cluster heads.

This section presents an improvement to LEACH-C protocol in two different forms [3, 4]. For the first form, density-aware clustering algorithm based on region density utilizes fuzzy clustering technique in cluster formation. For the second form, the selection method of cluster head depends on intra and inter-communication distances in addition to residual energy. It is a cluster-based, centralized and single-hop routing method. The algorithm does not require the number of cluster head to be predefined and it depends mainly on clustering radius. The density-aware algorithm effectively extends the network lifetime by taking the advantage of balanced energy consumption and optimized communication distances.

2.4.1 Density-Aware Clustering Protocol

Considering a single-hop cluster-based WSN that consists of thousands of sensors scattered in a sensing field, the base station is located in the middle of the field. The sensing field consists of several clusters, each of which has one cluster head that controls the coordination between members and transmits data to the base station. The energy radio model is same as LEACH. The system components are based on the assumptions stated as follows.

1. Sensor nodes are location-aware, energy-constrained and unattended after dispersion.
2. The sensors are randomly deployed. It is further assumed that the traffic load is lighter since the network is not so large.

3. All sensors can transmit with enough power to reach the base station if needed, and the nodes can vary their transmission power.
4. The base station is located at the center of the sensing field.
5. Nodes and the base station are both stationary.
6. The clustering is of single hop.
7. The base station has unlimited resources and wide transmission range coverage, hence it can use a single broadcast to reach all the sensor nodes within the sensing field.

Following the assumptions above, a density-aware clustering algorithm for WSNs using fuzzy clustering technique in cluster formation is proposed. The selection method of the cluster head depends on intra and inter-communication distances in addition to residual energy. The following sections discuss this protocol and its algorithm in detail. As same with LEACH-C, the algorithm is divided into two steps named as setup phase and steady state phase, respectively. The setup phase is the stage in which the network prepares itself to start the process of data transmission. It includes cluster formation, cluster head selection and TDMA schedule assignment for member nodes. In the steady state phase, the cluster head node uses the time slot allocated to each node by TDMA algorithm for data transmission.

2.4.1.1 Setup Phase

Clustering formation: Balanced energy consumption has direct effect on the network life time, since energy load is evenly distributed among nodes. For that reason, the region density is considered in cluster formation. The subtractive clustering method is a well known fuzzy clustering technique. It assumes each node is a potential cluster center. Also, it calculates a measure of the probability that each node would define the cluster center based on the density of neighboring nodes as stated below

$$D_i = \sum_{j=1}^{n} e^{\left(\|x_i - x_j\|^2/(r_a/2)^2\right)} \tag{2.54}$$

where D_i is the density function of node i, x_i and x_j are the coordinates of the node i and node j. r_a is radius which is employed to define a neighbor area or neighborhood.

It is reasonable to assume that the nodes have no influence on the density of node i when they are outside of neighbor area. After the density measure of each node has been calculated, the node with the highest density measure is selected as the first cluster center. Let x_{c1} be the node selected and D_{c1} its density measure. The density measure for each node x_i is revised by the formula

$$D_i <= D_i - D_{c1} e^{-\|x_i - x_c\|^2 / r_b^2 / 4} \tag{2.55}$$

where r_b defines neighborhood where the density function of node is reduced. Therefore the nodes near the first cluster center x_{c1} have significantly reduced density measure, thereby making the nodes unlikely to be selected as the next cluster center. The constant r_b is normally larger than r_a to prevent overlapping of adjacent clusters. After the density measure of each node is revised, the next cluster center x_{c2} is selected. Then the density measures of all nodes are revised again. This process is repeated until all or most of the nodes (above 80%) are within radius of cluster centers.

Cluster Head Selection: Considering each cluster, the base station selects the suitable cluster head according to a factor that maximizes the network lifetime. The factor is a combination of the residual energy of each node and the required energy for each node to play the role of cluster head. Nodes to the base station inform the residual energy while the second part is calculated by the base station depending on the distances of intra and inter-communications. The energy dissipation is proportional to the squared distance between transmitting and receiving nodes. The procedure of selection of cluster head can be expressed as

$$CH = n_1^* = \arg\max LT\left[n_i\left(Er_i / \sum_{k=1}^{n} d_{i,k}^2 + d_{i,BS}^2\right)\right] \tag{2.56}$$

where n_i denotes the i-th alternative sensor node, T is the lifetime of the WSN. Er_i is the residual energy of the i-th node. $d_{i,\ k}$ is the distance between node i and node k in the same cluster. $d_{i,BS}$ is the distance between node i and the base station.

The node with highest factor CH is preferred to be the cluster head. Then with high residual energy and small summation of communication distances, the node has high CH criteria and high priority to be cluster head. Therefore the equation can be rewritten as

$$CH = \arg\max\left(Er_i / \sum_{k=1}^{n} d_{i,k}^2 + d_{i,BS}^2\right) \tag{2.57}$$

TDMA Schedule Assignment: After finishing cluster head selection, the base station broadcasts notification message including clusters and cluster head IDs of each cluster. The cluster heads send TDMA schedules to their members and the network goes to the steady state phase. In this process, the non-member nodes send their data to the base station directly.

2.4.1.2 Steady State Phase

In this phase, the normal nodes collect and send data to cluster heads or base station. Cluster heads always keep their receivers turning on to receive data from their members, while the members wake up only during their assigned time-slots according to the TDMA schedule. Cluster heads aggregate data received from their own members and send to the base station. When any cluster loses its cluster head, the network goes to the setup phase to select new cluster heads for the clusters.

2.4.2 Simulation and Results

In the simulation environment the network consists of 100 sensor nodes, randomly deployed in 100 × 100 meters monitoring area. The transmitting power of each node is adjustable and the propagation radius is 18 meters while the base station is fixed and located in the center of the sensing field. Message exchanging scenario is expressed in the setup phase and steady state phases. The metrics to evaluate the network lifetime of WSN are time taken or number of rounds until all nodes die, packets sent to the base station within this time and stability period, which is the time before the first node dies.

Figure 2.18 shows the number of live nodes over the simulation time that is measured by the number of rounds. As we can see from the figure, the algorithm increases the number of rounds by approximately 40% than the normal LEACH-C which means increasing the life time of the network. The reason is that in the density-aware technique we optimize the distances from nodes to cluster heads and the distances from cluster heads to the base station. Therefore, by minimizing of the traversed distances we decrease the energy consumption, which lets more nodes to stay alive for longer time. In fact, as we minimized the traversed distances by the data, the clustering radius is an important factor since it determines the number of the clusters, hence the number of cluster heads. Small clustering radius

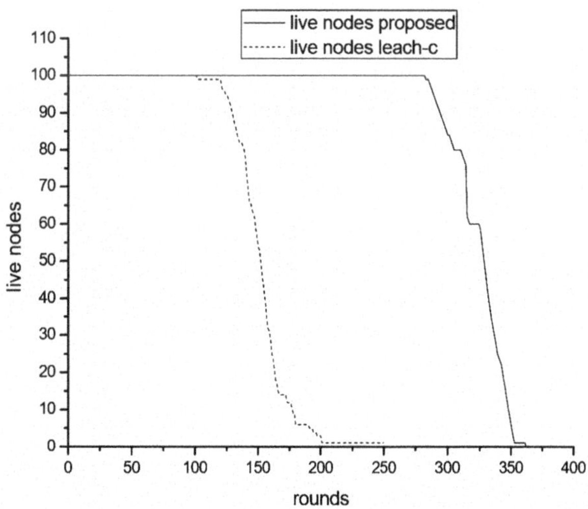

Fig. 2.18 Live nodes over simulation time

means more clusters and vice versa although clustering radius mainly can be controlled and defined according to the dimensions of sensing area. The cluster head number has a great impact on the energy consumption in the network.

Figure 2.19 shows the packets sent to the base station over the simulation time. It is clear that the total number of packets received by the base station in the algorithm is nearly twice the total packets received by the base station in the normal LEACH-C. This demonstrates the dramatic improvement in the network performance with the proposed clustering algorithm.

Figures 2.20 and 2.21 show the energy variation over the number of packets received at the base station and over the number of rounds respectively. It is obvious that new algorithm does not exhaust the network suddenly and quickly, because the density-aware clustering technique minimizes the energy consumption within the cluster and evenly distributes the energy consumption over all cluster nodes.

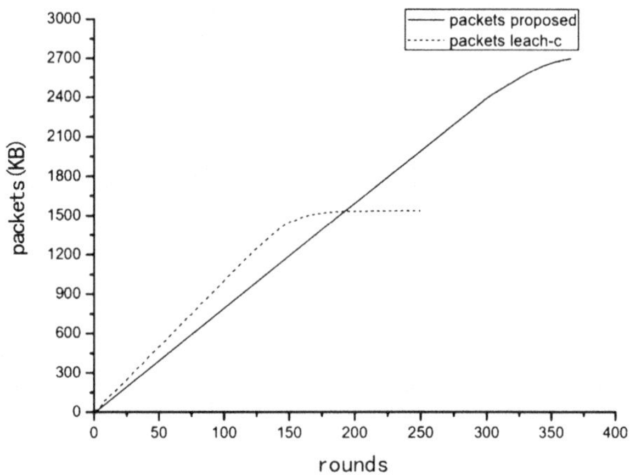

Fig. 2.19 Packets sent to base station over simulation time

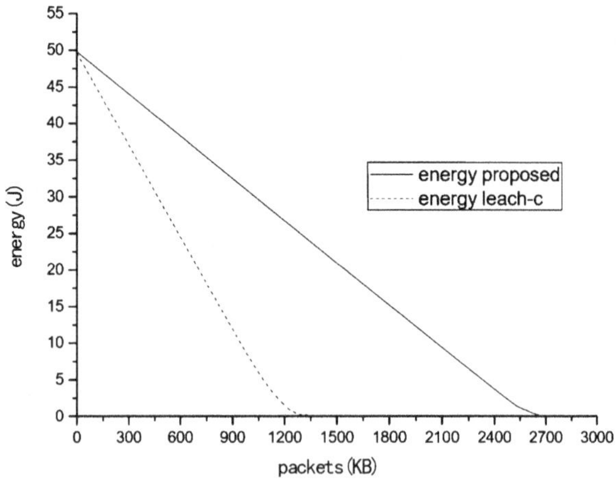

Fig. 2.20 Energy over packets sent to base station

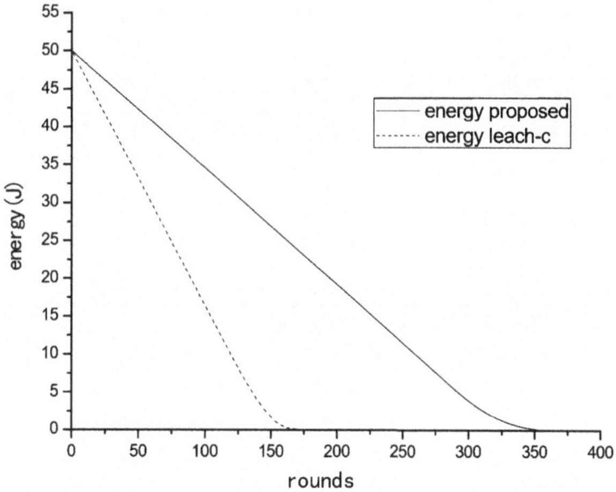

Fig. 2.21 Energy over the number of rounds

Moreover, the selection method of the cluster head is energy-efficient, because it is based on the energy required for each node when it is assumed the cluster head, which leads to the selection of the most suitable node.

From the above results, the density-aware method prolongs the stability period and lifetime of the wireless network, thereby improves the efficiency of the network.

The simulation results show that the algorithm can balance the load between nodes, reduce energy consumption and prolong the stability period and lifetime of the network compared with LEACH-C. The algorithm does not require the number of cluster head to be predefined and it depends mainly on clustering radius.

In fact, the density-aware algorithm effectively extends the network lifetime by taking advantage of balanced energy consumption and optimizing communication distances. Our simulation results show the effectiveness of the density-aware algorithm. It can achieve an increase by 40% in lifetime and about 90% in total packets sent to the base station compared with the normal LEACH-C. However, further studies are required including irregular density, excluded nodes, sudden death of cluster head, etc. Better performance is achievable by increasing the number of nodes due to density increment in sensing field. The algorithm can combine with multi-hop, shortest pass or negotiation-based routing to reduce energy consumption much more and increase scalability.

2.5 The Hierarchical AODV Protocol

In this section, the classical on demand routing protocol (AODV) is firstly presented in order to proposed the Hierarchical-AODV (HAODV) [5]. HAODV is based on AODV and mainly utilizes the hierarchical thought to generate a hierarchical tree-based protocol. First, the clustering algorithm is employed to avoid network congestion and generate a tree-based topology structure. Then, the gateway-auxiliary algorithm works to avoid the cluster heads from becoming the bottlenecks. The simulation results demonstrate that it works well in large-scale WSNs.

2.5.1 Working Procedure of AODV

WSNs are the wireless networks composed of Ad-Hoc sensors with limited sensing, computing and communication capabilities. AODV, which is a simple and efficient routing protocol, is quite practical and easy to implement in WSNs. The working procedures of AODV are simply described in Fig. 2.22 [6].

The source node would broadcast Route Request (RREQ) message when it does not have a route to the destination node. After receiving RREQ message, a node returns a Route Reply (RREP) message to the source node if it has a route to the destination node, and then the routing is established successfully. Otherwise, it forwards the broadcasting message to its neighbor nodes. The above process repeats until the destination node receives this message. If there are any errors from communication channel, one Route Error (RERR) message is sent to the neighbor nodes and finally return to the source node. Consequently, the corresponding route built along the way is removed and the route discovery process is restarted. This routing protocol is a kind of on demand routing, which establishes the route only when it needs to send data. However, the main disadvantage of AODV is that the flooding mechanism is used to seek the destination node, which is not efficient.

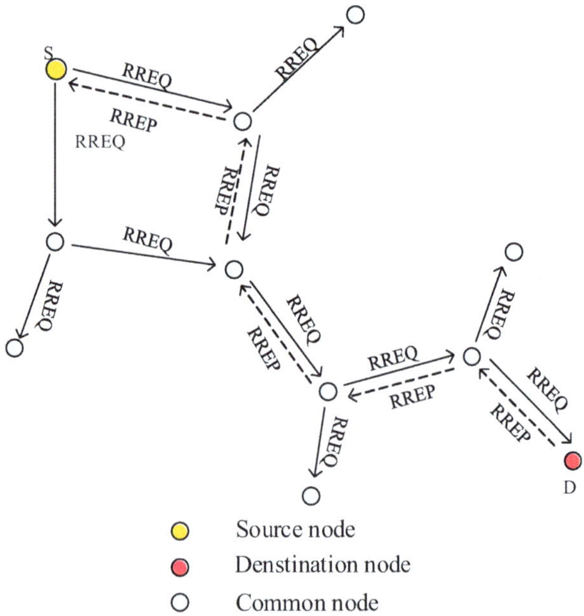

Fig. 2.22 Route discovery of AODV

2.5.2 *The Hierarchical AODV Protocol*

In this protocol, the clustering algorithm is employed to control flooding. A new election rule called ‘first declaration wins’ is used to select cluster heads and gateways. Meanwhile, gateways can be used to link two different clusters and share the task of cluster heads. The gateway-auxiliary algorithm avoids cluster heads from being bottlenecks and reduces transmission delay. In this protocol, the energy model uses the radio energy model.

In this algorithm, some assumptions for the protocol are made as follows.

1. Nodes are randomly distributed in a certain detecting region.
2. The source nodes and destination nodes are selected randomly.
3. Every node has the ability to forward and process data packets.
4. Nodes move to the targets at a speed randomly selected from 5 m/s to 20 m/s.
5. Nodes are isomorphic with a certain initial energy.

Similar to LEACH algorithm, we divide the protocol into four stages, which are selecting cluster head phase, selecting gateway phase, routed discovery phase and route maintenance phase, respectively, as shown in Fig. 2.23.

2.5.2.1 Selecting Cluster Head Phase

A new rule is used to select the cluster head, called ‘first declaration wins’. The node becomes the cluster head and ‘rules’ the rest of nodes in its clustered area or radio coverage, which sends a packet first. The specific process is depicted as follows. Each node exchanges information with the neighborhood nodes to check whether there is a cluster head. If one cluster head exists among its neighbors, the node joins the cluster. Otherwise, the node becomes the cluster head and ‘rules’ the rest nodes in its clustered area or radio coverage.

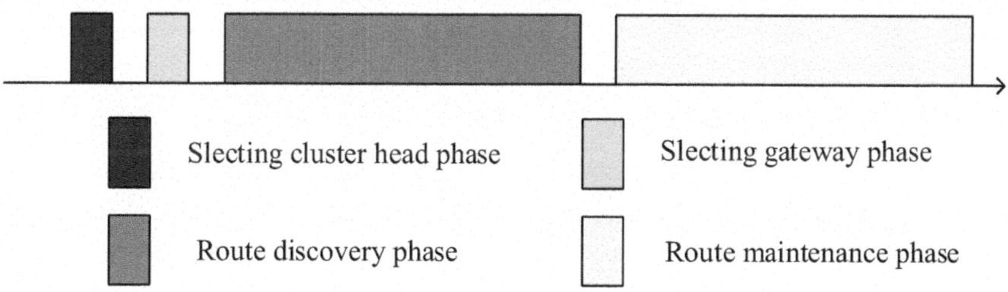

Fig. 2.23 Four parts in the proposed protocol

2.5.2.2 Selecting Gateway Phase

In the phase, only one gateway is chosen between two neighboring clusters, so that the proposed protocol can effectively reduce the flooding of control packets. Generally, two neighboring cluster heads can be connected directly through a common gateway by one hop. When two neighboring cluster heads cannot communicate directly, they have to be reached through their neighboring cluster by three hops. However, if there are two nodes, which are neighbors and separately belong to two clusters, the two cluster heads can be connected through the two nodes by two hops.

1. The selection of common gateway. As shown in Fig. 2.24, node *N* belongs to cluster 1 and cluster 2, and it has the shortest average distance to all its neighbors, so it becomes the common gateway. Then, it sends Gateway Message to all its neighbors and the two cluster heads. The neighbors will not become common gateway of the two clusters 1 and 2, which also belong to the two clusters, but have the longer average distance to all of its neighbors.
2. The selection of distributed gateway. It is shown that the clusters 2 and 3 cannot be directly connect in Fig. 2.24. Generally, they have to be connected through cluster 1. But, there are two neighboring nodes A and B, which respectively have the shortest average distance to all of its neighbors. Moreover, nodes A and B

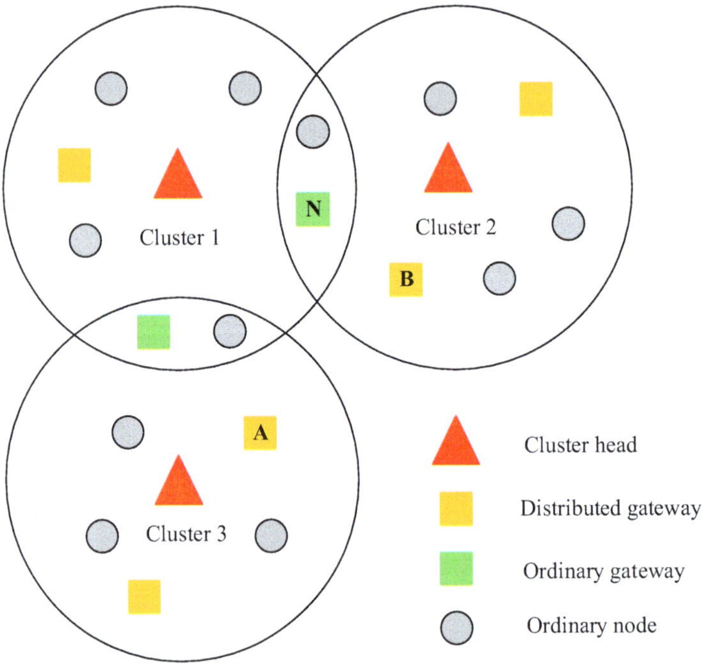

Fig. 2.24 The selection of gateways

separately belong to clusters 3 and 2. So, nodes A and B become the distributed gateways of clusters 2 and 3. Then, they send Distributed Gateway Message to the two cluster heads (2, 3) and all their neighbors. The neighbors will not become distributed gateways of the two clusters 2 and 3, which also meet the requirements, but have the longer average distance to all of its neighbors.

2.5.2.3 Route Discovery Phase

When a source node has data packets to send to the destination node, it first checks the destination node in its neighbor list. If the destination node is its neighbor, it directly sends data packets to the destination node. Otherwise, it checks its routing table to determine whether it has an active route to the destination node. If that is the case, it sends data packets through the route. Otherwise, the source node must perform a route discovery procedure to acquire a route to the destination node.

The source node first creates its RREQ packet and sends the RREQ packet to its cluster head. After the cluster head receives RREQ packet, it checks the destination node in its neighbor list, cluster member list and gateway list. If the cluster head finds the destination node in these lists, it responds RREP packet to the source node. Otherwise, the cluster head would insert some necessary information (ID) to RREQ packet, and send the RREQ packet to the gateways. These gateways check the destination node in their neighbor list and cluster head list. If the gateways find the destination node in these lists, they respond RREP packet to the source node. Otherwise, they would insert some necessary information (ID) to RREQ packet, and send it to their cluster heads. After these cluster heads receive RREQ packet, the cluster heads check the destination node in their neighbor lists, cluster member lists and gateway lists. If they find the destination node in these lists, they would respond RREP packet to source node. Otherwise, they would insert some necessary information (ID) to RREQ packet, and send it to their other gateways (except the above gateway). By that analogy, the destination node can be found. The nodes update their routing tables, when they receive RREQ and RREP packets. As shown in Fig. 2.25, an example is presented to show how the nodes generate and update their routing tables.

From the example, it is found that the backward route is established during the process of transmitting RREQ packet, and the forward route is established during the process of responding RREP packet. In this phase, nodes forward RREQ packets only through cluster heads and gateways. If cluster heads and gateways receive them at the first time, they process RREQ packets. Otherwise, they drop them out. When the source node receives many RREP packets, it gets many routes to the destination node. It chooses the route generated by gateways, when its hops are less than that generated by cluster heads. The above mechanism makes gateways share the task of cluster heads.

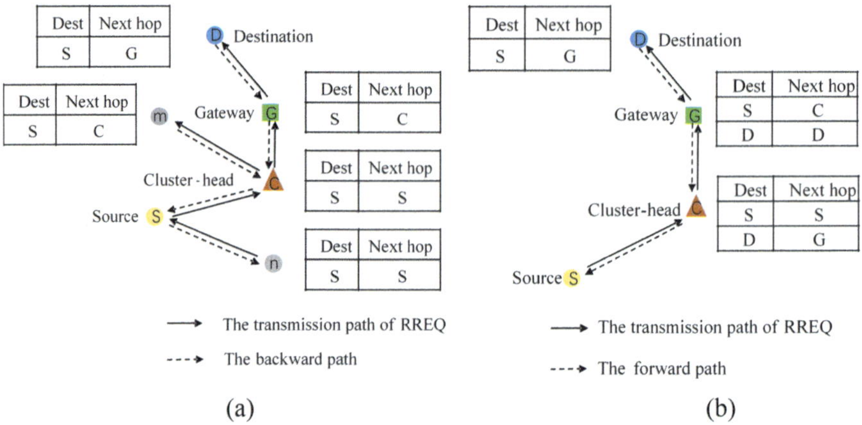

Fig. 2.25 How to generate and update the routing tables. (**a**) receive the RREQ packet. (**b**) receive the RREP packet

2.5.2.4 Route Maintenance Phase

Each node periodically sends HELLO message to confirm the existence of its neighbors. In this way, they can maintain the route periodically. Route maintenance mechanism includes four parts: routing table maintenance, cluster head maintenance, gateway maintenance, and cluster member maintenance.

1. Routing table maintenance. In the proposed protocol, the way of maintaining routing table is same as AODV. Nodes only maintain the active routes. The specific process is depicted as follows.
   Every node checks the validity of routes through sending HELLO message to its neighbors. When it does not receive response from a neighbor in a certain time, it gets that the neighbor has already moved or been dead. Then, the broken link is marked. The node checks the cluster state of its neighbor, and take corresponding measures in the following sections. When the node finds that the route to send data packet contains the broken link, it removes the routing information related to the broken link and send RRER packet to the source node. RRER packet includes the ID of unreachable destination nodes. Nodes would remove the corresponding routing information, which forward RRER packet.
2. Cluster head maintenance. When the cluster head moves, the number of its cluster members changes, and the cluster head should update the cluster member list. When there are no members in a cluster, the cluster head sets its cluster state to NULL. When its gateway, or cluster member is dead or moves to outside of its clustered area or radio coverage, the cluster head should reselect the gateway, or update the cluster member list.
3. Gateway maintenance. The gateway updates its cluster head list, when the cluster state of its cluster head becomes NULL or ORDINARY, or its cluster head has

been dead or moves to outside of its radio range. Gateway node may become ordinary cluster member when there is only one cluster head left in its cluster head list. Specially, the distributed gateway becomes the ordinary cluster member, when it has 0 joint gateway.

4. Cluster member maintenance. When the ordinary cluster member has no cluster head caused by the movement or the death of its cluster head, it should send message to its neighbors for joining a new cluster. If its neighbor responses to it, the ordinary cluster member joins the cluster. Otherwise, the ordinary cluster member becomes the cluster head. When the ordinary cluster member receives the packets from other cluster head, it joins the different cluster and becomes a gateway. The ordinary cluster member becomes a distributed gateway, after it receives the distributed gateway discovery control packets from its neighbor, which has a different cluster head compared with it.

2.5.3 Optimal Number of Cluster Heads

In the protocol, we assume that the cluster heads have longer distances to transmit the data, while non-cluster heads have relatively shorter distances to transmit. So the radio energy dissipation model can be redefined.

Data transmission energy cost of a cluster head is

$$E_{CH-TX} = lE_{elec} + l\varepsilon_{amp}d^4 \tag{2.58}$$

Data transmission energy cost of a non-cluster head is

$$E_{non-CH-TX} = lE_{elec} + l\varepsilon_{fs}d^2_{toCH} \tag{2.59}$$

Data receiving energy cost of a cluster head node or a non-cluster head node is calculate as

$$E_{CH-RX} = E_{non-CH-RX} = lE_{elec} \tag{2.60}$$

It is supposed that there are N nodes uniformly distributed in the $M \times M$ area. There are k clusters in the network. So, each cluster occupies M^2/k region and has N/k nodes, which include a cluster head and $N/(k-1)$ cluster members. From the radio energy dissipation model, the total energy cost of cluster heads and non-cluster heads can be achieved as follows.

The total energy cost of a cluster head is calculated by

$$E_{CH} = E_{CH-RX} + E_{CH-TX} = 2lE_{elec} + l\varepsilon_{mp}d^4 \tag{2.61}$$

The total energy cost of a non-cluster head is given by

$$E_{non-CH} = E_{non-CH-RX} + E_{non-CH-TX} = 2lE_{elec} + l\varepsilon_{fs}d_{toCH}^2 \tag{2.62}$$

Assume that the cluster head lies in the center of the cluster area and the distribution of nodes is $\rho(x, y)$, the expected square of the distance between a non-cluster head and its cluster head can be described as follows

$$\begin{aligned} E\left[d_{toCH}^2\right] &= \iint (x^2 + y^2)\rho(x, y)dxdy \\ &= \iint r^2\rho(r, \theta)drd\theta \end{aligned} \tag{2.63}$$

Assume that the region of a cluster is a circle with radius R. Obviously, the value of radius R is $M/(\pi k)^{1/2}$ As $\rho(r, \theta)$ is constant to r and θ the equation can be integrated as follows

$$E\left[d_{toCH}^2\right] = \rho\int_0^{2\pi}\int_{r=0}^{M} r^3 drd\theta = \frac{\rho}{2\pi}\frac{M^4}{k^2} \tag{2.64}$$

Assume that the node density is uniform, so the value of $\rho(r, \theta)$ is equal to $1/M^2/k$. Substitute it to the above formula, we can get

$$E\left[d_{toCH}^2\right] = \frac{1}{2\pi}\frac{M^2}{k} \tag{2.65}$$

Substitute Eq. (2.65) to Eq. (2.62)

$$E_{non-CH} = 2lE_{elec} + l\varepsilon_{fs}\frac{1}{2\pi}\frac{M^2}{k} \tag{2.66}$$

The total energy cost of a cluster is achieved by

$$\begin{aligned} E_{\text{cluster}} &= E_{CH} + \left(\frac{N}{k} - 1\right)E_{non-CH} \approx E_{CH} + \frac{N}{k}E_{non-CH} \\ &\approx 2lE_{elec} + l\varepsilon_{mp}d^4 + \frac{N}{k}\left(2lE_{elec} + l\varepsilon_{fs}\frac{1}{2\pi}\frac{M^2}{k}\right) \end{aligned} \tag{2.67}$$

The total energy cost of the network is given by

$$E_{total} = kE_{\text{cluster}} = k\left(2lE_{elec} + l\varepsilon_{mp}d^4\right) + N\left(2lE_{elec} + l\varepsilon_{fs}\frac{1}{2\pi}\frac{M^2}{k}\right) \tag{2.68}$$

Derivate the total energy cost of network with respect to k

$$\frac{dE_{total}}{dk} = 2lE_{elec} + l\varepsilon_{amp}d^4 - Nl\varepsilon_{fs}\frac{1}{2\pi}\frac{M^2}{k^2} = 0 \tag{2.69}$$

The optimal number of cluster heads is achieved by

$$n_{CH} = \frac{M\sqrt{N\varepsilon_{fs}}}{\sqrt{2\pi\left(2E_{elec} + \varepsilon_{mp}d^4\right)}} \tag{2.70}$$

The number of cluster heads can be calculated by the above equation. When the nodes are distributed uniformly in the network, the number of cluster heads is set closely as the optimal expectation value, and the whole network can be well connected and cost less energy. However, when the nodes are distributed randomly, it is difficult to determine the right number of cluster heads in the network. Under such circumstances, the network may include two parts. One has high density of node, which possesses several cluster heads. The other one has low density of node, which includes many cluster heads. Sometimes, a cluster includes only one node. Therefore, when the number of cluster heads is set as a certain value, some necessary cluster heads cannot be selected as the number of cluster heads is reached to its maximum. It can result in the network unconnected and unusable.

2.5.4 Simulation and Results

The simulation evaluates the effectiveness of HAODV protocol by comparing with AODV in terms of control overload, transmission delay and network lifetime.

The simulation environment is OMNET++ 4.0. All experiments are conducted in four randomized networks. n is 25 and area is 100 m × 100 m; n is 50 and area is 400 m × 400 m; n is 100 and area is 800 m × 800 m; n is 200 and area is 1600 m × 1600 m. All nodes are distributed randomly and the number of cluster heads is not set as a certain value in the simulation. Table 2.3 gives the simulation parameters. It is assumed that the node dies when the remaining energy is under 0.5 J. The network lifetime is defined as the time when the first node dies. Each group of sample data is with an average of 5 simulations.

Figures 2.26, 2.27, and 2.28 give the comparison of control overload, transmission delay and network lifetime, when the pause time is 15 s and 20 s. The purpose is

Table 2.3 The simulation parameters

Packet length l (bit)	Radius R (m)	E_{elec} (nJ/bit)	ε_{fs}(pJ/bit/m^2)
800	50	50	10
Initial energy (J)	ε_{mp}(pJ/bit/m^2)	Pause time (s)	Speed (m/s)
4	0.013	15,20	5 ~ 20

Fig. 2.26 The comparison of the control overload

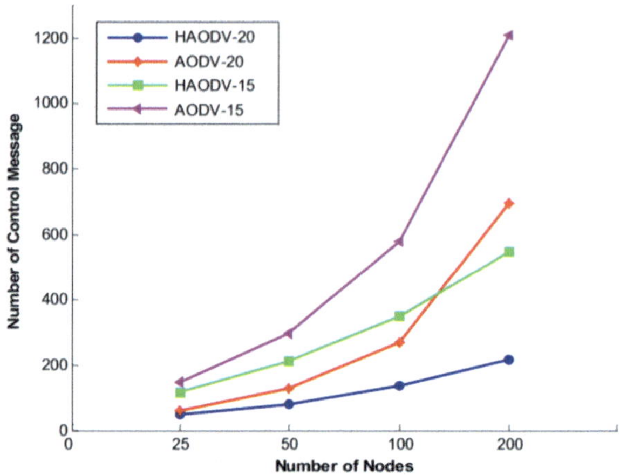

Fig. 2.27 The comparison of transmission delay

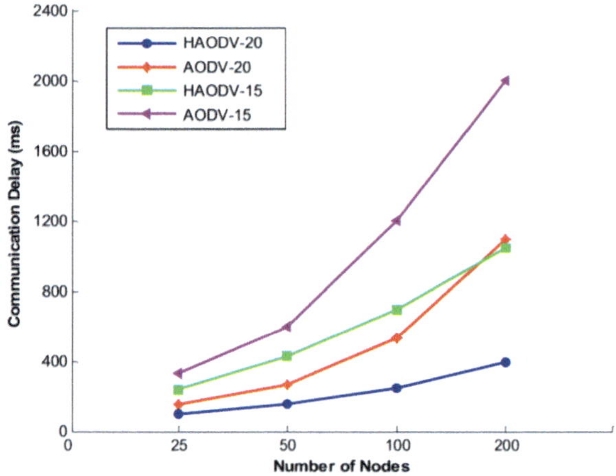

Fig. 2.28 The comparison of network lifetime

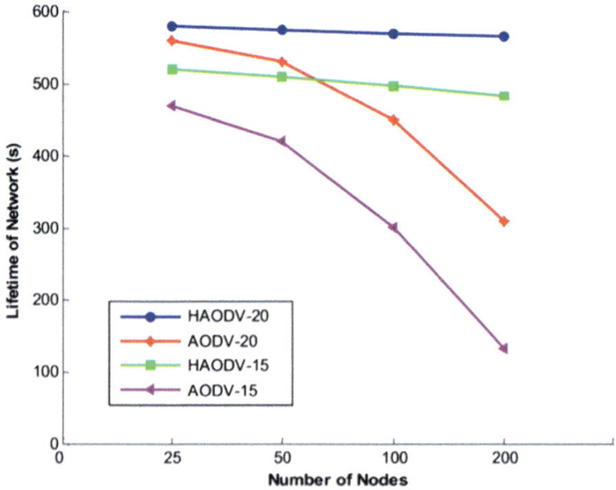

to evaluate the performance of the proposed protocol under different conditions and parameters.

2.5.4.1 The Comparison of Control Overload and Delay

Figures 2.26 and 2.27 give the comparison of control overload and transmission delay. The control overload and transmission delay of HAODV is less than that of AODV. Especially when the number of nodes increases to 200, the effect is more pronounced. The control overload and transmission delay of HAODV have respectively reduced up to 70% and 65% compared to AODV. The main reason for the difference is that HAODV employs the clustering algorithm. The algorithm controls the flooding significantly. Therefore, it can reduce the number of flooding control packets. At the same time, the transmission delay is deceased, which is caused by transmitted and received control packets. In the proposed protocol, the "first declaration wins" mechanism can fast select cluster head. And the data packets can be quickly transmitted to the destination node on the route from source node, cluster head, gateway, cluster head, gateway to destination node. This decreases the transmission delay and energy consumption. In the large-scale network, network congestion is avoided in HAODV, so that HAODV can work stably. But, the control overload and transmission delay of AODV increase rapidly, when the nodes number is close to 200. The reason is AODV uses flooding as the basic mechanism to propagate route control messages and discover routes, which can cause network congestion and make the network unavailable. It is seen that HAODV works better than AODV in the large-scale network.

2.5.4.2 The Comparison of Network Lifetime

HAODV protocol outperforms AODV in control overload and transmission delay. Further, the network lifetime in HAODV outperforms AODV. Figure 2.28 shows the comparison of the network life under different conditions. The network lifetime is longer than AODV. Especially when the node number increases to 200, the effect is more obvious. The lifetime of HAODV is up to 75% longer than AODV. One reason for the dramatic difference can be explained as that more control overload results in more energy cost. AODV has more control overload, so it costs more energy and has shorter network lifetime. Another primary reason is that the gateways share the task of cluster heads in HAODV. In this way, the cluster heads cannot cost so much energy and become the network bottlenecks and die after a short time. Therefore in HAODV, the energy consumption is balanced and reduced. It means that the lifetime of the first dead node is prolonged. When the number of nodes in the network is more than 200, it is easy to cause network congestion in AODV, which can make the nodes consume much energy and the network impossible to use. In other words, the network lifetime of AODV decreases rapidly, when the nodes number increases to 200.

2.6 Conclusion

The routing protocol of WSNs plays an important role in network performance such as control messages, network residual energy and network life time. In this chapter, we systematically study the routing protocols of WSNs. According to the characteristics of different protocols and energy imbalance, we propose some improved protocols ODCL, CECS and DACPL for the energy imbalance and connectivity in the existing LEACH protocol. In addition, we also propose HAODV, an improved ADOV protocol to improve energy efficiency and reduce the time delay in WSNs.

References

1. Khamiss A A, Chai S, Zhang B, et al. Energy-balanced on demand clustering improved LEACH protocol for wireless sensor networks[J]. Journal of Beijing Institute of Technology, 2016(3): 353-364.
2. Wen L, Zhang B, Cui L, et al. A connectivity-guaranteed and energy-efficient routing protocol for wireless sensor networks[C]. International Conference on Wireless Sensor Network, 2010: 349-354.
3. Khamiss A A, Energy-balanced clustering in wireless sensor networks[D]. Beijing Institute of Technology, 2015.
4. Khamiss A A, Chai S, Zhang B, et al. Combined metrics-clustering algorithm based on LEACH-C[C]. Chinese Control and Decision Conference, 2015: 5252-5257
5. Han H, Zhang B, Li Q. The improved routing protocol of wireless sensor network based on AODV[C]. Chinese Control Conference, 2013: 7410-7415.
6. Feng Y, Zhang B, Chai S, et al. An optimized AODV protocol based on clustering for WSNs[C]. International Conference on Computer Science and Network Technology, 2017: 410-414.

Chapter 3
Localization Technology for Wireless Sensor Networks

Network localization is a promising technology for providing high-accuracy positional information in GPS-challenged scenarios. The information is crucial in many location-based applications such as autonomous logistics, building security, as well as search-and-rescue. In this chapter, we discuss some effective localization methods of WSNs. In static sensor networks (SSNs), we present hop-count-based expectation of distance algorithm (HCED). In mobile sensor networks (MSNs), we present range-free Voronoi-based Monte Carlo localization algorithm (VMCL) and an optimal region selection strategy of Voronoi diagram based on VMCL (ORSS-VMCL). In anisotropic SSNs with holes, we present Heuristic Multi-dimensional Scaling (HMDS) and an Extended Kalman Filter Multidimensional Scaling (EKF-MDS) localization method. To solve the problem of Voronoi diagrams based localization scheme (VBLS), an optimal region selection strategy based on VBLS (ORSS-VBLS) is also proposed. Finally, We present a Delaunay triangulation based localization scheme (DBLS) and neighbor constraint assisted distributed localization (NCA-DL), which are effective in refining the distances required for localization.

3.1 Hop-Count-Based Expectation of Distance Localization

Traditional hop-count-based localization methods only take on integer value and have same distance estimation to all node's one-hop neighbors. In this section, we present hop-count-based expectation of distance (HCED) algorithm [1] in SSNs. The HCED algorithm applies the hop-count information of the unknown node with its neighbors, and the hop-count information of the reference node with its neighbors to establish a quantitative relationship between the distance from unknown node to anchor and the expectation of that.

S. Chai et al., *Wireless Sensor Networks*, Wireless Networks,
https://doi.org/10.1007/978-981-15-5757-6_3

3.1.1 System Model

In unit disk communication model, all nodes with same communication radius R are randomly deployed and formed a connected network. Let N and M denote the number of nodes and the number of anchors in a network area A respectively. Suppose the area of the network field is much larger than the area of node's communication disk, i.e., $A \geq \pi R^2$. The hop-counts h between all anchor-anchor and node-anchor pairs are known. Figure 3.1 illustrates a network of 1000 nodes randomly deployed in a region. We assume an anchor j locates at the center of the distribution disk. The hop-counts of all nodes with respect to the anchor j are represented by different markers. Let d_{wj} and d_{kj} denote the distance between node w and anchor j, the distance between node k and anchor j respectively. Both nodes w and k have the same hop-count, namely $h = 2$. If hop-count information of node w and node k is only applied to estimate d_{wj} and d_{kj}, the estimations of d_{wj} and d_{kj} should be the same, namely $d_{wj} < d_{kj}$. However, it can be observed obviously from Fig. 3.1 that $d_{wj} = d_{kj}$. Furthermore, we can also observe that the node w has more neighbors with hop-count $h = 1$ than the node k and the node k has more neighbors with hop-count $h = 3$ than the node w. According to the analysis above, there are some correlations between the distance from the anchor to the unknown node and the distribution of neighbors of the unknown nodes.

For a node with hop-count h, its communication disk intersects with three hop rings, namely hop ring $h - 1$, h and $h + 1$, then forms three disjoint regions, as illustrated in Fig. 3.2 where anchor j is represented by the triangle and the hop boundary distance is b. Let a_w^{h-1}, a_w^h and a_w^{h+1} denote the intersection regions respectively. Obviously, it can be seen from Fig. 3.2 that the areas of three regions

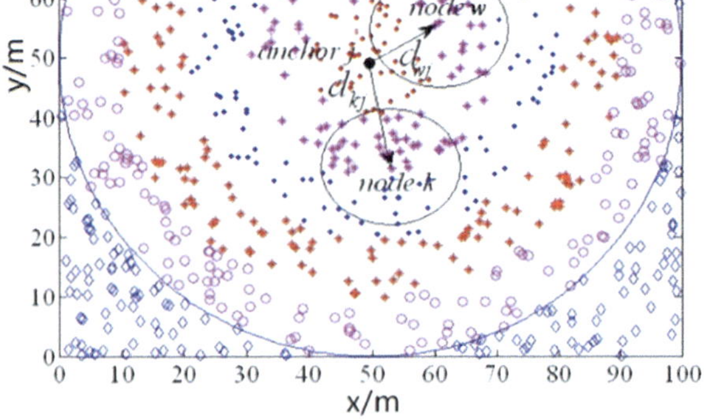

Fig. 3.1 Node distribution of different number of hop-counts

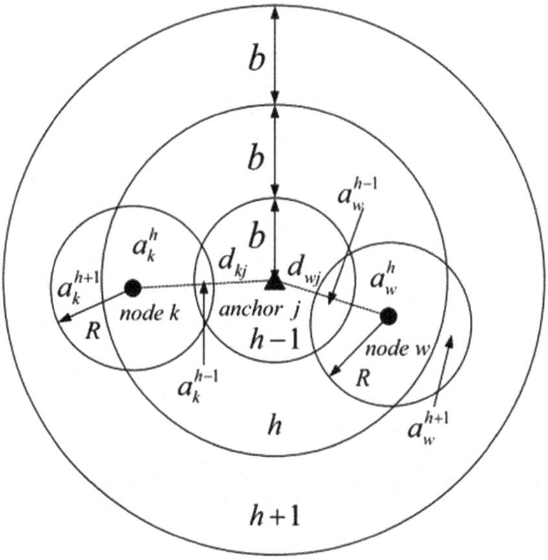

Fig. 3.2 Intersection area estimation

are dependent on the distance between the node and the anchor. If the areas of three intersection regions can be computed, we can obtain the distance between the anchor and the node using geometric equations.

3.1.2 HCED Localization

We approximate the area of an intersection region by the number of nodes falling into that region. This resembles the Hit-or-Miss Monte-Carlo integration method of given the uniform distribution assumption. It becomes more accurate when the node density increases.

Let n_i denotes the number of neighbors of anchor i, and n_i^{h-1}, n_i^h and n_i^{h+1} denote the number of nodes that falls in the intersection region a_i^{h-1} , a_i^h and a_i^{h+1} respectively. Let A_i^{h-1}, A_i^h and A_i^{h+1} denotes the true areas of the three intersection regions.

It is important to note that when the node density tends to be finite, the hop boundary distance b is unknown but satisfied $b^1 \neq b^2 \neq \ldots b^h$ in Fig. 3.3. As the areas of the three regions a_i^{h-1} , a_i^h and a_i^{h+1} are known, if we can obtain the expectation of the distance from anchor j to the region a_i^h, the inner hop boundary distance of hop ring h and the outer hop boundary distance of hop ring h can be uniquely determined. We can compute the distance between the anchor and the node. We define the distance between an anchor and a node that falls in the region of hop ring h as AN-distance (h). The expectation of the AN-distance (h) can be obtained by means of the reference node i of the unknown node k.

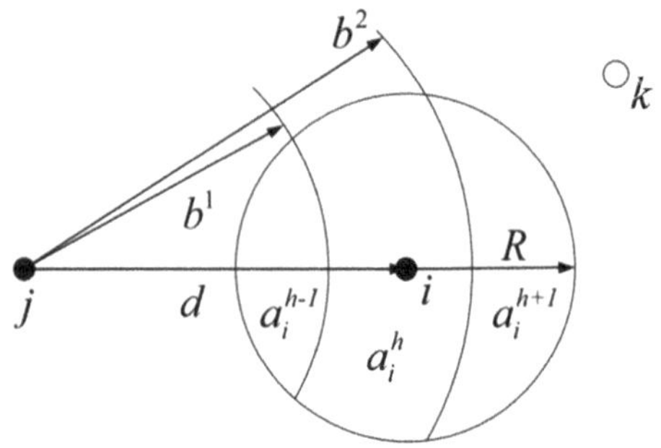

Fig. 3.3 Neighbors distribution of anchor *i*

Since N nodes are uniformly distributed in the filed with area A, the areas of the three intersection regions are given by

$$A_i^{h-1} = \frac{n_i^{h-1}}{N} \times A \tag{3.1}$$

$$A_i^{h} = \frac{n_i^{h}+1}{N} \times A \tag{3.2}$$

$$A_i^{h+1} = \frac{n_i^{h+1}}{N} \times A \tag{3.3}$$

Let d represents the distance between anchor j and anchor i. The intersection area is given by

$$\begin{aligned} A_i^{h-1} &= \left(b^1\right)^2 \cos^{-1}\left(\frac{d^2+\left(b^1\right)^2-R^2}{2db^1}\right) \\ &+R^2 \cos^{-1}\left(\frac{d^2+R^2-\left(b^1\right)^2}{2dR}\right) \\ &-\frac{1}{2}\sqrt{4d^2R^2-\left(R^2+d^2-\left(b^1\right)^2\right)^2} \end{aligned} \tag{3.4}$$

$$\begin{aligned} A_i^{h}+A_i^{h-1} &= \left(b^2\right)^2 \cos^{-1}\left(\frac{d^2+\left(b^2\right)^2-R^2}{2db^2}\right) \\ &-R^2 \cos^{-1}\left(\frac{\left(b^2\right)^2-d^2-R^2}{2dR}\right)+\pi R^2 \\ &-\frac{1}{2}\sqrt{4d^2R^2-\left(\left(b^2\right)^2-d^2-R^2\right)^2} \end{aligned} \tag{3.5}$$

We can apply the dichotomy to obtain values b^1 and b^2.

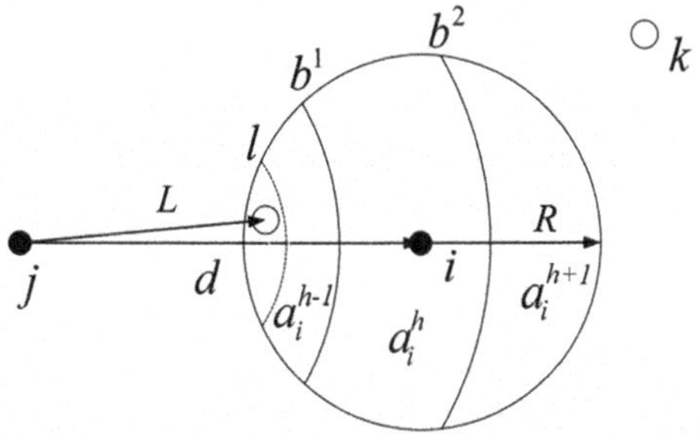

Fig. 3.4 The expectation of the AN-distance $(h - 1)$

Let L denotes the AN-distance $(h - 1)$ as illustrated in Fig. 3.4. We have

$$
\begin{aligned}
P_{h-1}[L \le l] = \Bigg[& l^2 \cos^{-1}\left(\frac{d^2 + l^2 - R^2}{2dl}\right) \\
& + R^2 \cos^{-1}\left(\frac{d^2 + R^2 + l^2}{2dR}\right) \\
& - \frac{1}{2}\sqrt{4d^2R^2 - \left(R^2 + d^2 - l^2\right)^2}\Bigg] \Big/ A_i^{h-1}
\end{aligned}
\tag{3.6}
$$

The probability density function of the AN-distance $(h - 1)$ is given by

$$
f_L^{h-1}(l) = \frac{d[P_{h-1}[L \le l]]}{dl} = \frac{2l \cos^{-1}\left(\frac{l^2 + d^2 - R^2}{2dl}\right)}{A_i^{h-1}} \tag{3.7}
$$

The expectation of the AN-distance $(h - 1)$ is given by

$$
d_i^{h-1} = E(L) = \int_{d-R}^{b^1} f_L^{h-1}(l)\, l\, dl = \int_{d-R}^{b^1} 2l^2 \cos^{-1}\left(\frac{l^2 + d^2 - R^2}{2dl}\right) dl / A_i^{h-1} \tag{3.8}
$$

Let L denote the AN-distance (h) as illustrated in Fig. 3.5. We have

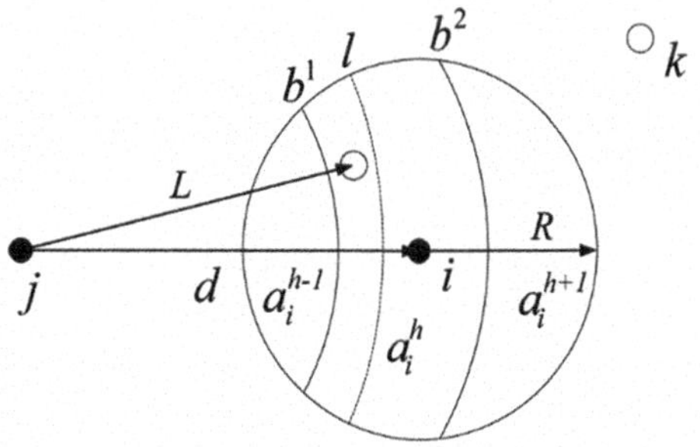

Fig. 3.5 The expectation of the AN-distance (h)

$$P_h[L \le l] = \left[l^2 \cos^{-1}\left(\frac{d^2 + l^2 - R^2}{2dl}\right) + R^2 \cos^{-1}\left(\frac{d^2 + R^2 - l^2}{2dR}\right) - \frac{1}{2}\sqrt{4d^2R^2 - \left(R^2 + d^2 - l^2\right)^2} - A_i^{h-1} \right] / A_i^h \tag{3.9}$$

where $l \le \sqrt{d^2 + R^2}$

$$P_h[L \le I] = \left[l^2 \cos^{-1}\left(\frac{d^2 + l^2 - R^2}{2dl}\right) + \pi R^2 - R^2 \cos^{-1}\left(\frac{l^2 - d^2 - R^2}{2dR}\right) - \frac{1}{2}\sqrt{4d^2R^2 - \left(l^2 - d^2 - R^2\right)^2} - A_i^{h-1} \right] / A_i^h \tag{3.10}$$

where $l \le \sqrt{d^2 + R^2}$

The probability density function of the AN-distance (h) is given by

$$f_L^h(l) = \frac{d[P_h[L \le l]]}{dl} = \frac{2l \cos^{-1}\left(\frac{l^2 + d^2 - R^2}{2dl}\right)}{A_i^h} \tag{3.11}$$

The expectation of the AN-distance (h) is given by

$$d_i^h = E(L) = \int_{b^1}^{b^2} f_L^h(l)dl = \int_{b^1}^{b^2} 2l^2 \cos^{-1}\left(\frac{l^2 + d^2 - R^2}{2dl}\right) dl / A_i^h \tag{3.12}$$

Let L denote the AN-distance (h + 1) as illustrated in Fig. 3.6. We have

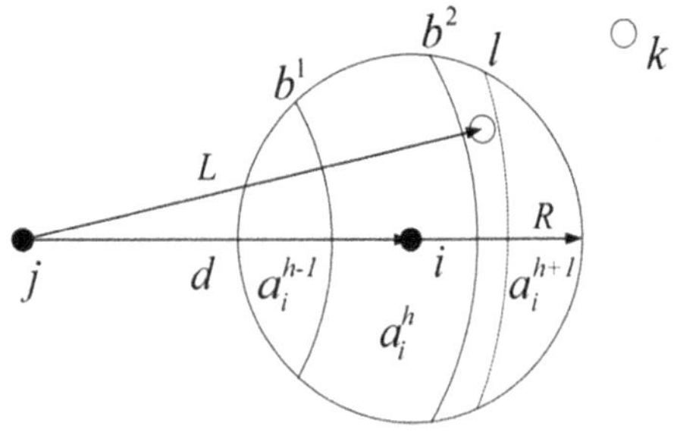

Fig. 3.6 The expectation of the AN-distance (h + 1)

$$P_{h+1}[L \leq l] = \left[l^2 \cos^{-1}\left(\frac{d^2 + l^2 - R^2}{2dl}\right) + \pi R^2 - R^2 \cos^{-1}\left(\frac{l^2 - d^2 - R^2}{2dR}\right) - \frac{1}{2}\sqrt{4d^2R^2 - \left(l^2 - d^2 - R^2\right)^2} - A_i^h \right] / A_i^{h+1} \tag{3.13}$$

The probability density function of the AN-distance (h + 1) is given by

$$f_L^{h+1}(l) = \frac{d[P_{h+1}[L \leq l]]}{dl} = \frac{2l \cos^{-1}\left(\frac{l^2 + d^2 - R^2}{2dl}\right)}{A_i^{h+1}} \tag{3.14}$$

The expectation of the AN-distance (h + 1) is given by

$$\begin{aligned} d_i^{h+1} = E(L) &= \int_{b^2}^{d+R} f_L^{h+1}(l)\, l\, dl \\ &= \int_{b^2}^{d+R} 2l^2 \cos^{-1}\left(\frac{l^2 + d^2 - R^2}{2dl}\right) dl / A_i^{h+1} \end{aligned} \tag{3.15}$$

We use the mean of d_i^{h-1}, d_i^h and d_i^{h+1} as the expectation of the AN-distance (h) is given by

$$E\left(a_i^h\right) = \frac{d_i^{h-1} + d_i^h + d_i^{h+1}}{3} \tag{3.16}$$

Let n denotes the hop ring of unknown node k with anchor j and unknown node k falls in the hop ring n, x denotes the distance between anchor j and unknown node k in Fig. 3.7. According to Eq. (3.16), we can calculate the expectation of the distance AN-distance (n), denoted by $E(x)$, as follows:

$$E(x) = E\left(a_k^n\right) \tag{3.17}$$

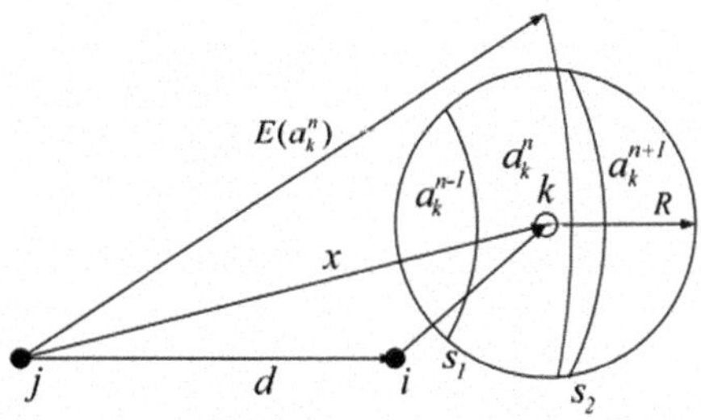

Fig. 3.7 The distance between anchor j and unknown node k

Here, $E(x)$ is an increasing function with respect to value x. We can apply the dichotomy to obtain value x. When the distances between unknown node k and three anchors are obtained, actual coordinates of the unknown nodes are calculated using the maximum likelihood algorithm.

3.1.3 Simulation Results

3.1.3.1 Simulation Setup

Our work is concerned with random topology. The basic network setup is a 100 m × 100 m square field. In our simulation study, the anchors are selected randomly.

The used parameters are as follows:

1. Nodes: 400, 600
2. Anchors: 3
3. Communication range error: 0

Let N and M denote the number of nodes and the number of anchors respectively, (x_i, y_i) and $(\widehat{x}_i, \widehat{y}_i)$ denote the true and estimated node location respectively. The localization error is given by

$$e = \frac{\sum_{i=1}^{N-M} \sqrt{(x_i - \widehat{x}_i)^2 + (y_i - \widehat{y}_i)^2}}{(N - M) \times R} \tag{3.18}$$

The system is subjected to the Distance Vector-Hop (DV-Hop) algorithm, the DV-regulated neighborhood distance (DV-RND) algorithm and the HCED algorithm to determine the locations of the unknown nodes.

3.1.3.2 Comparison with Other Methods

Figures 3.8 and 3.9 show that the results of the HCED algorithm for random topology with 3 anchors, 400 nodes and 600 nodes respectively. The green stars denote the true node location. The red circles denote the estimated node location and the blue lines denote the estimation errors. Figure 3.8 shows that the maximum localization error is 15 m for the HCED algorithm. Figure 3.9 shows that the maximum localization error is 8 m for the HCED algorithm. Increasing the number of the nodes improves the performance of localization. The reason is that the HCED algorithm obtains more precise distances among anchor and unknown nodes. The effect of the HCED algorithm is significantly improved.

We compare the HCED algorithm with the DV-Hop algorithm and the DV-RND algorithm, using different number of nodes. Figure 3.10 depicts the effect of different number of nodes on localization accuracy. These three curves have similar trend. Specifically, the DV-Hop algorithm and the DV-RND algorithm suffer larger

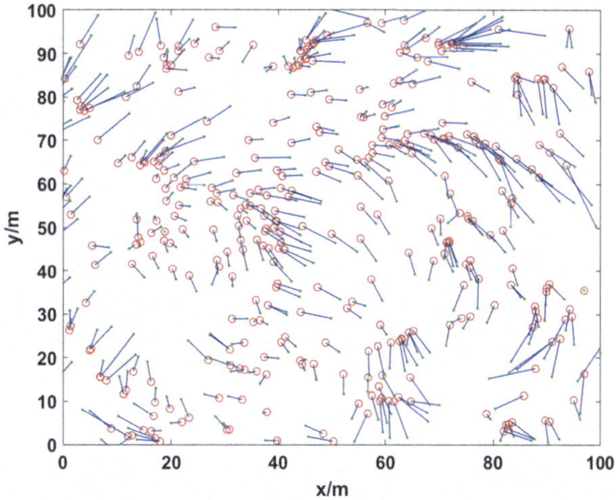

Fig. 3.8 Visualization of localization error of the HCED algorithm, N = 400

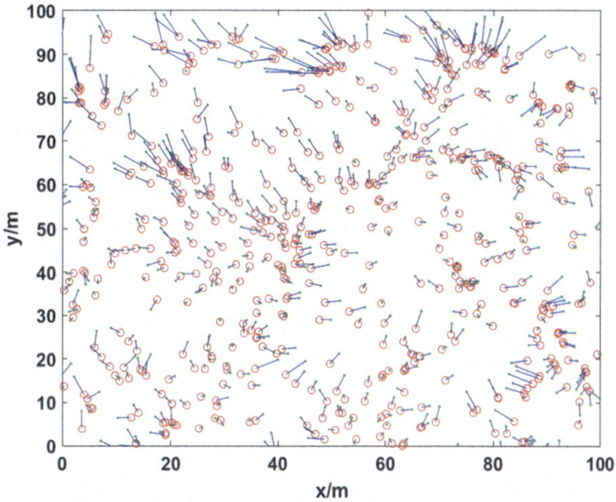

Fig. 3.9 Visualization of localization error of the HCED algorithm, N = 600

localization errors. The HCED algorithm achieves a much higher accuracy and has the minimum localization error.

3.2 Voronoi-Based Localization in MSNs

A Monte Carlo localization (MCL) scheme is specifically designed for MSNs. Aiming at the problem that sample point filtration condition of MCL localization algorithm is rather weak, we firstly present a range-free Voronoi-based Monte Carlo localization algorithm (VMCL) for MSNs. However, VMCL employs RSSI as basis for measuring distance and the large errors of RSSI technique. The localization precision of VMCL is negatively affected. To further reduce localization error, we

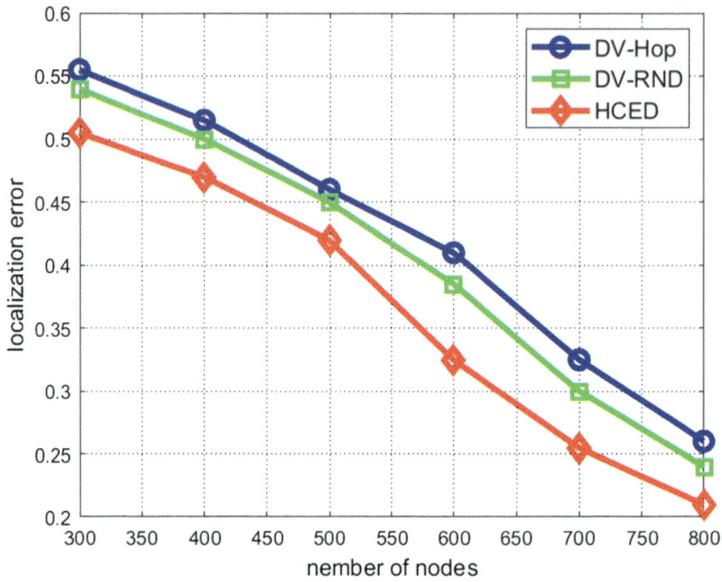

Fig. 3.10 Comparison of localization error in the case of different number of nodes

present an optimal region selection strategy of Voronoi diagram based on VMCL, called ORSS-VMCL [2]. This method increases the efficiency and accuracy for VMCL by adapting the size of Voronoi area during the filtering process.

3.2.1 Voronoi Diagram

Assume there are n points on the plane, $P = \{p_1, p_2, \ldots, p_n\}$, $2 < n \leq \infty$. Then on this plane, the Euclidean distance of any point $X(x, y)$ and $p_i(x_i, y_i)$ is

$$d(p_i, X) = \sqrt{\left((x_i - x)^2 + (y_i - y)^2\right)} \tag{3.19}$$

The plane is separated into n Voronoi areas by centering on each p_i, which makes any point X on $V(p_i)$ Voronoi area centering on p_i satisfy the condition of

$$V(p_i, X) = \left(\left\{x | d(p_i, X) \leq d\left(p_j, X\right), \forall i \neq j \wedge i, j \in I\right\}\right) \tag{3.20}$$

The n distinguish points on the plane construct n small areas according to the proximity principle.

Properties of the Voronoi diagram:

1. The dual graph for a Voronoi diagram (in the case of a Euclidean space with point sites) corresponds to the Delaunay triangulation for the same set of points.

2. The closest pair of points corresponds to two adjacent cells in the Voronoi diagram.
3. Assume the setting is the Euclidean plane and a group of different points are given. Then two points are adjacent on the convex hull if and only if their Voronoi cells share an infinitely long side.
4. If the space is a normed space and the distance to each site is attained (e.g. when a site is a compact set or a closed ball), then each Voronoi cell can be represented as a union of line segments emanating from the sites.
5. Under relatively general conditions (the space is a possibly infinite-dimensional uniformly convex space, there can be infinitely many sites of a general form, etc.), Voronoi cells enjoy a certain stability property. A small change in the shapes of the sites, e.g. a change caused by some translation or distortion, yields a small change in the shape of the Voronoi cells. This is the geometric stability of Voronoi diagrams.

3.2.2 RSSI Ranging Model

There are mainly three signal transmission models for RSSI, free-space model, two-ray ground reflection model and shadowing model. The free-space model and two-ray ground reflection model take the received signal as a fixed function of distance, so that their signal transmission model is an ideal circle.

The free-space model can be expressed as follows:

$$P_{r-\mathrm{FS}}(d) = \frac{P_t G_t G_r \lambda^2}{(4\pi)^2 d^2 L^2} \tag{3.21}$$

where P_t is the transmitted signal power. G_t and G_r are the antenna gains of the transmitter and the receiver, respectively. $L(L \geq 1)$ is the system loss, and λ is the wave length.

The two-ray ground reflection can be expressed as follows:

$$P_{r-\mathrm{RRG}}(d) = \begin{cases} P_{r-\mathrm{FS}}(d) & d \leq d_c \\ P_{r-\mathrm{FS}}(d)\left(\frac{d_c}{d}\right)^2 & d > d_c \end{cases} \tag{3.22}$$

where $d_c = 4\pi h_t h_r / \lambda$. h_t and h_r are the heights of the transmit and receive antennas, respectively.

However, due to multipath reflection and obstacles, the real signal transmission is anisotropic. Therefore, we choose a more general model, the shadowing model. As Eq. (3.23) shows

$$\left[\frac{P_r(d)}{P_r(d_0)}\right]^2 = -10\beta \lg\left(\frac{d}{d_0}\right) + X_{db} \tag{3.23}$$

where $X_{db} - N(0, \sigma^2)$, β is path loss exponent, $P_r(d_0)$ and d_0 represent reference energy and reference distance, respectively.

3.2.3 MCL Localization

The MCL method assumes that time is divided into discrete intervals and all nodes can move. Since a sensor node moves away from its previous location, it needs to re-localize in each time interval. In each time interval, a set of samples presenting the probabilistic distribution of a node's possible locations should be obtained based on possible movements and new observations.

In the initialization step of the algorithm, the nodes have no knowledge of their locations. There are no nodes presenting possible locations, so a location set of nodes $L_0 = \{l_0^1, l_0^2, \ldots, l_0^{N-1}\}$ are produced randomly. After initialization, there are two iterative steps, prediction and filtering. During the prediction step at time t, a new possible location set L_t is generated based on the possible movement and the possible location set L_{t-1} at the previous time. o_t denotes the observations from seed nodes received between time $t - 1$ and time t. A transition equation $p(l_t | l_{t-1})$ describes the prediction of nodes' current position based on previous position, and an observation equation $p(l_t | o_t)$ describes the likelihood of the node being at the location l_t given the observations.

In the prediction step, a node starts from the set of possible locations computed in the previous step, L_{t-1}, and applies the mobility model to each sample to get a set of new samples, L_t. We assume a node is unaware of its moving speed and direction, other than knowing its speed is less than v_{max}. So, if in previous step l_{t-1}^i is one possible position of a node, the possible current positions are contained in the circular region with origin l_{t-1}^i and radius v_{max}. $d(l_1, l_2)$ is used to denote the Euclidean distance between the two points l_1 and l_2. If speeds are distributed uniformly in the interval $[0, v_{max}]$, the probability of current location based on previous location estimate is given by a uniform distribution.

$$f(x) = \begin{cases} 1/\pi v_{max}^2 & d(l_t, l_{t-1}) < v_{\max} \\ 0 & d(l_t, l_{t-1}) \geq v_{\max} \end{cases} \tag{3.24}$$

In the filtering step, the node filters the impossible locations based on new observations. For simplification of presentation and analysis, we assume that time is discrete and all messages are received instantly. Hence, at time t, every node within radio range of a seed will hear a location announcement from that seed. In a realistic deployment, it would be necessary to deal with network collisions and account for lost messages. Let S denotes the set of one-hop anchor node neighbors of the sensor node, T denotes the set of two-hop anchor node neighbors of the sensor node and r stands for communication range. Every candidate samples l_t is filtered by the following formula.

$$filter\ (l_t) = \forall s \in S, d(l_t, s) \le r \wedge \forall s \in T, r < d(l_t, s) \le 2r \tag{3.25}$$

The prediction step and the filtering step are repeated to update the sample set presenting the distribution of nodes' possible locations. Taking use of these samples, nodes' location can be estimated by the weighted average method.

The MCL algorithm is addressed in detail as follows.

Initialization: initially the node has no knowledge of its location. N is a constant that denotes the number of samples to maintain,

$$L_0 = \{\text{set of } N \text{ random locations in the deployment area}\}$$

Step: compute a new possible location set L_t based on L_{t-1} and the new observations, o_t.

$$L_t = \{\}$$

while (size(L_t) $< N$) do

Prediction: $R = \{l_t^i | l_t^i \text{ is selected from } p(l_t\ | l_{t-1}^i), l_{t-1}^i \in L_{t-1} \text{ for all } 1 < i < N\}$

Filtering: $R_{filtered} = l_t^i\ |\ l_t^i, \text{ where } l_t^i \in R \text{ and } p(l_t^i|0,) > 0$

$$L_t = \text{choose}\left(L_t \cup R_{filtered}, N\right)$$

The above chart shows an overview of the algorithm. Initially, the node has no knowledge about its position, so the initial samples are selected randomly from all possible locations. At each time step, the location set is updated based on possible movements and new observations. The location of the node is estimated by computing the average location of all possible locations in L_t.

An example for choosing valid samples is shown in Figs. 3.11 and 3.12.

3.2.4 VMCL Localization

In the filtering step of MCL, one-hop and two-hop anchor nodes are obtained to predict the possible location of nodes. But the accuracy is not satisfactory when the number of anchor nodes is not enough. We find that Voronoi diagram can be employed in the filtering step of MCL.

Let *RSSI*(*ij*) be the RSSI value from node j to node i. We have the following assumption.

Assumption 3.1 Suppose that $RSSI(i_1,j_1) > RSSI(i_2,j_2)$. Then $d(i_1,j_1) < d(i_2,j_2)$.

Obviously, the Assumption 3.1 holds only if the absolute value of difference of $RSSI(i_1,j_1) > RSSI(i_2,j_2)$ is large enough.

Figure 3.13 illustrates the VMCL algorithm in three cases. According to the characteristics of the Voronoi diagram, l is the perpendicular bisector of line A_1A_2 in Fig. 3.13a. Obviously, node N can tell whether it is closer to anchor node A_1 or A_2

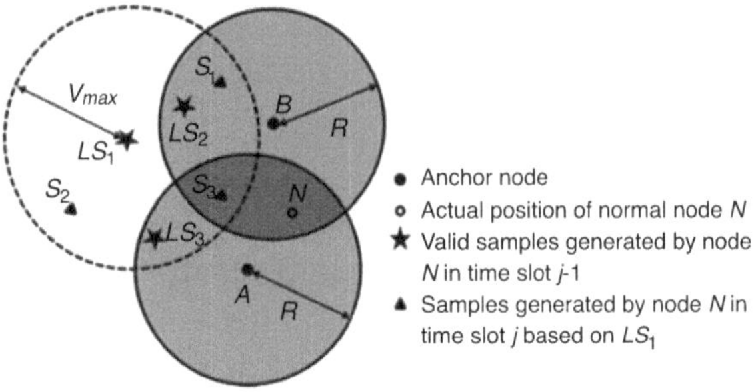

Fig. 3.11 An example for choosing valid samples

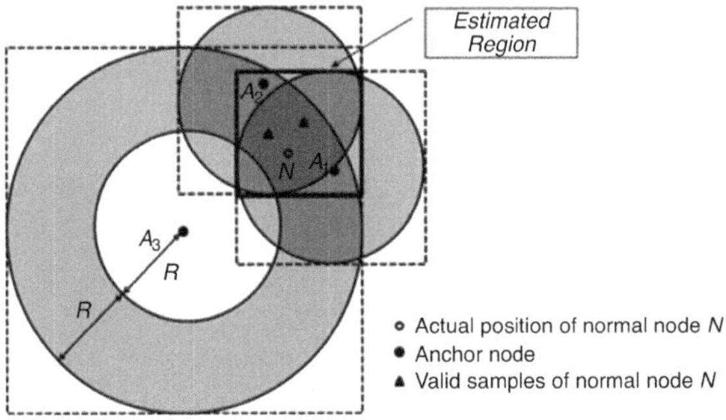

Fig. 3.12 An example for choosing valid samples

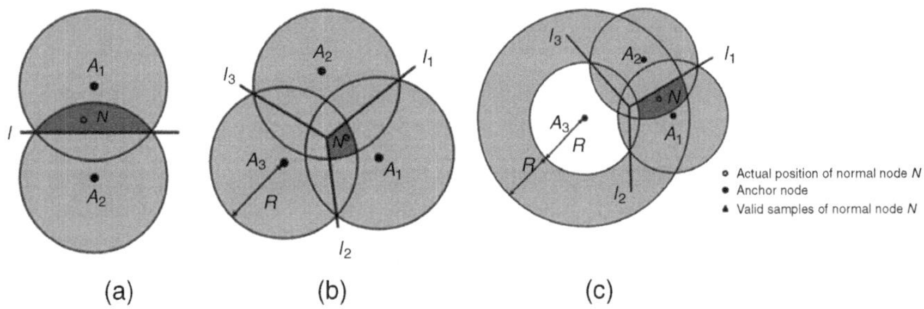

Fig. 3.13 Voronoi-based Monte Carlo localization algorithm. (**a**) The first case of VMCL (**b**) The second case of VMCL (**c**) The third case of VMCL

based on the RSSI. Thus, the filtering area is further shrinking. Similarly, l_1, l_2 and l_3 are the perpendicular bisector of line A_1A_2, A_1A_3 and A_2A_3, respectively in Fig. 3.13b and c. The probability distribution $p(l_t| o_t)$ is zero if the filter condition is false, and evenly distributed otherwise. Thus, we eliminate locations that are inconsistent with observations from the possible location set. After filtering, there may be fewer than N possible locations remaining. Repeat the prediction and filtering processes and union the possible points until at least N possible locations have been acquired.

3.2.5 ORSS-VMCL Localization

As mentioned in Sect. 3.2.2, there are mainly three signal transmission models. However, due to multipath reflection and obstacles, the real signal transmission is anisotropic. Therefore, we choose a more general model: the shadowing model.

Before we proceed further, let us identify an assumption.

Assumption 3.2 The maximum noise is *MaxNoise*, which means the maximum value of X_{db} is *MaxNoise*.

We assume that d_A and d_B are the distance from the unknown node to anchor nodes A and B. According to Eq. (3.23), we have

$$\left[\frac{P_r(d_A)}{P_r(d_0)}\right]_{db} = -10\beta \lg\left(\frac{d_A}{d_0}\right) + X_{A-db} \tag{3.26}$$

$$\left[\frac{P_r(d_B)}{P_r(d_0)}\right]_{db} = -10\beta \lg\left(\frac{d_B}{d_0}\right) + X_{B-db} \tag{3.27}$$

While Eqs. (3.26) and (3.27) can be expressed as

$$\left[\frac{P_r(d_A)}{P_r(d_0)}\right]_{db} - X_{A-db} + 10\beta \lg(d_0) = -10\beta \lg(d_A) \tag{3.28}$$

$$\left[\frac{P_r(d_B)}{P_r(d_0)}\right]_{db} - X_{B-db} + 10\beta \lg(d_0) = -10\beta \lg(d_B) \tag{3.29}$$

We subtract Eq. (3.29) from Eq. (3.28) as follows:

$$\left[\frac{P_r(d_A)}{P_r(d_0)}\right]_{db} - X_{A-db} - \left[\frac{P_r(d_B)}{P_r(d_0)}\right]_{db} + X_{B-db} = -10\beta \lg\left(\frac{d_A}{d_B}\right) \tag{3.30}$$

If Eq. (3.30) satisfies

$$\left[\frac{P_r(d_A)}{P_r(d_0)}\right]_{db} - X_{A-db} - \left[\frac{P_r(d_B)}{P_r(d_0)}\right]_{db} + X_{B-db} < 0 \tag{3.31}$$

Then we can get the conclusion that $d_A > d_B$. Furthermore, Eq. (3.32) can be obtained by the aforementioned assumption and Eq. (3.31) as follows.

$$\left[\frac{P_r(d_B)}{P_r(d_0)}\right]_{db} - \left[\frac{P_r(d_A)}{P_r(d_0)}\right]_{db} > 2\text{MaxNoise} \tag{3.32}$$

In the VMCL mentioned above, if the subtraction of Eq. (3.32) is smaller than ε (ε is determined by the noise), the rank of RSSI will not reflect the rank of distance accurately. As a result, the unknown node will be estimated in the wrong direction. There are some improved algorithms. For example, the original Voronoi region can be replaced by the new one which is η times bigger than the original one in order to contain some fuzzy region, but the localization accuracy is reduced. Another improved method is proposed in Huang and Wu [11], the maximum likelihood estimation method is used to obtain a pre-estimate. To some extent, this method optimizes the rank of distance and takes full use of the information obtained from all anchor nodes, but obtaining neighbor anchor nodes' information table will largely increase energy consumption.

Because of the drawback mentioned above, we present an ORSS to deal with this issue.

Figure 3.14a shows a Voronoi diagram. There are mainly three kinds of situations to consider.

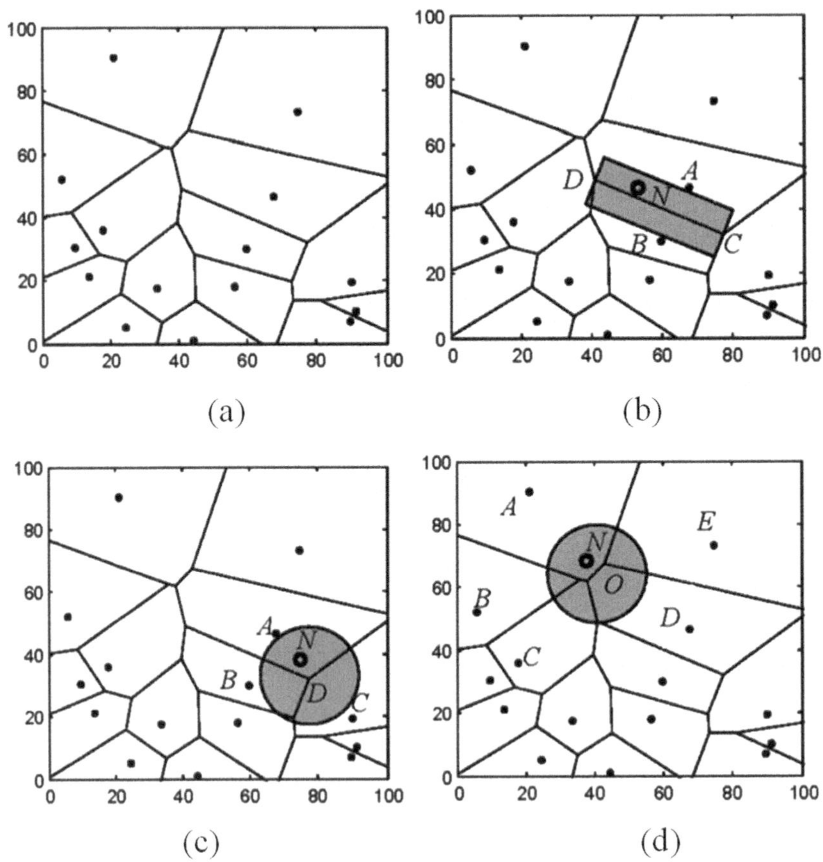

Fig. 3.14 Voronoi-based Monte Carlo localization algorithm. (**a**) Voronoi diagram (**b**) Situation 1 of Voronoi diagram (**c**) Situation 2 of Voronoi diagram (**d**) Situation 3 of Voronoi

Situation 1: as shown in Fig. 3.14b, there are two anchor nodes which almost have the equal distance to the unknown node. According to the characteristics of the Voronoi diagram, *CD* is the perpendicular bisector of *AB*, so nodes around *CD* satisfy the above condition. Therefore, we take *CD* as the mid-line to draw a rectangle. The length of the rectangle is the length of *CD* and the width is related to the noise signal. The greater the noise signal is, the greater the width is. Finally, the Voronoi region is replaced with the rectangle region, which is shown as the shadow in Fig. 3.14b.

Situation 2: as shown in Fig. 3.14c, there are three anchor nodes which almost have the equal distance to the unknown node. According to the characteristics of the Voronoi diagram, *D* is the circumcenter of nodes *A*, *B* and *C*, so nodes around *D* satisfy the above condition. Therefore, we take *D* as the center to draw a circle. The radius of the circle is related to the noise signal. The greater the noise signal is, the greater the radius is. Finally, the Voronoi region is replaced with the circle region, indicated by the shadow in Fig. 3.14c.

Situation 3: as shown in Fig. 3.14d, there are $n(n > 3)$ anchor nodes which almost have the equal distance to the unknown node. According to the characteristics of the Voronoi diagram, there will be several circumcenters which are like node *D* in Fig. 3.14c. Centroid of these *D*s is named *O*, as can be seen in Fig. 3.14c. So, nodes around *O* satisfy the above condition. Therefore, we take *O* as the center to draw a circle. The radius of the circle is related to the noise signal. The greater the noise signal is, the greater the radius is. Finally, the Voronoi region is replaced with the circle region, illustrated by the shadow in Fig. 3.14d.

We have talked about the ORSS above. It is applied to VMCL, as depicted in Fig. 3.15. Based on this algorithm, the filtering area is further shrinking.

The region selection strategy mentioned above cannot match a dynamic noise. Therefore, we have to find the relationship between the noise and the selected region. A noise observer is applied to find the environment noise (EN).

We assume that the maximum transient noise is ε. For situation 1, the width of the rectangle is width. According to Eq. (3.32) we can get the conclusion that,

$$\text{width} \geq 2\varepsilon \tag{3.33}$$

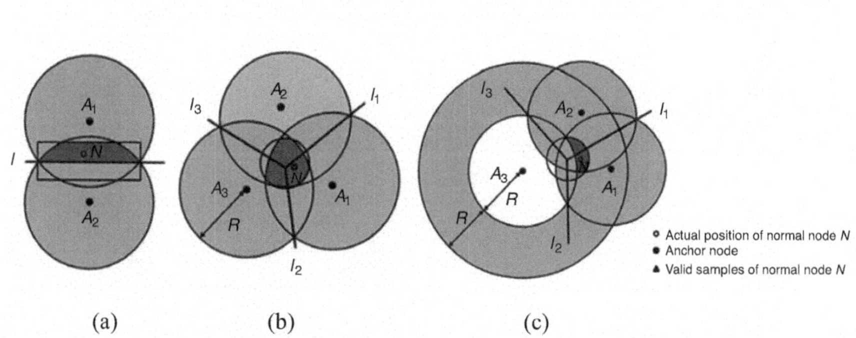

Fig. 3.15 ORSS-VMCL algorithm. (**a**) The first case of ORSS-VMCL (**b**) The second case of ORSS-VMCL (**c**) The third case of ORSS-VMCL

For situation 2 and situation 3, the radius of the circle is radius. According to Eq. (3.32), the following conclusion can be obtained:

$$\text{radius} \geq 2\varepsilon \tag{3.34}$$

3.2.6 Simulation Results

3.2.6.1 Simulation Setup

Simulations are on a 500 m × 500 m square field with $EN \sim N(\mu, \sigma^2)$, where $\mu = 0$ and $\sigma = 0.1$. The probability density function is as follows:

$$f(x) = \left(\frac{1}{\sqrt{2\pi\sigma}}\right) \exp\left(-\frac{(x-\mu)^2}{2\sigma^2}\right) \tag{3.35}$$

Figure 3.16 shows the signal transmission, the circumference stands for RSSI without EN.

Scenario defined for simulation environment is:

1. The simulation area is 500 m × 500 m.
2. The communication radius $r = 100$m.
3. The max speed of nodes $v_{max} = 0.5$, $r = 50\, m/s$ and this speed has been changed for specific tests.
4. s_d is the average number of anchor nodes within one hop.
5. n_d is the average number of nodes within one hop.

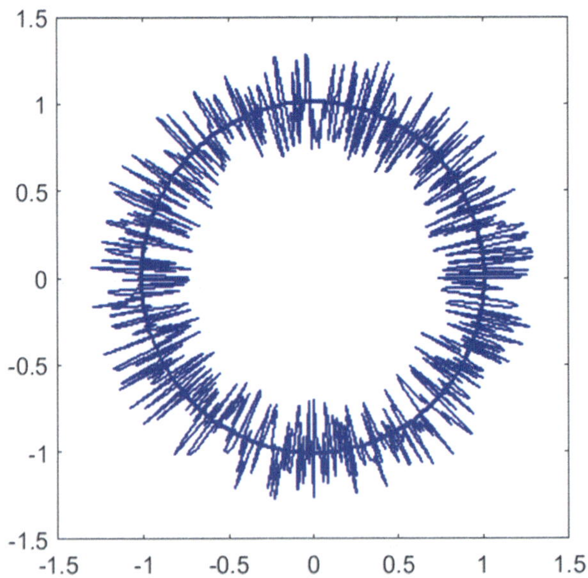

Fig. 3.16 Simulation for environment noise

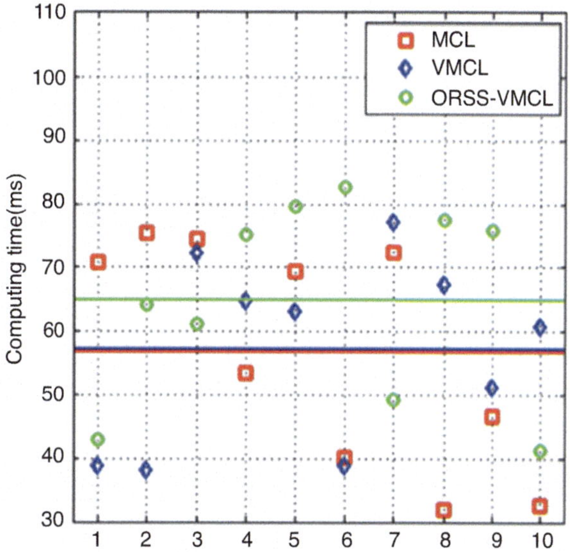

Fig. 3.17 Computing time for localization

3.2.6.2 Comparison with Other Methods

Time consumption: Fig. 3.17 shows average time consumption for localization used by MCL, VMCL and ORSS-VMCL. This experiment has been tested for 10 groups. Each group indicates the average time consumption of 10 location calculations. VMCL has almost the same time consumption with MCL, and ORSS-VMCL has a higher value than them.

Localization accuracy: Fig. 3.18 compares the localization accuracy for the three algorithms. This experiment has also been tested for 10 groups. Each group indicates the average localization accuracy of 10 location calculations. The average localization error of ORSS-VMCL is 17.5 meters which is slightly below the MCL and VMCL.

Convergence speed: Fig. 3.19 illustrates the convergence speed for MCL, MCB, WMCL and ORSS-VMCL. Four algorithms have roughly the same convergence speed. After five iterations, all these four algorithms have generally convergent.

Anchor density: Fig. 3.20 shows the effect of anchor node density on localization accuracy. Anchor density is a critical parameter of anchor-based localization schemes. These four algorithms have the same tendency. With the increase of s_d, more and more location information can be obtained for each node. However, when $s_d > 2.5$, redundant location information has little impact on localization accuracy.

Maximum velocity: Fig. 3.21 displays the effect of maximum velocity on localization accuracy at different s_d. These four curves have the similar trend. While v_{max} is changing from $0.2r$ to $0.4r$, the localization accuracy has been increased. Because the moving node can obtain more information from anchor nodes. On the contrary, when v_{max} is more than $0.4r$, the uncertainty of nodes is increasing as well. So, the localization errors are getting larger.

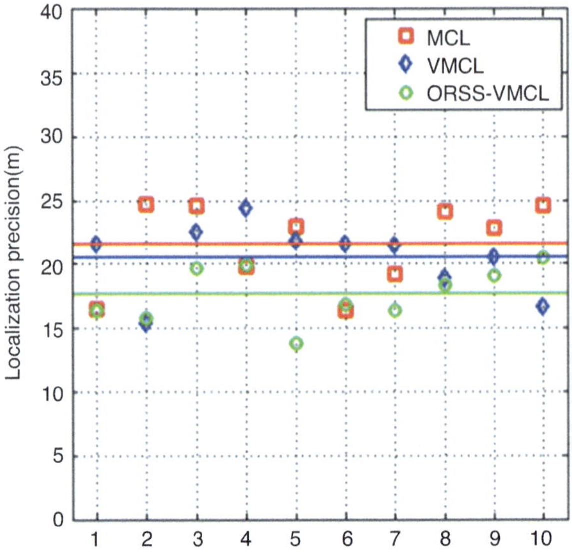

Fig. 3.18 Comparison in localization accuracy

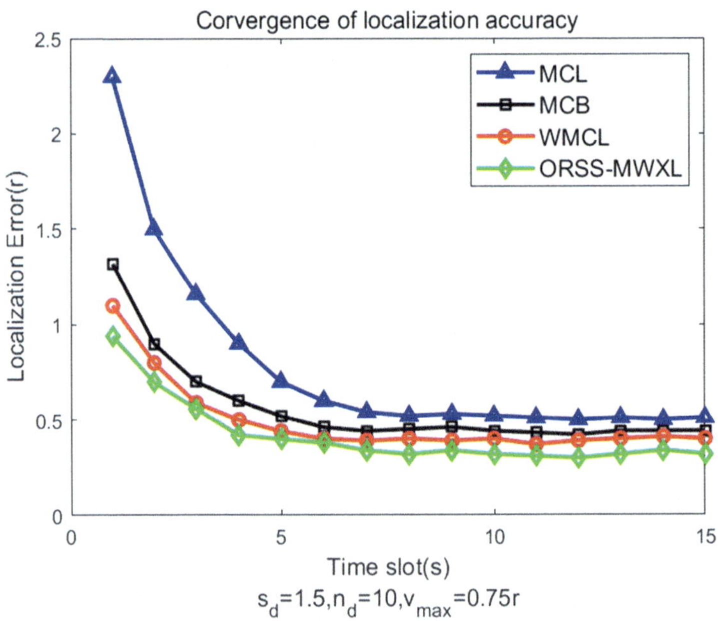

Fig. 3.19 Localization accuracy change with time

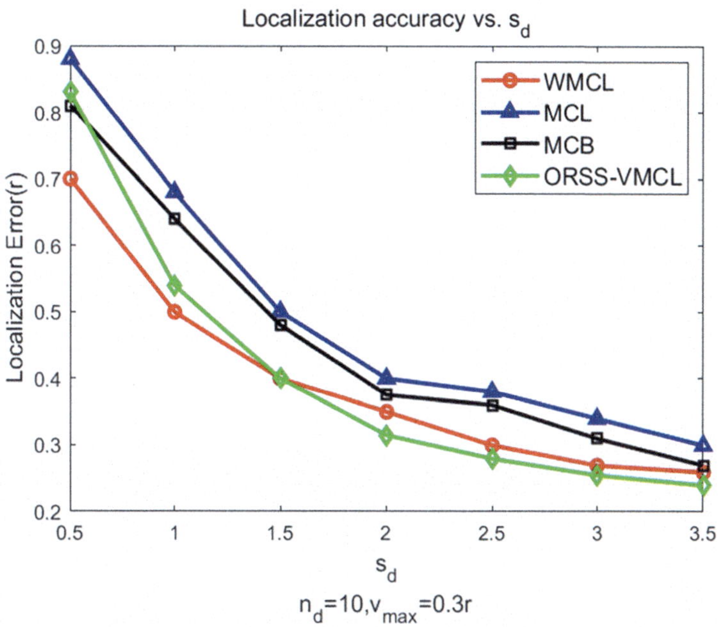

Fig. 3.20 The effect of anchor density on localization accuracy

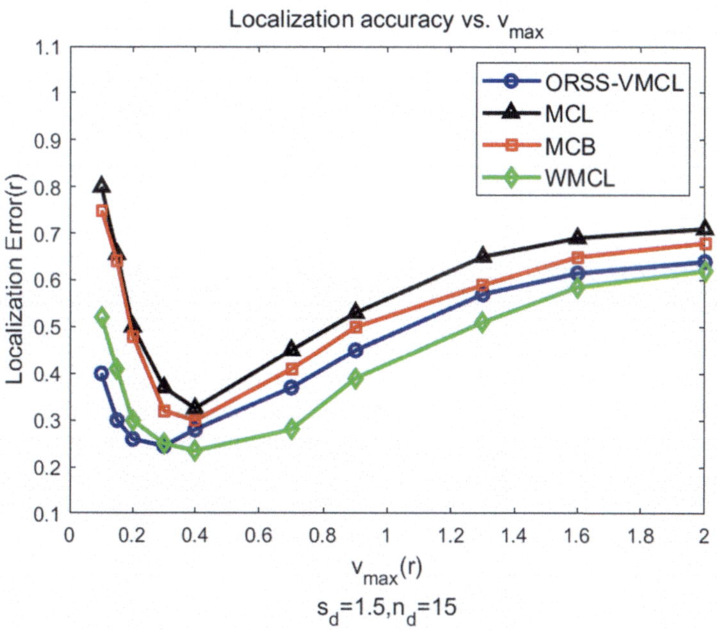

Fig. 3.21 The effect of maximum velocity on localization accuracy

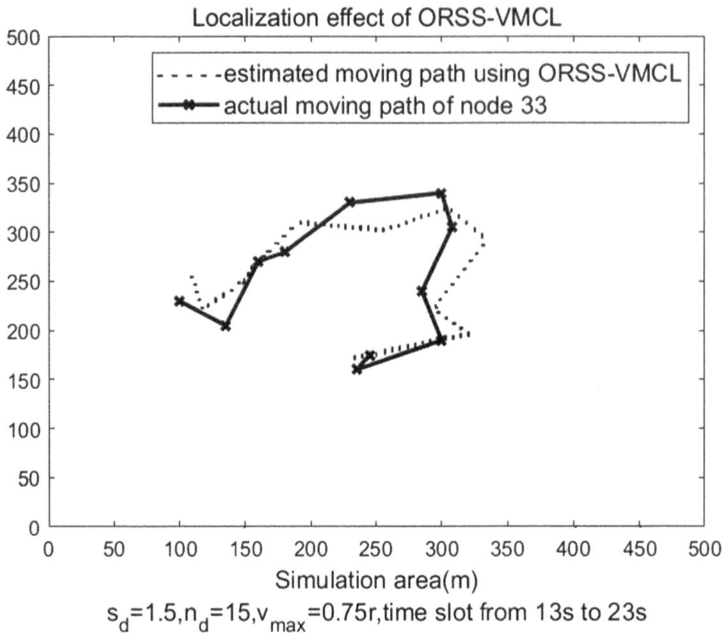

Fig. 3.22 The effect of ORSS-VMCL

Tracking effects: Fig. 3.22 shows the estimated moving path using ORSS-VMCL and an actual moving path of a generic node. Obviously, the estimated location can roughly express the real nodes' location, especially after 10 time slots, when the algorithm has been stability.

3.3 Heuristic Multi-Dimensional Scaling Localization

In this section, a novel algorithm termed the Heuristic Multi-dimensional Scaling (HMDS) is presented in anisotropic SSNs with holes [3]. The HMDS algorithm has the advantage of low complexity in communication and computation. Simulation results demonstrate that the HMDS algorithm is suitable for four different types of topologies, namely the semi-C-shape, the O-shape, the multiple O-shape and the concave-shape topologies, and is more accurate and efficient comparing with state-of-the-art methods in anisotropic SSNs with holes.

3.3.1 MDS Localization

The MDS algorithm has been widely used in many fields such as information processing and exploratory data analysis. The MDS algorithm can be used in WSNs where only distances between nodes are known.

Let $X_{m \times n}$ be a space coordinate matrix of n nodes. Each row of matrix X denotes $m-$dimensional coordinates of a node. Let $D^2(X)$ represent the matrix of squared distances, which is given by

$$D^2(X) = \begin{bmatrix} 0 & d_{12}^2 & \cdots & d_{1n}^2 \\ d_{21}^2 & 0 & \cdots & d_{2n}^2 \\ \vdots & \vdots & \ddots & \vdots \\ d_{n1}^2 & d_{n2}^2 & \cdots & 0 \end{bmatrix} = ce^T + ec^T - 2B \tag{3.36}$$

where $c = \left[\sum_{k=1}^{m} x_{1k}^2 \sum_{k=1}^{m} x_{2k}^2 \cdots \sum_{k=1}^{m} x_{nk}^2 \right]^T$, $e = (1,\ 1, \ldots,\ 1)^T$ and the matrix $B = XX^T$.

Multiplying the left and the right sides of Eq. (3.36) by the centering matrix $J = (I - n^{-1} ee^T)$ and by the factor $-1/2$ results in

$$\begin{aligned} B &= -\frac{1}{2} JD^2(X)J = -\frac{1}{2} J\left(ce^T + ec^T - 2B\right)J \\ &= -\frac{1}{2} Jce^T J - \frac{1}{2} Jec^T J + \frac{1}{2} J(2B)J = -\frac{1}{2} Jc0^T - \frac{1}{2} 0c^T + JBJ \end{aligned} \tag{3.37}$$

Applying the singular value decomposition on matrix B results in

$$B = Q\Lambda Q^T = \left(Q\Lambda^{1/2}\right)\left(Q\Lambda^{1/2}\right)^T \tag{3.38}$$

where the coordinate matrix is defined as $X = Q\Lambda^{1/2}$.

We use MDS to obtain relative coordinates of each node in the $m-$dimensional space. The relative coordinates of the nodes are transformed into absolute coordinates of the nodes to complete the localization process by using anchors.

The distance information is obtained by the shortest path algorithm in order to achieve high localization accuracy. If the nodes are deployed uniformly in WSNs, as shown in Fig. 3.23a, the shortest paths are close to the distances between nodes. However, if the nodes are deployed in anisotropic WSNs with a hole, as shown in Fig. 3.23b, using the shortest path algorithm to estimate the distances between nodes will generate larger errors.

The MDS algorithm is a precise algorithm for dimensionality reduction. If the Euclidean distances between each pair of nodes are given, the accurate distance matrix is achieved. The MDS algorithm obtains the node locations without error. When the Euclidean distances between some of the nodes are not able to be measured in the network, they will be replaced by using the Dijkstra shortest path

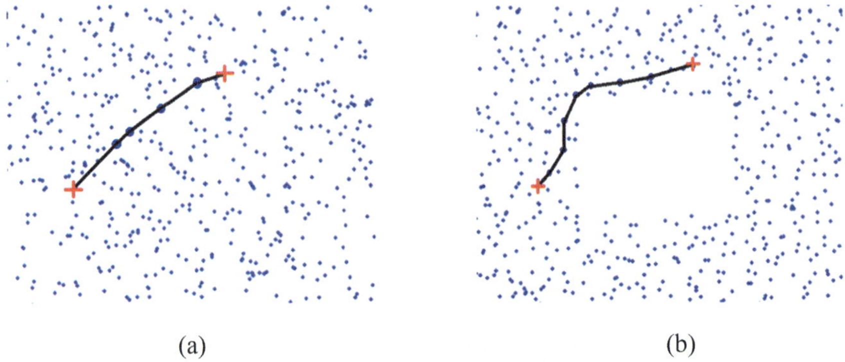

Fig. 3.23 The shortest paths between nodes. (**a**) Uniform distribution. (**b**) Non-uniform distribution

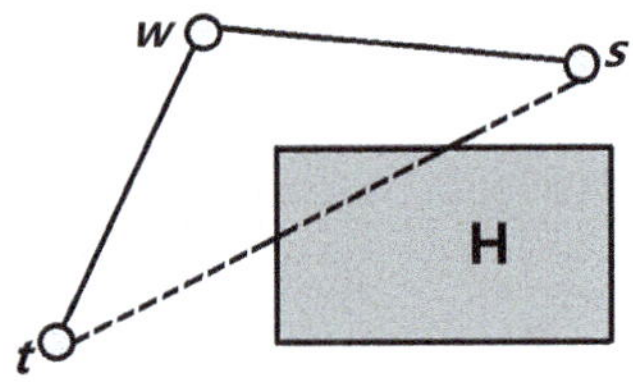

Fig. 3.24 Distance calculation using Dijkstra algorithm

algorithm. But this approximation method produces larger localization errors when the nodes are deployed in anisotropic WSNs with holes.

The process of Dijkstra algorithm is illustrated in Fig. 3.24. Because the presence of hole H obstructs the line of sight, the two nodes *s* and *t* cannot communicate directly to each other although they are extremely close. Using the Dijkstra algorithm and considering another node *w* outside of hole H, the distance *st* between nodes *s* and *t* can be approximated as $st = sw + wt$, which is longer than the Euclidean distance between nodes *s* and *t*. In this case, the approximated distance will generate larger localization errors. To solve this problem, the HMDS algorithm is proposed to calculate precise Euclidean distances between nodes in anisotropic WSNs with holes.

3.3.2 HMDS Localization

3.3.2.1 Mathematical Model

We can explore a border node *o* which intersects with hole H using algorithm presented in [4]. As Fig. 3.25 shows, the shortest path between unknown nodes

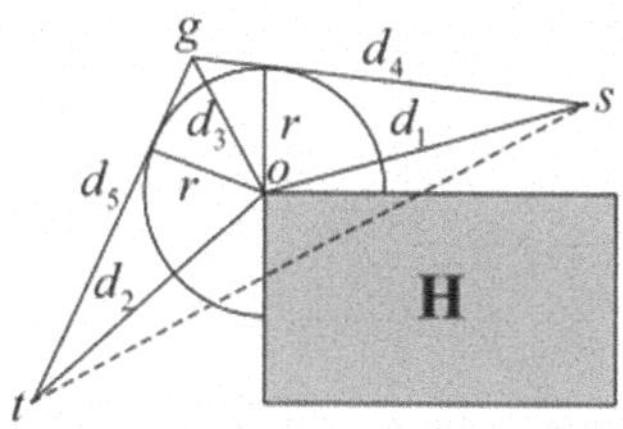

Fig. 3.25 Explore the virtual node g

s and t is segmented into the length of so and ot. We assume that $|so| = d_1$ and $|ot| = d_2$. From trigonometry, we have $|st|^2 = |so|^2 + |ot|^2 - 2\,|so| \cdot |ot| \cdot \cos \angle sot$ and the distance d_{st} is given by

$$d_{st} = \sqrt{d_1^2 + d_2^2 - 2|d_1||d_2| \cdot \cos \angle sot} \tag{3.39}$$

To find the angle $\angle sot$, we can explore a virtual node g which is close to the border node o and is the intersection of the lines sg and tg. The lines sg and tg are tangent to the virtual circle. The radius r is determined by Li and Liu [5].

As illustrated in Fig. 3.25, the new shortest path between nodes s and t is segmented into the length of sg and gt. Now, assume that $|og| = d_3$, $|sg| = d_4$ and $|gt| = d_5$. The angle $\angle sot$ can be derived from Fig. 3.25 as follows:

$$\angle \mathrm{sot} = 2\pi - \angle \mathrm{sog} - \angle \mathrm{got} = 2\pi - \arccos \frac{d_1^2 + d_3^2 - d_4^2}{2d_1 d_3} - \arccos \frac{d_2^2 + d_3^2 - d_5^2}{2d_2 d_3} \tag{3.40}$$

Using Eq. (3.40), we are able to calculate the distance d_{st} between nodes s and t, and solve the problem of the non-line of sight. The distance d_{st} is more accurately than the approximation distance by using the Dijkstra algorithm.

It can be seen from above results that we assume d_1 and d_2 are shorter than the maximum communication range of node in the network. However, the nodes are often deployed non-uniformly over a large-scale network. The distances d_1 and d_2 are longer than R which is the communication radius, and d_1, d_2 cannot be obtained directly.

As Fig. 3.26 shows, we can explore a virtual node w' using the aforementioned method. However, if we use the virtual node w' to calculate the distance d_{st}, the value of d_{st} is not calculated accurately. The reason is that the distance $d_{sw'}$ is longer than R which is the communication radius. To accurately calculate the distance d_{st}, we need to explore a new virtual node w which lies in the lines s_2o and sw'. We also can explore a virtual node g' which lies in the reverse extension line so. Furthermore, we assume node s lies exactly in the middle of the curve s_1s_2 using the HMDS

Fig. 3.26 Explore the virtual node w

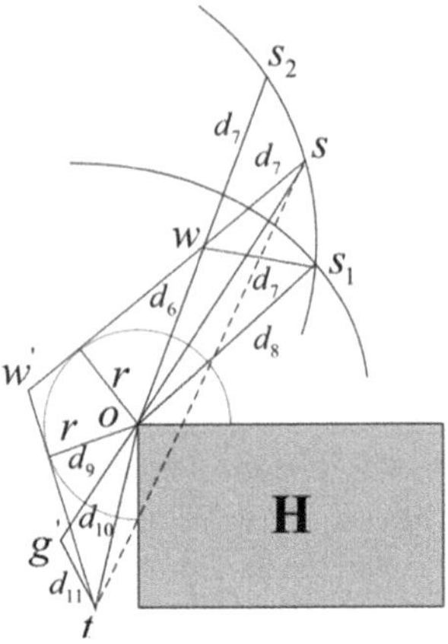

algorithm. We assume that $|ow| = d_6$, $|s_1w| = |s_2w| = |sw| = d_7$, $|os_1| = d_8$, $|og'| = d_9$, $|ot| = d_{10}$ and $|g't| = d_{11}$. Using trigonometry, the distance d_{so} is given by

$$d_{so}^2 = d_6^2 + d_7^2 - 2d_6d_7\cos(\angle ows) \tag{3.41}$$

In order to achieve the distance d_{so}^2, the angle $\angle ows$ needs to be found as follows:

$$\angle ows = \angle ows_1 + \angle s_1ws \tag{3.42}$$

where angle $\angle ows_1$ can be found as follows:

$$\angle ows_1 = \arccos\left(\frac{d_6^2 + d_7^2 - d_8^2}{2d_6d_7}\right) \tag{3.43}$$

Since $\angle s_1ws = \angle sws_2$, we have

$$\angle s_1ws = \frac{1}{2}\angle s_1ws_2 = \frac{1}{2}(\pi - \angle ows_1) \tag{3.44}$$

$$\angle ows = \angle ows_1 + \frac{1}{2}(\pi - \angle ows_1) = \frac{\pi}{2} + \frac{1}{2}\angle ows_1 \tag{3.45}$$

$$\begin{aligned} d_{so}^2 &= d_6^2 + d_7^2 - 2d_6d_7\cos(\angle ows) = d_6^2 + d_7^2 - 2d_6d_7\cos\left(\frac{\pi}{2} + \frac{1}{2}\angle ows_1\right) \\ &= d_6^2 + d_7^2 + 2d_6d_7\sin\frac{1}{2}\angle ows_1 \end{aligned} \tag{3.46}$$

$$\angle g'ot = \arccos\left(\frac{d_9^2 + d_{10}^2 - d_{11}^2}{2d_9d_{10}}\right) \tag{3.47}$$

$$\angle sot = \pi - \angle g'ot \tag{3.48}$$

$$d_{st} = \sqrt{d_{so}^2 + d_{ot}^2 - 2 \mid d_{so} \mid\mid d_{ot} \mid \cdot \cos \angle sot} \tag{3.49}$$

and

$$\angle sot = \pi \; \arccos\left(\frac{d_9^2 + d_{10}^2 - d_{11}^2}{2d_9d_{10}}\right) \tag{3.50}$$

It follows from the above formulas that the distance d_{st} can be calculated more accurately. If the distance d_{ot} is longer than R, the same method can be applied. After all the Euclidean distances between nodes are obtained, we can apply the MDS algorithm to the distance matrix and obtain relative coordinates of the nodes.

3.3.2.2 Coordinate Calculation

The locations of the nodes are absolute coordinates of the nodes that are transformed from relative coordinates. Obtained by the MDS algorithm, the relative coordinates of the nodes, are shifted, rotated and reflected with respect to the anchors to compute the locations of the nodes.

Suppose that $R = (R_1, R_2, R_3, \ldots, R_n)$ denotes relative coordinates of n nodes in the two-dimensional space and $T = (T_1, T_2, T_3, \ldots, T_n)$ denote absolute coordinates of those nodes in two-dimensional space. A vector R_i can be shifted to $R_i^{(a)}$ by $R_i^{(a)} = R_i + X$ or rotated counter clockwise through an angle α to $R_i^{(b)}$ by $R_i^{(b)} = Q_a R_i$, where

$$Q_a = \begin{bmatrix} \cos\alpha & -\sin\alpha \\ \sin\alpha & \cos\alpha \end{bmatrix} \tag{3.51}$$

Also, the vector R_i can be reflected across the line L to $R_i^{(c)}$ by $R_i^{(c)} = Q_b R_i$, where

$$L = \begin{pmatrix} \cos(\beta/2) \\ \sin(\beta/2) \end{pmatrix} \tag{3.52}$$

$$Q_b = \begin{bmatrix} \cos\beta & \sin\beta \\ \sin\beta & -\cos\beta \end{bmatrix} \tag{3.53}$$

The available data for the alignment process are matrix R from the MDS algorithm and relative coordinates of n nodes $(R_1, R_2, R_3, \ldots, R_n)$ and absolute coordinates of three anchors (T_1, T_2, T_3). Using the available data, relative coordinates of other nodes $(T_4, T_5, \ldots, T_n)$ will be calculated. We can write

$$(T_1 - T_1, T_2 - T_1, T_3 - T_1) = Q_a Q_b (R_1 - R_1, R_2 - R_1, R_3 - R_1) \quad (3.54)$$

The values for T_1, T_2, T_3 and R_1, R_2, R_3 are known. Therefore,

$$Q = Q_a Q_b = (T_1 - T_1, T_2 - T_1, T_3 - T_1)(R_1 - R_1, R_2 - R_1, R_3 - R_1)^{-1} \quad (3.55)$$

After calculating the value Q, values of $(T_4, T_5, \ldots, T_n)$ can be evaluated as follows:

$$(T_4 - T_1, T_5 - T_1, \ldots, T_n - T_1) = Q(R_4 - R_1, R_5 - R_1, \ldots, R_n - R_1) \quad (3.56)$$

$$(T_4, T_5, \ldots, T_n) = Q(R_4 - R_1, R_5 - R_1, \ldots, R_n - R_1) + (T_1, T_1, \ldots, T_1) \quad (3.57)$$

Locations of the nodes in the network are computed in this way.

3.3.2.3 The HMDS Algorithm

The HMDS algorithm concentrates on how to geometrically calculate the Euclidean distances between nodes. It consists of five steps:

1. The border node of the hole is explored and marked using the algorithm presented in [4].
2. The Dijkstra algorithm is used to calculate the shortest paths between every pair of nodes as well as to construct the distance matrix.
3. The Euclidean distances between nodes are achieved in accordance with the following principles:
   (a) If there are no marked nodes in the shortest paths between nodes, the communication between nodes is not across the holes. The Dijkstra shortest paths are regarded as the Euclidean distances between nodes.
   (b) If there are marked nodes in the shortest paths between nodes, the communication between nodes is across the holes. We explore a virtual node and recalculate the Euclidean distances between nodes. If the distance between a virtual node and an unknown node is shorter than communication range R, the Euclidean distances between unknown nodes can be achieved by using Eqs. (3.39) and (3.40). If the distance between a virtual node and an unknown node is larger than communication range R, the Euclidean distances between unknown nodes can be achieved by using Eqs. (3.41)–(3.49).
4. The recalculated Euclidean distances between nodes replace the shortest paths of the marked nodes in the distance matrix and form a new distance matrix.
5. The new distance matrix is applied to the MDS algorithm and the node locations are obtained.

Table 3.1 Complexity comparison

Algorithm	Communication cost	Computation cost
DV-Hop	$O(N^2)$	$O(NL)$
MDS-MAP	$O(N \log N)$	$O(N^3L)$
REP	$O(N^2)$	$O(NL)$
HMDS	$O(N^2)$	$O(NL)$

3.3.2.4 Complexity Analysis

The HMDS algorithm can use as few as three anchors to obtain the locations of the node in anisotropic networks. Each anchor is assumed to have the same communication capability as the node. To calculate the locations of the node, each node needs several rounds of query flooding to find different shortest paths and accordingly calculate the Euclidean distances to the anchors. We assume that N denotes the number of nodes and L denotes the number of holes within the network. The communication cost is bounded by $O(N^2)$. The anchors bear most of the computational cost. Each anchor processes distance queries from entire network and for each query, the anchor does at most $O(L)$ computations to calculate the shortest path. For each anchor, the computation cost is $O(NL)$.

Table 3.1 compares HMDS with the three state-of-the-art methods: DV-Hop, MDS-MAP and REP. DV-Hop presumes anisotropic networks and triangulates the node locations with its network distances to the three anchors. Each node floods the network for computing the hop count, so the communication cost of DV-Hop is $O(N^2)$. Each anchor accepts requests from entire network and sends out feedbacks with $O(NL)$ computation cost. The core of MDS-MAP is singular value decomposition, which has communication cost of $O(N \log N)$. MDS-MAP has computation cost $O(N^3L)$. The communication cost of REP is $O(N^2)$ and the computation cost of REP is $O(NL)$.

According to the aforementioned analysis, the proposed HMDS algorithm has low complexity in communication and computation compared with the MDS-MAP algorithm and has the same complexity compared with the DV-Hop algorithm and the REP algorithm.

3.3.3 *Simulation Results*

3.3.3.1 Simulation Setup

We have considered four random topologies including semi-C-shape, O-shape, multiple O-shape and concave-shape topologies to evaluate the proposed algorithm. We assume a network composed of thousands of sensor nodes deployed randomly over a large-scale region. The semi-C-shape topology setup is a 100 m × 100 m square field with a 60 m × 60 m hole. The O-shape topology setup is a 100 m × 100 m square field with a 40 m × 40 m hole. The multiple O-shape topology setup is a 100 m × 100 m square field with two 60 m × 20 m holes. The concave-shape topology setup is a 100 m × 100 m square field with an irregular hole. In our simulation study, anchors are selected randomly.

The used parameters are as follows:

1. Different number of nodes: 600, 800, 1000, 1200
2. Different number of anchors: 3, 5, 8, 10, 12
3. Different communication range error: 0, 0.05R, 0.1R

The system is subjected to many localization algorithms to determine the locations of the nodes in different topologies.

3.3.3.2 Comparison with Other Methods

Simulation results of the MDS-MAP algorithm and the HMDS algorithm with 1200 nodes, 5 anchors and without communication range error in a semi-C-shape topology and an O-shape topology are shown in Figs. 3.27, 3.28, 3.29, and 3.30. The blue dots denote the randomly deployed nodes whose locations are to be determined. The red crosses denote the anchors. The pink diamonds denote the localization result of the algorithm and the green lines denote the estimation errors. Figures 3.27 and 3.28 show that the maximum localization error in a semi-C-shape topology is 20 m for the MDS-MAP algorithm and 9 m for the HMDS algorithm. Figures 3.29 and 3.30 show that the maximum localization error in an O-shape topology is 18 m for the MDS-MAP algorithm and 7 m for the HMDS algorithm. Figures 3.27 and 3.29 indicate that larger localization errors are generated by the MDS-MAP algorithm in such environments. Figures 3.28 and 3.30 demonstrate that localization errors are significantly improved in the HMDS algorithm. Simulation results show that the estimation error of the HMDS algorithm is much smaller than that of the MDS-MAP

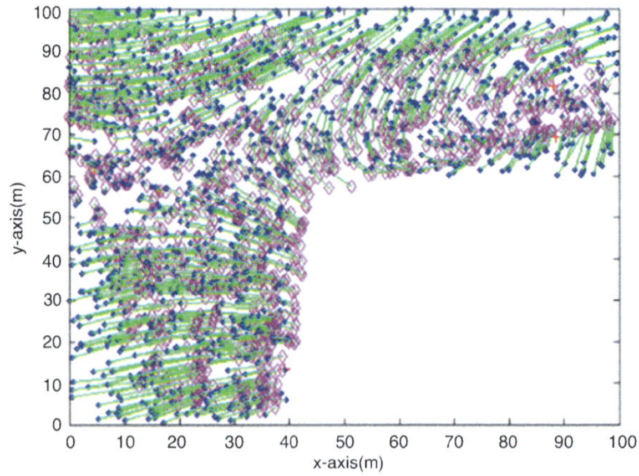

Fig. 3.27 Localization performance of the MDS-MAP algorithm in a semi-C-shape topology

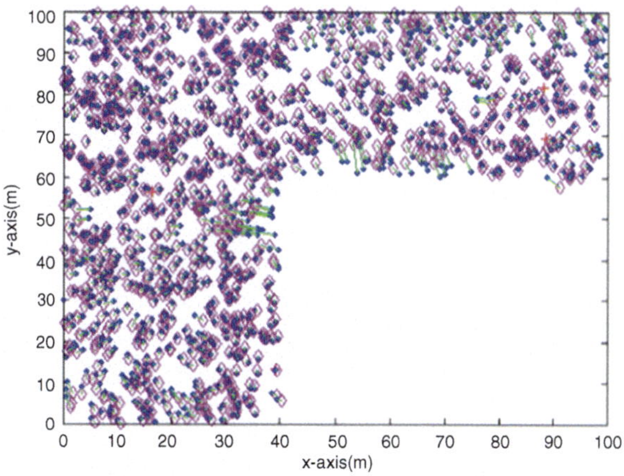

Fig. 3.28 Localization performance of the HMDS algorithm in a semi-C-shape topology

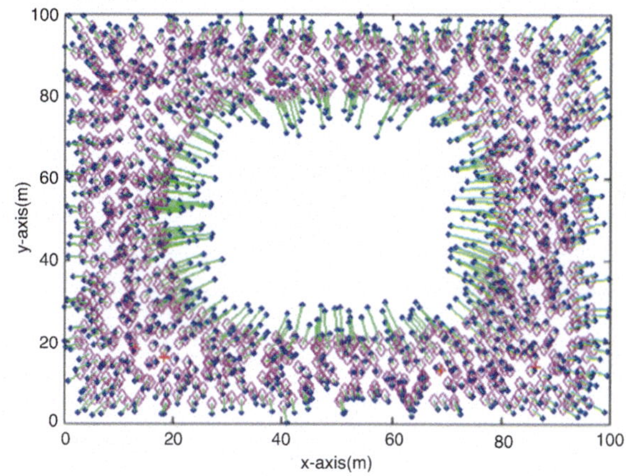

Fig. 3.29 Localization performance of the MDS-MAP algorithm in an O-shape topology

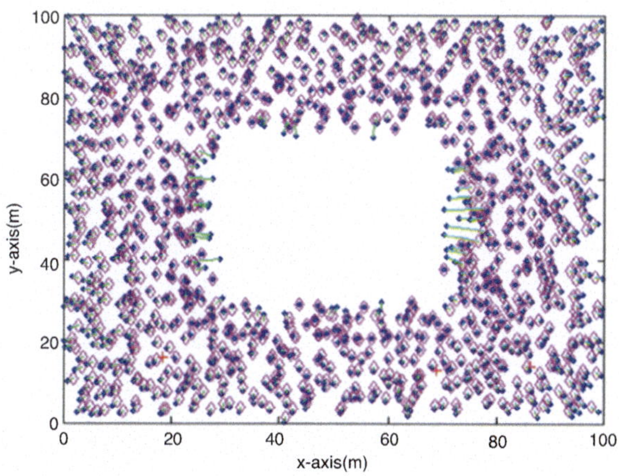

Fig. 3.30 Localization performance of the HMDS algorithm in an O-shape topology

algorithm. The reason is that the HMDS algorithm obtains more precise Euclidean distances between nodes.

Simulation results of the HMDS algorithm with 600 nodes, 5 anchors and without communication range error in a multiple O-shape topology and a concave-shape topology are shown in Figs. 3.31 and 3.32. The figures demonstrate that the maximum localization error is 10 m for the multiple O-shape topology and 11 m for the concave-shape topology. Simulation results show that the number of nodes is decreased from 1200 to 600 and the topology is more complex than the semi-C-shape and the O-shape topologies. However, the localization error of the HMDS algorithm achieves the same accuracy. The reason is that the node localization performance of the HMDS algorithm does not depend on the number of nodes and the HMDS algorithm is able to obtain more precise Euclidean distances. The simulations in Figs. 3.27, 3.28, 3.29, 3.30, 3.31, and 3.32 confirm that the HMDS algorithm improves the estimation results in all of the four topologies mentioned.

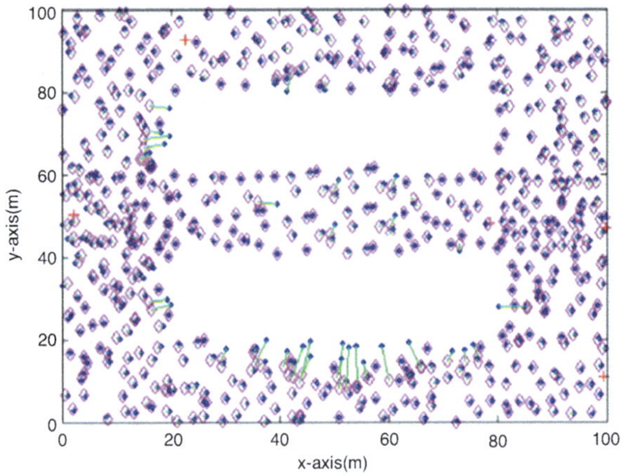

Fig. 3.31 Localization performance of the HMDS algorithm in a multiple O-shape topology

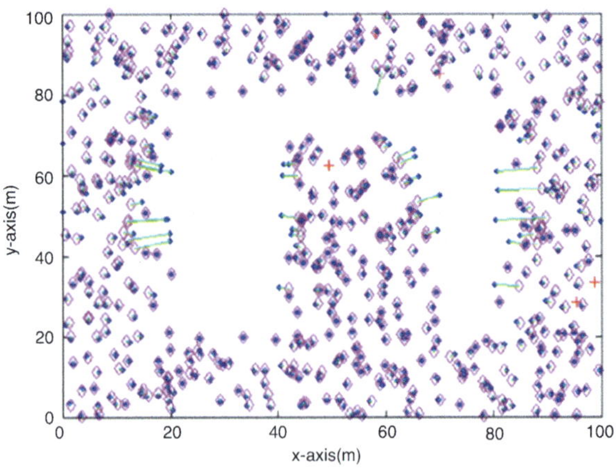

Fig. 3.32 Localization performance of the HMDS algorithm in a concave-shape topology

Table 3.2 Average error of distance estimation

Number of nodes (average connectivity)	Ave, localization error (m)	
	MDS-Map algorithm	HMDS algorithm
n = 800 (10.7)	19	8
n = 1000 (14.5)	16	6
n = 1200 (18.9)	12	4

In Table 3.2, the MDS-MAP and the HMDS algorithms of different parameters are compared in a semi-C-shape topology. Each experiment includes 20 runs and the average localization error for different numbers of nodes and connectivity are presented in the Table 3.2. The localization error is reduced as the number of nodes is increased for both the MDS-MAP and HMDS algorithms. The average localization error of the HMDS algorithm is decreased about 70% compared with the MDS-MAP algorithm in the case of high connectivity. The average localization error of the HMDS algorithm is much smaller than the MDS-MAP algorithm for all cases.

Figures 3.33, 3.34, and 3.35 show that the simulation results of the HMDS and the MDS-MAP algorithms for random semi-C-shape topology with 1200 nodes and 5 anchors. We have considered three cases for the communication range error as no the communication range error, 0.05R and 0.1R. Increasing the communication range error deteriorates the performance of both algorithms and the MDS-MAP algorithm yields larger localization errors. Simulation results show that the HMDS algorithm can improve accuracy of node localization in all cases.

We compare the HMDS algorithm with the REP algorithm, the MDS-MAP algorithm and the DV-Hop algorithm in a semi-C-shape topology with 1200 nodes

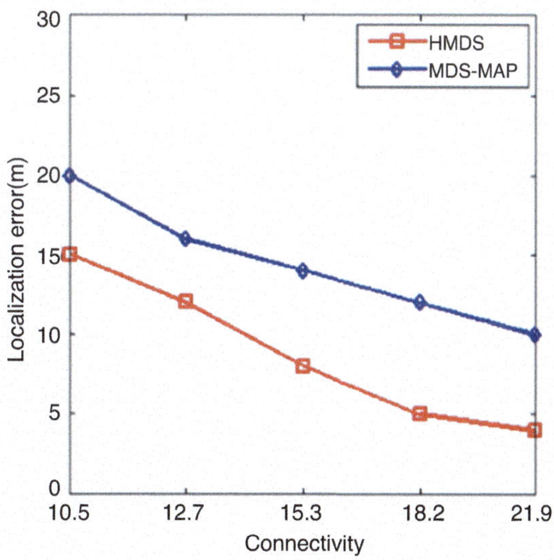

Fig. 3.33 Comparison of the MDS-MAP and HMDS without communication range error

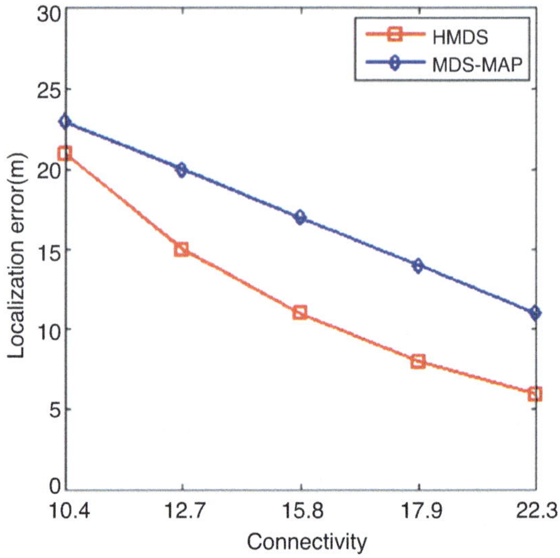

Fig. 3.34 Comparison of the MDS-MAP and HMDS with communication range error of 0.05R

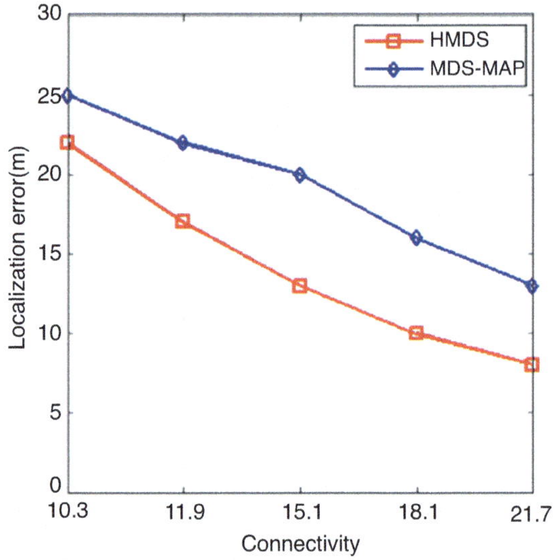

Fig. 3.35 Comparison of the MDS-MAP and HMDS with communication range error of 0.1R

and without the communication range error. Figure 3.36 illustrates that the increase of anchor number improves the localization accuracy for all four methods. Specifically, the MDS-MAP algorithm and the DV-Hop algorithm suffer larger localization errors. The REP algorithm depends on a large number of anchors within the network to assist localization. The HMDS algorithm achieves a much higher accuracy and has the minimum localization error in anisotropic WSNs with holes.

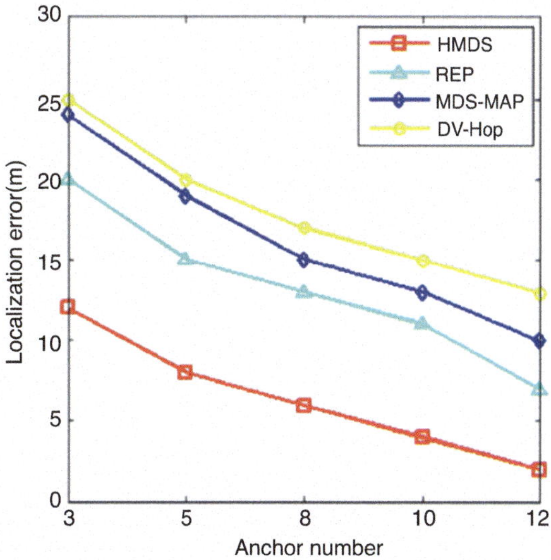

Fig. 3.36 Localization error against anchor number

3.4 Voronoi-Based Localization in SSNs

Voronoi diagrams based localization scheme (VBLS) is a widely used algorithm among computational-geometry based localization algorithms. However, due to the fact that VBLS employs RSSI as basis for measuring distance and the large errors of RSSI technique, the localization precision of VBLS is negatively affected. To solve this problem, an optimal region selection strategy based on VBLS (ORSS-VBLS) is presented in this section [6]. This algorithm settles the problem that the intersection of Voronoi areas is empty, which is caused by error of RSSI in traditional VBLS.

3.4.1 VBLS Localization

The whole WSNs could be modeled as a connected undirected graph $\mathbf{G}(\mathbf{V}, \mathbf{E})$. $\mathbf{V}$ stands for a collection of all sensors' nodes; $\mathbf{E}$ is a collection of all lines that connect neighbor sensors. The collection $\mathbf{A}$ stands for a collection of all anchor nodes, $\mathbf{S} = \mathbf{V} - \mathbf{A}$ represents for a collection of unknown nodes.

When an unknown node S_i needs to be located, it first gains the RSSI value of its k neighboring anchor nodes and rank them. Before we proceed further, let us identify some assumptions that are used throughout this article.

Assumption 3.3

$$RSSI_{i,j_1} \geq \mathrm{RSSI}_{i,j_2} \geq \ldots \geq RSSI_{i,j_k}.$$

Assumption 3.4 The maximum noise is MaxNoise.

A point collection, among which each point $X(x, y)$ meets up with the following condition, needs to be searched $0 \leq d_{i,j_1} \leq d_{i,j_2} \leq \ldots \leq d_{i,j_k} \leq R$, in which d is defined in Sect. 3.2.1 and R stands for communication radius.

Then Voronoi area $V_1(j_1)$ which contains the unknown node S_i is recorded. After removing j_1 from collection **A**, a new Voronoi diagram can be obtained. In the new Voronoi diagram, j_2 is nearest anchor node to S_i, so $V_2(j_2)$ is recorded. This process is repeated k times, until the $V_k(j_k)$ is collected. Finally, the intersection of these Voronoi areas $p(m)$ is calculated.

$$p(m) = V_1(x_1) \cap V_2(x_2) \cap \ldots \cap V_k(x_k) \tag{3.58}$$

The centroid of the intersection area $p(m)$ is the estimated location of S_i. Figure 3.37 shows the example and phases of VBLS.

There are mainly three signal transmission models: Free-space model, Two-ray ground reflection model and Shadowing model. Free-space model and Two-ray ground reflection model take the received signal as a fixed function of distance, so that their signal transmission model is an ideal circle.

However, due to multipath reflection and obstacles, the real signal transmission is anisotropic. So, we chose a more general model: Shadowing model. As Eq. (3.59) shows:

$$\left[\frac{P_r(d)}{P_r(d_0)}\right]_{dB} = -10\beta \log\left(\frac{d}{d_0}\right) + X_{db} \tag{3.59}$$

where $X_{db} \sim N(0, \sigma^2)$, β is a path loss exponent, $P_r(d_0)$ and d_0 represent reference energy and reference distance respectively.

We assume that d_A and d_B is the distance from the unknown node to anchor nodes A and B. According to Eq. (3.59), we have

$$\left[\frac{P_r(d_A)}{P_r(d_0)}\right]_{dB} = -10\beta \log\left(\frac{d_A}{d_0}\right) + X_{A-db} \tag{3.60}$$

$$\left[\frac{P_r(d_B)}{P_r(d_0)}\right]_{dB} = -10\beta \log\left(\frac{d_B}{d_0}\right) + X_{B-db} \tag{3.61}$$

While Eqs. (3.60) and (3.61) can be expressed as:

$$\left[\frac{P_r(d_A)}{P_r(d_0)}\right]_{dB} - X_{A-db} - 10\beta \log(d_0) = -10\beta \log(d_A) \tag{3.62}$$

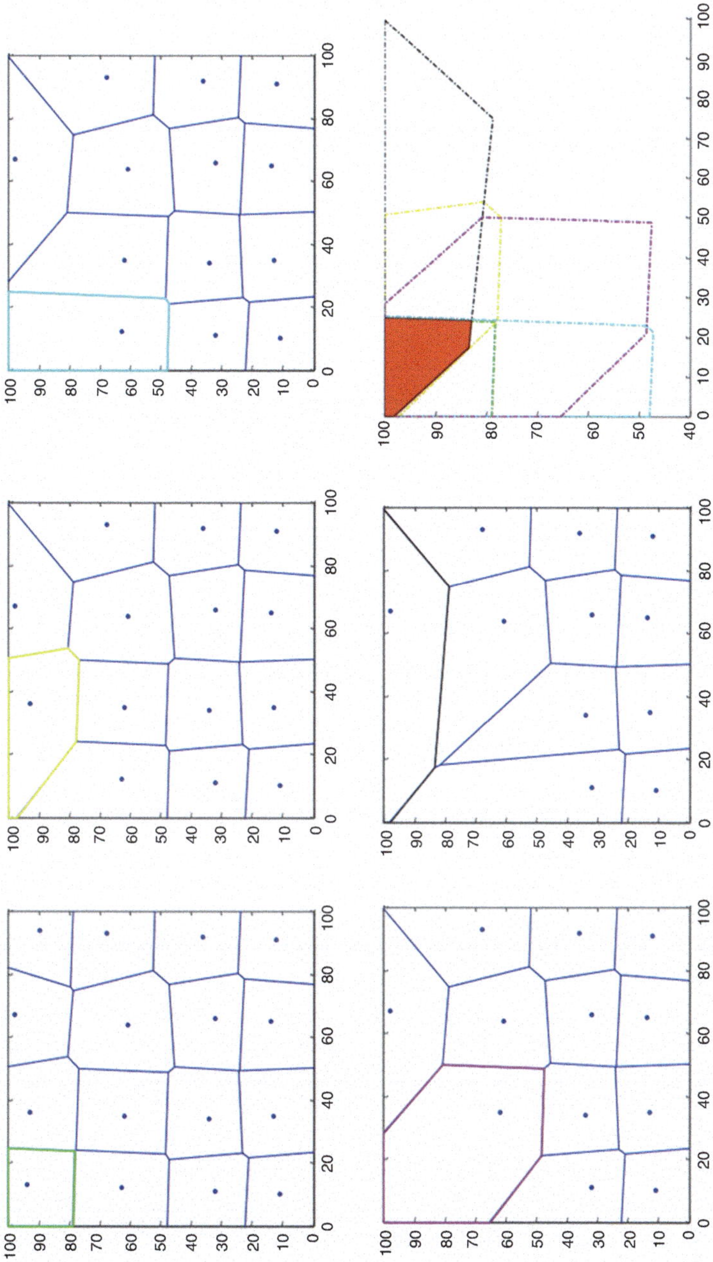

Fig. 3.37 Example and phases of VBLS

$$\left[\frac{P_r(d_B)}{P_r(d_0)}\right]_{dB} - X_{B\cdot db} - 10\beta\log(d_0) = -10\beta\log(d_B) \tag{3.63}$$

We subtract Eq. (3.63) from Eq. (3.62) as follows:

$$\left[\frac{P_r(d_A)}{P_r(d_0)}\right]_{dB} - X_{A-db} - \left[\frac{P_r(d_B)}{P_r(d_0)}\right]_{dB} + X_{B-db} = -10\beta\log\left(\frac{d_A}{d_B}\right) \tag{3.64}$$

If Eq. (3.64) satisfies

$$\left[\frac{P_r(d_A)}{P_r(d_0)}\right]_{dB} - X_{A-db} - \left[\frac{P_r(d_B)}{P_r(d_0)}\right]_{dB} + X_{B-db} < 0$$

Then we can get the conclusion that $d_A > d_B$. Furthermore, Eq. (3.65) can be obtained by Assumption 3.4 and Eq. (3.64) as follows:

$$\left[\frac{P_r(d_B)}{P_r(d_0)}\right]_{dB} - \left[\frac{P_r(d_A)}{P_r(d_0)}\right]_{dB} > 2\ \text{MaxNoise} \tag{3.65}$$

In the traditional VBLS, if the subtraction of $P_r(d_A)$ and $P_r(d_B)$ is smaller than θ (θ is determined by the noise), the rank of RSSI will not reflects the rank of distance accurately. As a result, $p(m)$ may be empty.

3.4.2 *ORSS-VBLS Localization*

3.4.2.1 Optimal Region Selection Strategy

Because of the drawback mentioned above, we present an optimal region selection strategy to deal with this issue.

Situation 1: as is shown in Fig. 3.38, there are two anchor nodes which almost have the equal distance to the unknown node. According to the characteristics of the Voronoi diagram, CD is the perpendicular bisector of AB, so nodes around CD satisfy the above condition. Therefore, we take CD as the midline to draw a rectangle. The length of the rectangle is the length of CD and the width is related to the noise signal. The greater the noise signal is, the greater the width is. Finally, the Voronoi region is replaced with the rectangle region. As the shadow in Fig. 3.38 shows, "*" is the unknown node and "·" are the anchor nodes.

Situation 2: as is shown in Fig. 3.39, there are three anchor nodes which almost have the equal distance to the unknown node. According to the characteristics of the Voronoi diagram, D is the circumcenter of nodes A, B and C, so nodes around D satisfy the above condition. Therefore, we take D as the center to draw a circle. The radius of the circle is related to the noise signal. The greater the noise signal is, the

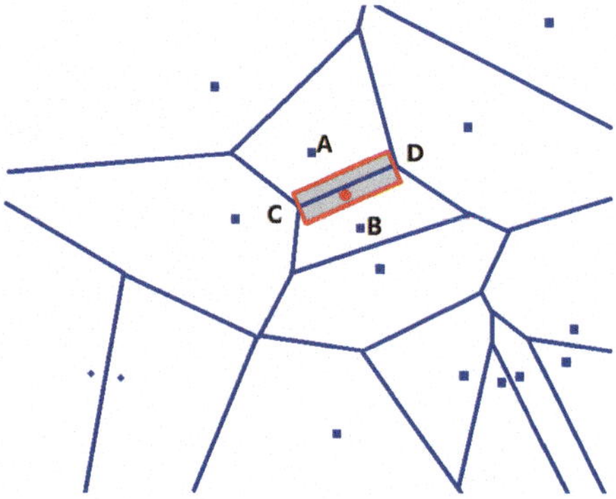

Fig. 3.38 "Two anchor nodes" situation

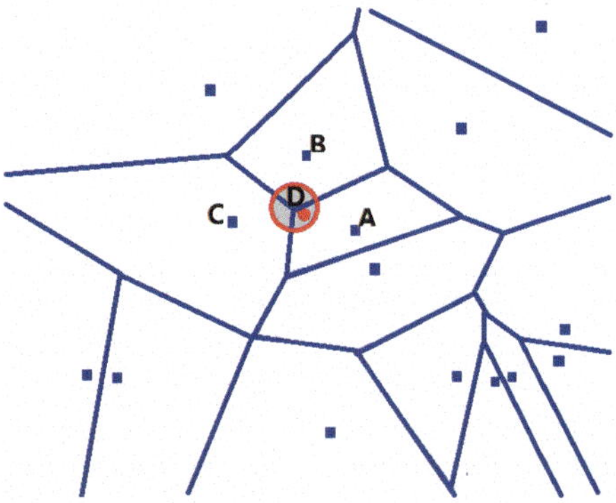

Fig. 3.39 "Three anchor nodes" situation

greater the radius is. Finally, the Voronoi region is replaced with the circle region, as the shadow in Fig. 3.39 shows.

Situation 3: as is shown in Fig. 3.40, there are $n(n > 3)$ anchor nodes which almost have the equal distance to the unknown node. According to the characteristics of the Voronoi diagram, there will be several circumcenters which are like node D in Fig. 3.39. Centroid of these Ds is named O, as can be seen in Fig. 3.40. So, nodes around O satisfy the above condition. Therefore, we take O as the center to draw a circle. The radius of the circle is related to the noise signal. The greater the noise signal is, the greater the radius is. Finally, the Voronoi region is replaced with the circle region. As the shadow in Fig. 3.40 shows.

The step is shown as follows.

1. The unknown nodes broadcast a request signal to anchor nodes within its communication radius.
2. The anchor nodes which obtained the signal will broadcast a reply signal.
3. Determine whether the number of received signal is more than zero. If the number is zero, which means there is no anchor node around, the localization failed, go to 6. If the number is not zero, then go to 4.

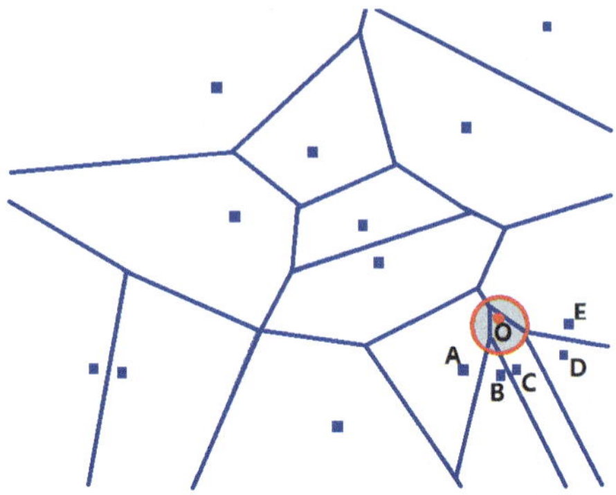

Fig. 3.40 "*n* anchor nodes" situation

4. Determine whether we can tell $d_A > d_B$ clearly. If we can, then we choose VBLS. If we can't, then the ORSS-VBLS is selected.
5. Remove the anchor node with the biggest RSSI and go to 1.
6. End.

3.4.2.2 Noise Estimation

An adaptive Kalman filtering method has the advantage of simple, real-time, and it can estimate the first moment and second moment of the system noise and observation noise. The Kalman filter model assumes the real state at time k evolved from the state at $k - 1$ according to

$$X_k = \Phi_{k,k-1}X_{k-1} + W_{k-1} \tag{3.66}$$

where $\Phi_{k,\ k\ -\ 1}$ is the state transition model which is applied to the previous state $X_{k\ -\ 1}$. W_k is the process noise which is assumed to be drawn from a zero mean multivariate normal distribution with covariance Q_k.

$$W_k - N(0, \mathrm{Q}_k)$$

At time k an observation (or measurement) Z_k of the true state X_k is made according to

$$Z_k = H_kX_k + V_k \tag{3.67}$$

where H_k is the observation model which maps the true state space into the observed space and V_k is the observation noise which is assumed to be zero mean Gaussian white noise with covariance R_k.

$$V_k - N(0, R_k)$$

The initial state, and the noise vectors at each step $\{x_0, w_1, \ldots, w_k, v_1, \ldots, v_k\}$ are all assumed to be mutually independent.

$$E[W_k] = q_k, E\left[(W_k - q_k)(W_j - q_j)^T\right] = Q_k\delta_{kj} \tag{3.68}$$

$$E[V_k] = r_k, E\left[(V_k - r_k)(V_j - r_j)^T\right] = R_k\delta_{kj} \tag{3.69}$$

$$E\left[(W_k - q_k)(V_j - r_j)^T\right] = 0 \tag{3.70}$$

An adaptive Kalman filtering method has the advantage of simple, real-time, and it can estimate the first moment and second moment of the system noise and observation noise. The method is as follows.

$$\widehat{X}_k = \widehat{X}_{k,k-1} + K_k\widetilde{Z}_k \tag{3.71}$$

$$\widehat{X}_{k,k-1} = \Phi_{k,k-1}\widehat{X}_{k-1} + \widehat{q}_k \tag{3.72}$$

$$\widetilde{Z}_k = Z_k - H_k\widehat{X}_{k,k-1} - \widehat{r}_k \tag{3.73}$$

$$K_k = P_{k,k-1}H_k^T\left[H_kP_{k,k-1}H_k^T + \widehat{R}_k\right]^{-1} \tag{3.74}$$

$$P_{k,k-1} = \Phi_{k,k-1}P_{k-1}\Phi_{k,k-1}^T + \widehat{Q}_{k-1} \tag{3.75}$$

$$P_k = [I - K_kH_k]P_{k,k-1} \tag{3.76}$$

$$\widehat{r}_{k+1} = (1 - d_k)\widehat{r}_k + d_k\left(Z_{k+1} - H_{k+1}\widehat{X}_{k+1,k}\right) \tag{3.77}$$

$$\widehat{R}_{k+1} = (1 - d_k)\widehat{R}_k + d_k\left(\widetilde{Z}_{k,1}\widetilde{Z}_{k+1}^T - H_{k+1}P_{k+1,k}H_{k+1}^T\right) \tag{3.78}$$

$$\widehat{q}_{k+1} = (1 - d_k)\widehat{q}_k + d_k\left(\widehat{X}_{k+1} - \Phi_{k+1,k}\widehat{X}_k\right) \tag{3.79}$$

$$\widehat{Q}_{k+1} = (1 - d_k)\widehat{Q}_k + d_k\left(K_{k+1}\widetilde{Z}_{k+1}\widetilde{Z}_{k+1}^T K_{k+1}^T + P_{k+1} - \Phi_{k+1,k}P_k\Phi_{k+1}^T\right) \tag{3.80}$$

$$d_i = \frac{1 - b}{1 - b^i} \quad i = 0, 1, \cdots k - 1 \tag{3.81}$$

where b is the forgetting factor.

3.4.2.3 Dynamic Region Selection

The region selection strategy mentioned above cannot match a dynamic noise. So, we have to find the relationship between the noise and the selected region. A noise observer is applied to find the environment noise.

We assume that the maximum transient noise is ε. For situation 1, the width of the rectangle is l. According to Eq. (3.65), we can get the conclusion that:

$$l \geq 2 \times \varepsilon \tag{3.82}$$

For situation 2 and situation 3, the radius of the circle is r. According to Eq. (3.65), the following conclusion can be obtained.

$$r \geq 2 \times \varepsilon \tag{3.83}$$

To guarantee the location accuracy, we choose to double the maximum transient error as l and r.

Figure 3.41 shows the example and phases of ORSS-VBLS. Obviously, in contrast to VBLS mentioned above, our algorithm has narrowed the selected region significantly.

3.4.3 Simulation Results

3.4.3.1 Simulation Setup

The basic network setup is a 100 m*100 m square field with an environment noise $EN \sim N(\mu, \sigma^2)$, where $\mu = 0$ and $\sigma = 0.1$. The probability density function is as follows

$$f(x) = \frac{1}{\sqrt{2\pi}\sigma} \exp\left(-\frac{(x-\mu)^2}{2\sigma^2}\right) \tag{3.84}$$

Figure 3.42 shows the signal transmission, the circumference stands for RSSI without environment noise.

3.4.3.2 Comparison with Other Methods

As is shown in Figs. 3.43 and 3.44, the localization error becomes smaller with more anchor nodes and bigger communication radius. It is because each node can communicate more anchor nodes in average, and enjoys more localization information.

Figure 3.43 shows the influence of the different number of anchor nodes to the average localization error. The data in chart shows that the average localization error of ORSS-VBLS is lower than that of I-VBLS and centroid algorithm, and will reduce with the increase of the proportion of anchor nodes.

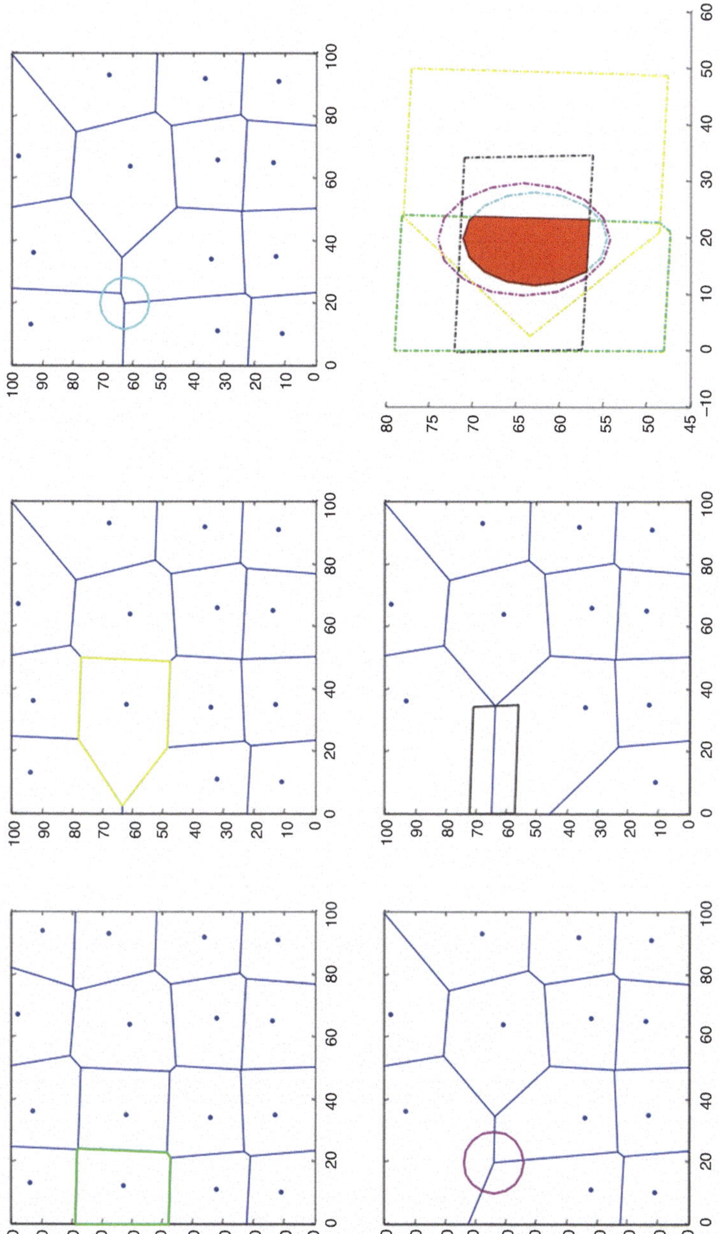

Fig. 3.41 Example and phases of ORSS-VBLS

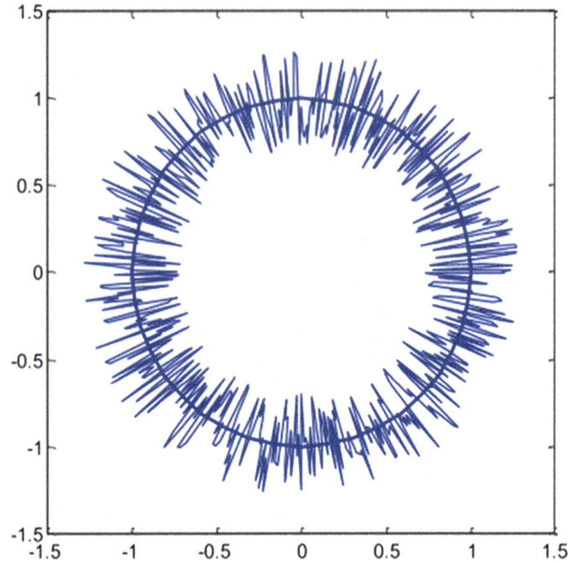

Fig. 3.42 Radio propagation pattern

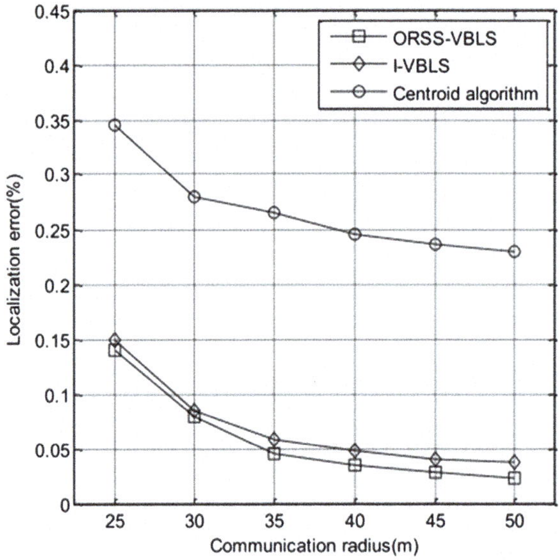

Fig. 3.43 The average localization error of different communication radius in environment

Figure 3.44 depicts the localization errors of the 3 approaches in networks with different numbers of anchor nodes. Centroid algorithm suffers large errors especially when anchor node number is small. ORSS-VBLS achieves the best performance among three methods. ORSS-VBLS and I-VBLS are not affected by the anchor node number when anchor node number is not small (more than 15).

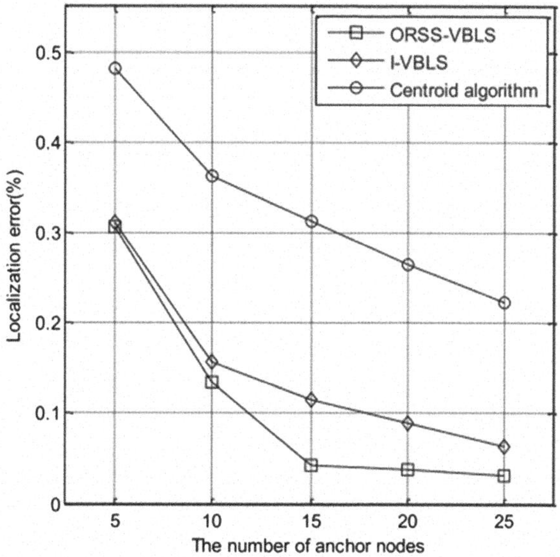

Fig. 3.44 The average localization error of the number of different anchor nodes in environment

3.5 Delaunay Triangulation Based Localization Scheme

In this section, we present a Delaunay triangulation based localization scheme [7] (DBLS) in SSNs. Based on RSSI, the Delaunay area is generated. The unknown node determines which Delaunay area it belongs to and then record it down. After several iteration, the overlapping region of all Delaunay areas is identified as the possible region where sensor resides in. The simulation result shows that DBLS has a better performance than VBLS algorithm and centroid algorithm. That's because the Delaunay triangulation divide the plane into more parts than Voronoi diagram.

3.5.1 Delaunay Triangulation

The definition of Delaunay triangulation is given as follows:

Definition A Delaunay triangulation for a set P of points in a plane is a triangulation $DT(P)$ such that no point in P is inside the circumcircle of any triangle in $DT(P)$.

There are some properties of Delaunay triangulation. Let n be the number of points.

1. In the plane, if there are b vertices on the convex hull, then any triangulation of the points has $2n - 2 - 2b$ triangles at most.
2. A circle circumscribing any Delaunay triangle does not contain any other input points in its interior.

3. If a circle passing through two of the input points doesn't contain any other of them in its interior, then the segment connecting the two points is an edge of a Delaunay triangulation of the given points.
4. The convex hull is unique.

3.5.2 DBLS Localization

3.5.2.1 Localization Scheme Without Noise

First, all the anchor nodes generate the global Delaunay triangulation. The unknown nodes estimate their distance from their neighbor anchor nodes by RSSI. After that, the unknown node U_i judges which Delaunay triangulation it belongs to and records it down by using cosine theorem. Then we delete the anchor node which has the minimum distance from U_i. We iterate the above steps (k is determined by accuracy requirement). At last, U_i must be in the overlapping area of all the recorded Delaunay triangulation.

As is shown in Fig. 3.45, the vertexes of triangle $\Delta A_1A_2A_3$ are $A_1(x_1, y_1)$, $A_2(x_2, y_2)$, $A_3(x_3, y_3)$. The distance of unknown node U_i to A_1, A_2 and $_{\mathrm{v}}$ are d_1, d_2 and d_3 respectively. The necessary and sufficient condition that U_i lies in $\Delta A_1A_2A_3$ is

$$\angle A_1U_iA_3 + \angle A_2U_iA_3 + \angle A_1U_iA_2 = 2\pi \tag{3.85}$$

According to cosine theorem,

$$\begin{aligned} \cos\angle A_1U_iA_2 &= \frac{d_1^2 + d_2^2 - A_1A_2^2}{2d_1d_2} \\ \cos\angle A_1U_iA_3 &= \frac{d_1^2 + d_3^2 - A_1A_3^2}{2d_1d_3} \\ \cos\angle A_2U_iA_3 &= \frac{d_2^2 + d_3^2 - A_2A_3^2}{2d_2d_3} \end{aligned} \tag{3.86}$$

Fig. 3.45 Relationship between the unknown node and the triangular

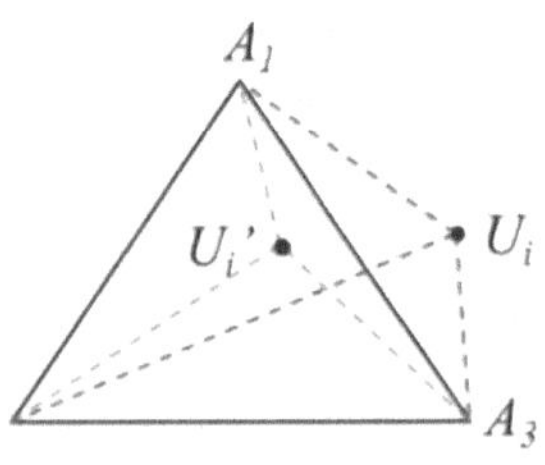

Thus if $\angle A_1U_iA_2 + \angle A_1U_iA_3 + \angle A_2U_iA_3 = 2\pi$, U_i lies in $\Delta A_1A_2A_3$. If $\angle A_1U_iA_2 + \angle A_1U_iA_3 + \angle A_2U_iA_3 < 2\pi$, U_i is outside $\Delta A_1A_2A_3$.

Without loss of generality, we assume that U_i lies in Delaunay area $\Delta A_1A_2A_3$. After that, we record the Delaunay area and delete the closest anchor node to U_i. Then, a new Delaunay triangulation is generated. We repeat the above steps and record all the Delaunay areas. At last, the centroid of overlapping area of all the Delaunay areas is the estimate location of U_i.

The flow chart of DBLS is as follows:

1. All the anchor nodes generate the global Delaunay triangulation. The unknown node U_i send a Request signal, to ask for localization from its neighbor anchor nodes.
2. All the anchor nodes which received the Request signal send a Reply signal which contains the anchor nodes' location and RSSI.
3. U_i estimates the distance from its neighbor anchor nodes by RSSI.
4. U_i determines which Delaunay area it belongs to.
5. Receive Reply messages from m anchors, the DBLS algorithm is shown in Algorithm 3.1.
6. Calculate the centroid of overlapping area of all the Delaunay areas.

Algorithm 3.1 The DBLS Algorithm

Step 1 For $j = 1$ to m (m is the number of neighbor anchor nodes.)
  Step 2 Calculate Delaunay area D_{ij} of anchor $A_i[j]$
  Step 3 Add Delaunay (D_{ij})
    Step 4 For $k = j + 1$ to m (k is the iteration number.)
    Step 5 Remove $A_i[j]$ from $A_i[k]$'s one-hop neighbors table
    Step 6 End for
  Step 7 End for

Figure 3.46 is an overview of DBLS. The blue∗ is anchor nodes, and the red∗ is the unknown node. After generating the global Delaunay triangulation, U_i begins to determine which Delaunay areas it belongs to according to Eq. (3.85). Figure 3.46a is the first iteration. The green area is the first Delaunay area. Then the nearest anchor node is deleted and a new global Delaunay triangulation is generated as Fig. 3.46b shows. After several iterate, U_i must be in the overlapping area of all the Delaunay areas.

3.5.2.2 Localization Scheme with Noise

In reality, DBLS is easily affected by measurement noise. We assume that there are three points A, B and O lying on the plan. d_A and d_B are the distance of OA and OB respectively. According to Eq. (3.23),

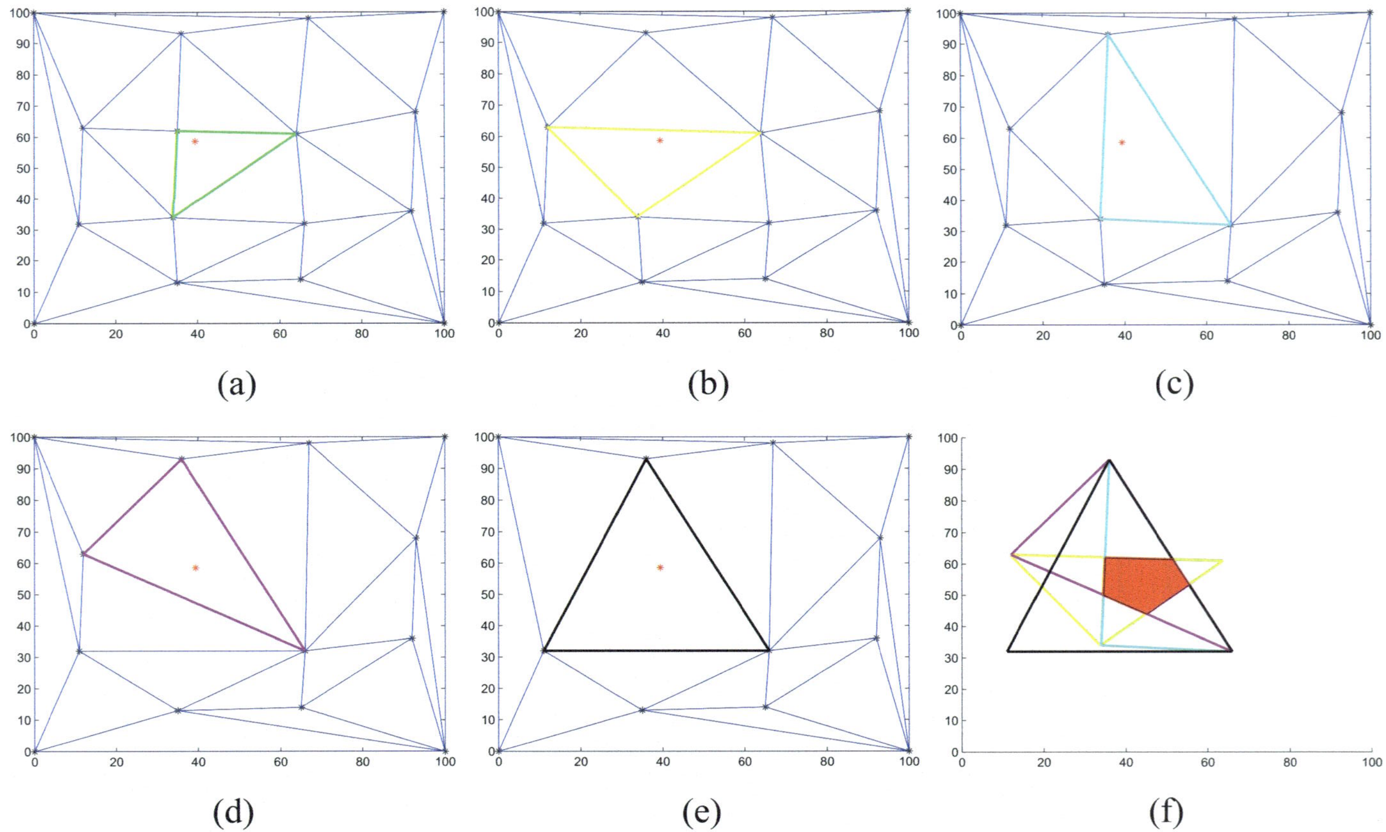

Fig. 3.46 DBLS Overview. (**a**) The First iteration of DBLS (**b**) The second iteration of DBLS (**c**) The third iteration of DBLS (**d**) The fourth iteration of DBLS (**e**) The fifth interation of DBLS (**f**) Finally result of DBLS

$$\left[\frac{P_r(d_A)}{P_r(d_0)}\right]_{dB} = -10\beta \log\left(\frac{d_A}{d_0}\right) + X_{A-db} \tag{3.87}$$

$$\left[\frac{P_r(d_B)}{P_r(d_0)}\right]_{dB} = -10\beta \log\left(\frac{d_B}{d_0}\right) + X_{B-db} \tag{3.88}$$

where X_{A-db} and X_{B-db} are the instantaneous measurement noise.

We subtract Eq. (3.87) from Eq. (3.88) as follows:

$$\left[\frac{P_r(d_A)}{P_r(d_0)}\right]_{dB} - X_{A-db} - \left[\frac{P_r(d_B)}{P_r(d_0)}\right]_{dB} + X_{B-db} = -10\beta \log\left(\frac{d_A}{d_B}\right) \tag{3.89}$$

If Eq. (3.89) satisfies

$$\left[\frac{P_r(d_A)}{P_r(d_0)}\right]_{dB} - X_{A-db} - \left[\frac{P_r(d_B)}{P_r(d_0)}\right]_{dB} + X_{B-db} < 0 \tag{3.90}$$

Then we can get the conclusion that $d_A > d_B$. We assume that the maximum instantaneous measurement noise is MaxNoise, which means $X <$ MaxNoise. Then

$$\left[\frac{P_r(d_B)}{P_r(d_0)}\right]_{dB} - \left[\frac{P_r(d_A)}{P_r(d_0)}\right]_{dB} > 2\ \text{MaxNoise} \tag{3.91}$$

In DBLS, if the subtraction of $P_r(d_A)$ and $P_r(d_B)$ is smaller than θ (θ is determined by the noise), the RSSI cannot reflect the distance exactly. That means the unknown node may wrongly determine the Delaunay area it belongs to. As a result, the overlapping area may be empty.

As shown in Fig. 3.47, while U_i is moving towards $\Delta A_1A_2A_3$ from outside, $\angle A_1U_iA_3 \rightarrow \pi$, while

$$\angle A_1U_iA_3 + \angle A_2U_iA_3 + \angle A_1U_iA_2 = 2\angle A_1U_iA_3 \rightarrow 2\pi \tag{3.92}$$

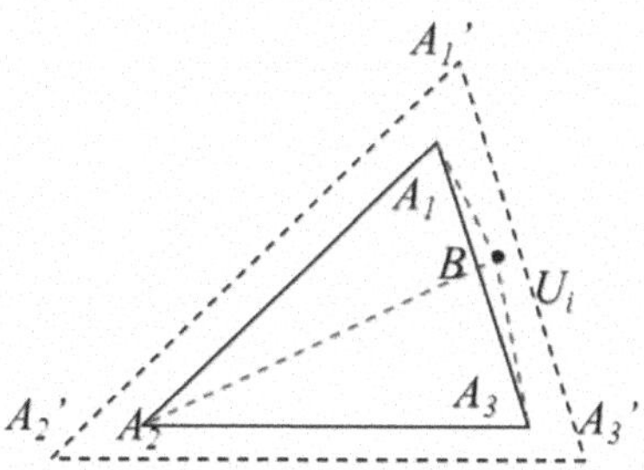

Fig. 3.47 The unknown node near the edge of triangular

In this case, we cannot tell whether U_i lies in $\Delta A_1A_2A_3$ or not with certainty. Fortunately, this case only occur when U_i is close to $\Delta A_1A_2A_3$. So, we can just enlarge the triangle η times (η is determined by maximum instantaneous measurement noise). As Fig. 3.47 shows, we take $\Delta A_1'A_2'A_3'$ as the Delaunay area instead of $\Delta A_1A_2A_3$. Thus the new Delaunay area must contains U_i .

3.5.3 *Simulation Results*

3.5.3.1 Simulation Setup

Several simulations are carried out to test the performance of DBLS. Simulations are on a 100 m × 100 m square field with an environment noise $EN{\sim}N(\mu, \sigma^2)$, where $\mu = 0$ and $\sigma = 0.1$. The probability density function is as follows Eq. (3.84).

3.5.3.2 Comparison with Other Methods

As shown in Figs. 3.48 and 3.49, with the increase of the communication radius or the number of anchor nodes, the localization error of the three algorithms decreases. That's because the unknown nodes can receive more information which can help them to localization. In our simulation, DBLS has a better performance to VBLS, for the Delaunay triangulation dividing the plan into $2n - b - 2$ parts (b is vertices on

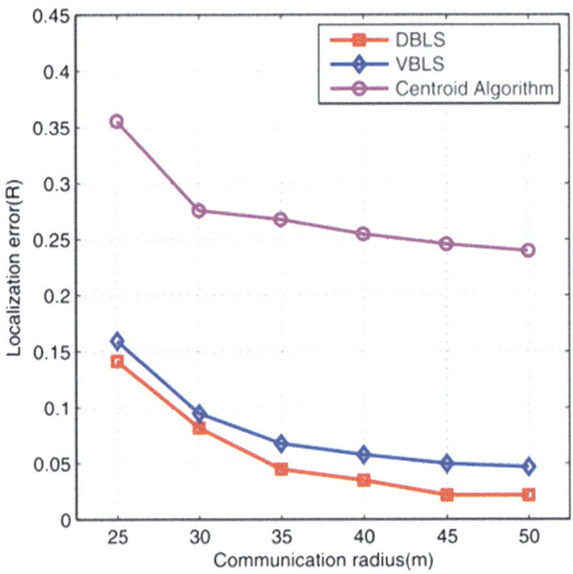

Fig. 3.48 Localization error varying communication range

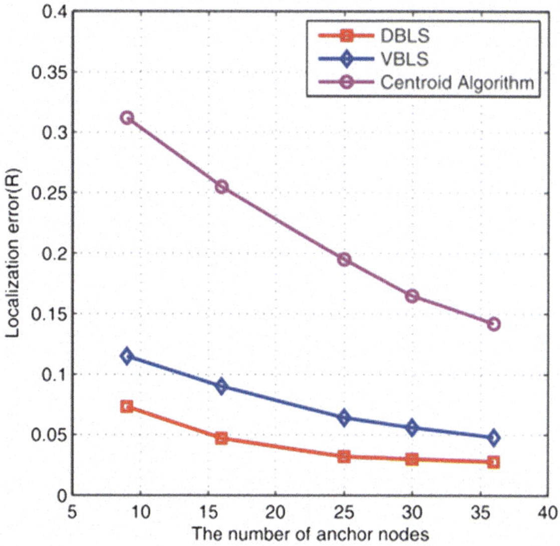

Fig. 3.49 Localization error varying anchor density

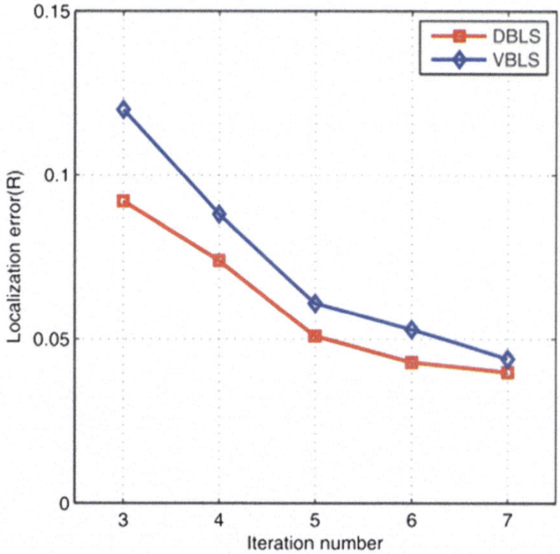

Fig. 3.50 Localization error varying iteration number

the convex hull). Furthermore, if we set four anchor nodes at the vertex of the square simulation area, the plan will be divided into $2n - 6$ parts. As to VBLS, the Voronoi diagram divides the plan into n parts. That means if $n > 6$, DBLS will have a better performance than VBLS.

Figure 3.50 displays that with the increase of iteration times, the overlapping area is getting smaller which leads to a better localization performance. In Fig. 3.51, with

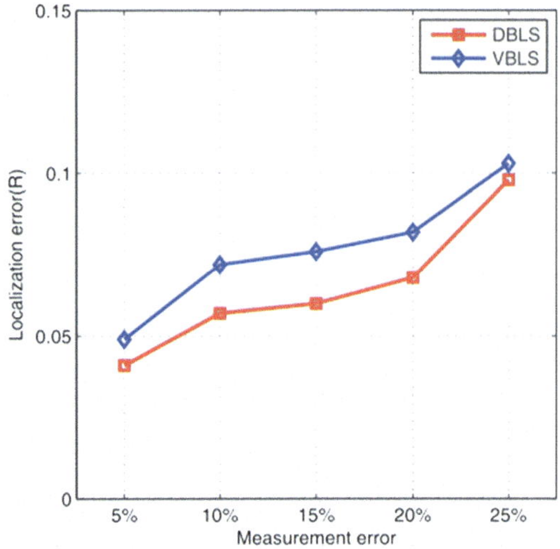

Fig. 3.51 Localization error varying measurement error

the increase of measurement error, the localization error becomes large. However, when the measurement error is between 10% and 20%, both two curves are gently. That is to say, both two schemes are not very sensitive to the measurement error in a certain range.

3.6 Neighbor Constraint Assisted Distributed Localization

In this section, we present a novel algorithm called neighbor constraint assisted distributed localization (NCA-DL) which is effective in refining the distances required for localization [8]. The core idea behind NCA-DL is to use the Cayley-Menger determinant. The major contribution of this algorithm is twofold. First, the proposed algorithm is distributed so that sensor nodes can estimate their locations by themselves. Second, it introduces the idea of geometric constraints and decreases the distance estimation errors with the help of an adjacent neighbor.

3.6.1 Cayley-Menger Determinant

Let us first consider $\{\mathbf{a}_1, \mathbf{a}_2, \ldots, \mathbf{a}_n\}$ and $\{\mathbf{b}_1, \mathbf{b}_2, \ldots, \mathbf{b}_n\}$ which represent a set of n distinct points, respectively. The Cayley-Menger matrix of these two sets can be defined as

$$C(\mathbf{a}_1, \mathbf{a}_2, \cdots, \mathbf{a}_n; \mathbf{b}_1, \mathbf{b}_2, \ldots, \mathbf{b}_n) \triangleq \begin{bmatrix} d^2(\mathbf{a}_1, \mathbf{b}_1) & d^2(\mathbf{a}_1, \mathbf{b}_2) & \cdots & d^2(\mathbf{a}_1, \mathbf{b}_n) & 1 \\ d^2(\mathbf{a}_2, \mathbf{b}_1) & d^2(\mathbf{a}_2, \mathbf{b}_2) & \cdots & d^2(\mathbf{a}_2, \mathbf{b}_n) & 1 \\ \vdots & \vdots & \ddots & \vdots & \vdots \\ d^2(\mathbf{a}_n, \mathbf{b}_1) & d^2(\mathbf{a}_n, \mathbf{b}_2) & \cdots & d^2(\mathbf{a}_n, \mathbf{b}_n) & 1 \\ 1 & 1 & \cdots & 1 & 0 \end{bmatrix} \tag{3.93}$$

where $d(\mathbf{a}_i, \mathbf{b}_j)$, $\forall\ (i,j) \in \{1, 2, \ldots, n\}$ denotes the Euclidean distance between the points $\mathbf{a}_i$ and $\mathbf{b}_j$.

The Cayley-Menger determinant of these two sequences of n points is defined as

$$D(\mathbf{a}_1, \mathbf{a}_2, \cdots, \mathbf{a}_n; \mathbf{b}_1, \mathbf{b}_2, \ldots, \mathbf{b}_n) \triangleq \det C(\mathbf{a}_1, \mathbf{a}_2, \cdots, \mathbf{a}_n; \mathbf{b}_1, \mathbf{b}_2, \ldots, \mathbf{b}_n)) \tag{3.94}$$

The above determinant is widely used in distance geometry theory. When the two sequences of points are the same, $D(\mathbf{a}_1, \mathbf{a}_2, \ldots, \mathbf{a}_n; \mathbf{a}_1, \mathbf{a}_2, \ldots, \mathbf{a}_n)$ is denoted for convenience by $D(\mathbf{a}_1, \mathbf{a}_2, \ldots, \mathbf{a}_n)$ which is simply called a Cayley-Menger determinant.

A brief summary of the Cayley-Menger determinant is generalized as follows.

Theorem 3.1 Consider an n-tuple of points $\mathbf{a}_1, \mathbf{a}_2, \ldots, \mathbf{a}_n$ in m-dimensional space. If $n \geq m + 2$, then the Cayley-Menger matrix $C(\mathbf{a}_1, \mathbf{a}_2, \ldots, \mathbf{a}_n)$ is singular, namely,

$$D(\mathbf{a}_1, \mathbf{a}_2, \ldots, \mathbf{a}_n) = 0 \tag{3.95}$$

Theorem 3.2 Consider an n-tuple of points $\mathbf{a}_1, \mathbf{a}_2, \ldots, \mathbf{a}_n$ in m-dimensional space. If $n \geq m + 1$, the rank of Cayley-Menger matrix $C(\mathbf{a}_1, \mathbf{a}_2, \ldots, \mathbf{a}_n)$ is at most $m + 1$.

In a 2-dimensional Euclidean space, each node has a set of coordinates (x_i, y_i). The study of the localization problem applied to WSNs first requires some basic terminology and concepts to be defined.

According to the above theorems and definitions, an interesting development of localization is how to use the Cayley-Menger determinant to reduce the impact of distance measurement errors. As shown in Fig. 3.52, let $d_{ij} = d(\mathrm{a}_i, \mathbf{a}_j)$ denote the accurate Euclidean distance between anchors a_i and a_j with $i \neq j$, $(i,j = 1,2,3)$, which can be inferred from known anchor positions. $d_{0i} = d(\mathbf{r}_0, \mathbf{a}_i)$ denotes the accurate distances between the regular node $\mathbf{r}_0$ and node a_i with $i = 1,2,3$, and $\overline{d}_{0i}$ denotes the inaccurate distances acquired by either noisy range measurement or computations. Then the following equation is defined:

$$\overline{d}_{0i}^2 = d_{0i}^2 - \varepsilon_i \tag{3.96}$$

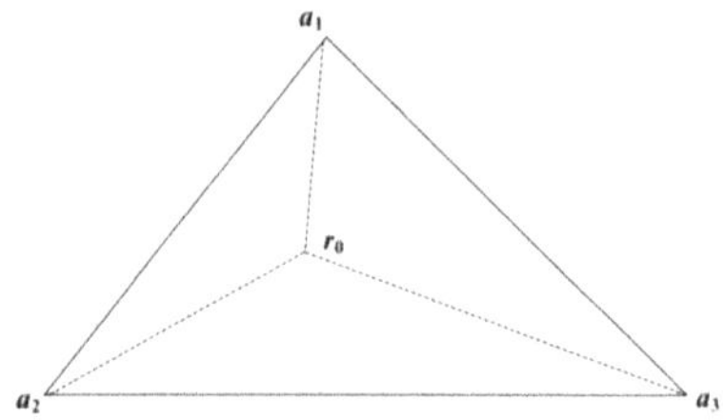

Fig. 3.52 A regular node and the anchors

Theorem 3.3 The errors ε_i for $i = 1,2,3$ as defined immediately above satisfy a single algebraic equality which is quadratic though not homogeneous in the ε_i's:

$$\boldsymbol{\varepsilon}^T \mathbf{A} \boldsymbol{\varepsilon} \;\; + \boldsymbol{\varepsilon}^T \mathbf{b} + c = 0 \tag{3.97}$$

where

$$\boldsymbol{\varepsilon} = [\varepsilon_1, \varepsilon_2, \varepsilon_3]^T$$

$$\mathbf{A} = \begin{pmatrix} 2d_{23}^2 & d_{12}^2 - d_{13}^2 - d_{23}^2 & d_{13}^2 - d_{23}^2 - d_{12}^2 \\ d_{12}^2 - d_{13}^2 - d_{23}^2 & 2d_{13}^2 & d_{23}^2 - d_{12}^2 - d_{13}^2 \\ d_{13}^2 - d_{23}^2 - d_{12}^2 & d_{23}^2 - d_{12}^2 - d_{13}^2 & 2d_{12}^2 \end{pmatrix}$$

$$b_1 = 4d_{23}^2\overline{d}_{01}^2 + 2\left(d_{12}^2 - d_{13}^2 - d_{23}^2\right)\overline{d}_{02}^2 + 2\left(d_{13}^2 - d_{12}^2 - d_{23}^2\right)\overline{d}_{03}^2 + 2d_{23}^2\left(d_{23}^2 - d_{12}^2 - d_{13}^2\right)$$

$$b_2 = 4d_{13}^2\overline{d}_{02}^2 + 2\left(d_{12}^2 - d_{13}^2 - d_{23}^2\right)\overline{d}_{01}^2 + 2\left(d_{23}^2 - d_{12}^2 - d_{13}^2\right)\overline{d}_{03}^2 + 2d_{13}^2\left(d_{13}^2 - d_{12}^2 - d_{23}^2\right)$$

$$b_3 = 4d_{12}^2\overline{d}_{03}^2 + 2\left(d_{13}^2 - d_{12}^2 - d_{23}^2\right)\overline{d}_{01}^2 + 2\left(d_{23}^2 - d_{12}^2 - d_{13}^2\right)\overline{d}_{02}^2 + 2d_{13}^2\left(d_{12}^2 - d_{13}^2 - d_{23}^2\right)$$

$$\begin{aligned} c = {} & 2d_{12}^2 d_{13}^2 d_{23}^2 + 2d_{23}^2\overline{d}_{01}^4 + 2d_{13}^2\overline{d}_{02}^4 + 2d_{12}^2\overline{d}_{03}^4 + 2\left(d_{12}^2 - d_{13}^2 - d_{23}^2\right)\overline{d}_{01}^2\overline{d}_{02}^2 \\ & + 2\left(d_{13}^2 - d_{12}^2 - d_{23}^2\right)\overline{d}_{01}^2\overline{d}_{03}^2 + 2\left(d_{23}^2 - d_{12}^2 - d_{13}^2\right)\overline{d}_{02}^2\overline{d}_{03}^2 + 2d_{23}^2\left(d_{23}^2 - d_{12}^2 - d_{13}^2\right)\overline{d}_{01}^2 \\ & + 2d_{13}^2\left(d_{13}^2 - d_{12}^2 - d_{23}^2\right)\overline{d}_{02}^2 + 2d_{12}^2\left(d_{12}^2 - d_{13}^2 - d_{23}^2\right)\overline{d}_{03}^2 \end{aligned}$$

Proof According to Theorem 3.1, we know that $D(\mathbf{r}_0, \mathbf{a}_1, \mathbf{a}_2, \mathbf{a}_3) = 0$, namely

$$\det \begin{pmatrix} 0 & d_{01}^2 & d_{02}^2 & d_{03}^2 & 1 \\ d_{01}^2 & 0 & d_{12}^2 & d_{13}^2 & 1 \\ d_{02}^2 & d_{12}^2 & 0 & d_{23}^2 & 1 \\ d_{03}^2 & d_{13}^2 & d_{23}^2 & 0 & 1 \\ 1 & 1 & 1 & 1 & 0 \end{pmatrix} = 0 \tag{3.98}$$

Suppose the anchors are nonlinear, $D(\mathbf{a}_1, \mathbf{a}_2, \mathbf{a}_3) \neq 0$. From Eq. (3.98), the following equation can be derived:

$$\begin{pmatrix} d_{01}^2 & d_{02}^2 & d_{03}^2 & 1 \end{pmatrix} \mathbf{E}^{-1} \begin{pmatrix} d_{01}^2 \\ d_{02}^2 \\ d_{03}^2 \\ 1 \end{pmatrix} = 0 \tag{3.99}$$

where

$$\mathbf{E} = \begin{pmatrix} 0 & d_{12}^2 & d_{13}^2 & 1 \\ d_{12}^2 & 0 & d_{23}^2 & 1 \\ d_{13}^2 & d_{23}^2 & 0 & 1 \\ 1 & 1 & 1 & 0 \end{pmatrix} \tag{3.100}$$

Then according to Eq. (3.96), we can obtain

$$\left(\overline{d}_{01}^2 + \varepsilon_1 \overline{d}_{02}^2 + \varepsilon_2 \overline{d}_{03}^2 + \varepsilon_3 1 \right) \mathbf{E}^{-1} \begin{pmatrix} \overline{d}_{01}^2 + \varepsilon_1 \\ \overline{d}_{02}^2 + \varepsilon_2 \\ \overline{d}_{03}^2 + \varepsilon_3 \\ 1 \end{pmatrix} = 0 \tag{3.101}$$

Multiplying both sides of Eq. (3.101) by the determinant of $\mathbf{E}^{-1}$, we can arrive at Eq. (3.97). This completes the proof.

3.6.2 Geometric Relations with Neighbor Constraint

We will focus on the geometric relations among the distances between nodes, which can be transformed to an algebraic constraint of the distance estimation errors. At first, we define another basic term.

As shown in Fig. 3.53, $\mathbf{r}(x_0, y_0)$ represents a regular node which needs to be located, $\mathbf{n}(x_1, y_1)$ represents a neighbor of node $\mathbf{r}'$ and $\mathbf{a}_i(x_i, y_i)$ represents the anchors with $i = 2{,}3{,}4$. Then let $d_{ij} = d(\mathbf{a}_i, \mathbf{a}_{i\cdot})$ denotes the accurate Euclidean distance between anchors $\mathbf{a}_i$ and $\mathbf{a}_j$ with $i \neq j$, $i, j = 2{,}3{,}4$, $d_{1i} = d(\mathbf{n}, \mathbf{a}_i)$ denotes the accurate distances between the neighbor node $\mathbf{n}$ and anchor $\mathbf{a}_i$ with $i = 2{,}3{,}4$, $d_{0i} = d(\mathbf{r}, \mathbf{a}_i)$ denotes the accurate distances between the regular node $\mathbf{r}$ and anchor $\mathbf{a}_i$ with $i = 2{,}3{,}4$, and d_{01} denote the accurate distance between regular node $\mathbf{r}$ and its neighbor node $\mathbf{n}$.

In this case, suppose we know the accurate distances $d_{1i}(i = 2{,}3{,}4)$ and the accurate distance d_{01} by refinement or setting node $\mathbf{n}$ as a mobile anchor. Then $\overline{d}_{0i}^2 = d_{0i}^2 - \varepsilon_i$ denotes the inaccurate distances squared between node $\mathbf{r}$ and anchor $\mathbf{a}_i$ with $i = 2{,}3{,}4$, for some error ε_i. That is to say, the true distances represented by the dotted line in Fig. 3.53 cannot be obtained. In this work, we aim to refine these inaccurate distances to trend toward actual values.

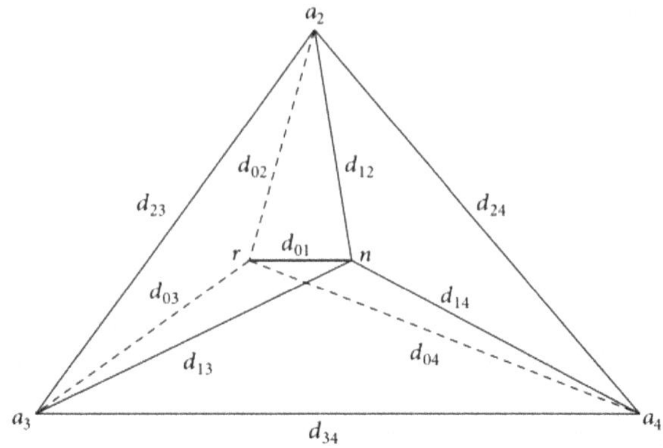

Fig. 3.53 Anchors, a regular node, and its neighbor node

Theorem 3.4 The errors ε_i for $i = 2,3,4$ as defined immediately above satisfy an algebraic equality in the ε_i's, and whose coefficients are determined by d_{01}, $\overline{d}_{0i}$ for $i = 2,3,4$ and d_{ij} for $i, j = 1,2,3,4$, $i \neq j$:

$$\alpha\varepsilon_2 + \beta\varepsilon_3 + \gamma\varepsilon_4 + \delta = 0 \tag{3.102}$$

where

$$\begin{aligned}
\alpha &= d_{34}^2\left(d_{23}^2 + d_{24}^2 - d_{34}^2\right) - 2d_{34}^2 d_{12}^2 + d_{13}^2\left(d_{24}^2 - d_{23}^2 + d_{34}^2\right) + d_{14}^2\left(d_{23}^2 - d_{24}^2 + d_{34}^2\right)\\
\beta &= d_{24}^2\left(d_{23}^2 - d_{24}^2 + d_{34}^2\right) - 2d_{24}^2 d_{13}^2 + d_{12}^2\left(d_{24}^2 - d_{23}^2 + d_{34}^2\right) + d_{14}^2\left(d_{23}^2 + d_{24}^2 - d_{34}^2\right)\\
\gamma &= d_{23}^2\left(d_{24}^2 - d_{23}^2 + d_{34}^2\right) - 2d_{23}^2 d_{14}^2 + d_{12}^2\left(d_{23}^2 - d_{24}^2 + d_{34}^2\right) + d_{13}^2\left(d_{23}^2 + d_{24}^2 - d_{34}^2\right)\\
\delta &= \overline{d}_{02}^2\left(d_{34}^2\left(d_{23}^2 + d_{24}^2 - d_{34}^2\right) - 2d_{34}^2 d_{12}^2 + d_{13}^2\left(d_{24}^2 - d_{23}^2 + d_{34}^2\right) + d_{14}^2\left(d_{23}^2 - d_{24}^2 + d_{34}^2\right)\right)\\
&\quad + \overline{d}_{03}^2\left(d_{24}^2\left(d_{23}^2 - d_{24}^2 + d_{34}^2\right) - 2d_{24}^2 d_{13}^2 + d_{12}^2\left(d_{24}^2 - d_{23}^2 + d_{34}^2\right) + d_{14}^2\left(d_{23}^2 + d_{24}^2 - d_{34}^2\right)\right)\\
&\quad + \overline{d}_{04}^2\left(d_{23}^2\left(d_{24}^2 - d_{23}^2 + d_{34}^2\right) - 2d_{23}^2 d_{14}^2 + d_{12}^2\left(d_{23}^2 - d_{24}^2 + d_{34}^2\right) + d_{13}^2\left(d_{23}^2 + d_{24}^2 - d_{34}^2\right)\right)\\
&\quad + d_{23}^2 d_{14}^2\left(d_{24}^2 - d_{23}^2 + d_{34}^2\right) + d_{24}^2 d_{13}^2\left(d_{23}^2 - d_{24}^2 + d_{34}^2\right) + d_{34}^2 d_{12}^2\left(d_{23}^2 + d_{24}^2 - d_{34}^2\right)\\
&\quad - 2d_{23}^2 d_{24}^2 d_{34}^2 n + d_{01}^2\left(d_{23}^4 + d_{24}^4 + d_{34}^4 - 2d_{23}^2 d_{24}^2 - 2d_{23}^2 d_{34}^2 - 2d_{24}^2 d_{34}^2\right)
\end{aligned}$$

Proof According to Theorem 3.1, we know that $D(\mathbf{r}, \mathbf{n}, \mathbf{a}_2, \mathbf{a}_3, \mathbf{a}_4) = 0$ namely,

$$\det\begin{pmatrix}
0 & d_{01}^2 & d_{02}^2 & d_{03}^2 & d_{04}^2 & 1\\
d_{01}^2 & 0 & d_{12}^2 & d_{13}^2 & d_{14}^2 & 1\\
d_{02}^2 & d_{12}^2 & 0 & d_{23}^2 & d_{24}^2 & 1\\
d_{03}^2 & d_{13}^2 & d_{23}^2 & 0 & d_{34}^2 & 1\\
d_{04}^2 & d_{14}^2 & d_{24}^2 & d_{34}^2 & 0 & 1\\
1 & 1 & 1 & 1 & 1 & 0
\end{pmatrix} = 0 \tag{3.103}$$

One can then obtain,

$$\det\begin{pmatrix} \mathbf{B}_{11} & \mathbf{B}_{12} \\ \mathbf{B}_{12}^T & \mathbf{B}_{22} \end{pmatrix} = 0 \tag{3.104}$$

where

$$\mathbf{B}_{11} = \begin{pmatrix} 0 & d_{01}^2 \\ d_{01}^2 & 0 \end{pmatrix}$$

$$\mathbf{B}_{12} = \begin{pmatrix} d_{02}^2 & d_{03}^2 & d_{04}^2 & 1 \\ d_{12}^2 & d_{13}^2 & d_{14}^2 & 1 \end{pmatrix}$$

$$\mathbf{B}_{22} = \begin{pmatrix} 0 & d_{23}^2 & d_{24}^2 & 1 \\ d_{23}^2 & 0 & d_{34}^2 & 1 \\ d_{24}^2 & d_{34}^2 & 0 & 1 \\ 1 & 1 & 1 & 0 \end{pmatrix}$$

Suppose the anchors are nonlinear, $D(\mathbf{a}_2, \mathbf{a}_3, \mathbf{a}_4) \neq 0$, so $\mathbf{B}_{22}$ is nonsingular. From Eq. (3.104), the following equation can be derived:

$$\det\left(\mathbf{B}_{11} - \mathbf{B}_{12}\mathbf{B}_{22}^{-1}\mathbf{B}_{12}^T\right)\det\mathbf{B}_{22} = 0 \tag{3.105}$$

That is,

$$\det\begin{pmatrix} 0 - \left(\mathbf{b}_{120}^T\mathbf{B}_{22}^{-1}\mathbf{b}_{120}\right) & d_{01}^2 - \left(\mathbf{b}_{120}^T\mathbf{B}_{22}^{-1}\mathbf{b}_{121}\right) \\ d_{01}^2 - \left(\mathbf{b}_{121}^T\mathbf{B}_{22}^{-1}\mathbf{b}_{120}\right) & 0 - \left(\mathbf{b}_{121}^T\mathbf{B}_{22}^{-1}\mathbf{b}_{121}\right)\right) \end{pmatrix} = 0 \tag{3.106}$$

where

$$\mathbf{b}_{120} = \begin{pmatrix} d_{02}^2 & d_{03}^2 & d_{04}^2 & 1 \end{pmatrix}^T$$
$$\mathbf{b}_{121} = \begin{pmatrix} d_{12}^2 & d_{13}^2 & d_{14}^2 & 1 \end{pmatrix}^T$$

According to the proof procedure of Theorem 3.3, we can obtain $\mathbf{b}_{120}^T\mathbf{B}_{22}^{-1}\mathbf{b}_{120} = \mathbf{b}_{121}^T\mathbf{B}_{22}^{-1}\mathbf{b}_{121} = 0$. Then from Eq. (3.106), we obtain

$$\mathbf{b}_{120}^T\mathbf{B}_{22}^{-1}\mathbf{b}_{121} = \mathbf{b}_{121}^T\mathbf{B}_{22}^{-1}\mathbf{b}_{120} = d_{01}^2 \tag{3.107}$$

This yield

$$\left(\overline{d}_{02}^{2} + \varepsilon_2 \overline{d}_{03}^{2} + \varepsilon_3 \overline{d}_{04}^{2} + \varepsilon_4 1\right) \mathbf{B}_{22}^{-1} \begin{pmatrix} d_{12}^2 \\ d_{13}^2 \\ d_{14}^2 \\ 1 \end{pmatrix} = d_{01}^2 \tag{3.108}$$

Multiplying both sides of Eq. (3.108) by the determinant of $\mathbf{B}_{22}$, we obtain

$$\left(\overline{d}_{02}^{2} + \varepsilon_2 \overline{d}_{03}^{2} + \varepsilon_3 \overline{d}_{04}^{2} + \varepsilon_4 1\right) \mathbf{B}_{22}^{*} \begin{bmatrix} d_{12}^2 \\ d_{13}^2 \\ d_{14}^2 \\ 1 \end{bmatrix} = d_{01}^2 \times \det(\mathbf{B}_{22}) \tag{3.109}$$

where

$$\mathrm{B}_{22}^{*} = \begin{pmatrix} 2d_{34}^2 & d_{23}^2 - d_{24}^2 - d_{34}^2 & d_{24}^2 - d_{23}^2 - d_{34}^2 & d_{34}^2\left(d_{34}^2 - d_{23}^2 - d_{24}^2\right) \\ d_{23}^2 - d_{24}^2 - d_{34}^2 & 2d_{24}^2 & d_{34}^2 - d_{23}^2 - d_{24}^2 & d_{24}^2\left(d_{24}^2 - d_{23}^2 - d_{34}^2\right) \\ d_{24}^2 - d_{23}^2 - d_{34}^2 & d_{34}^2 - d_{23}^2 - d_{24}^2 & 2d_{23}^2 & d_{24}^2\left(d_{24}^2 - d_{23}^2 - d_{34}^2\right) \\ d_{34}^2\left(d_{34}^2 - d_{23}^2 - d_{24}^2\right) & d_{24}^2\left(d_{24}^2 - d_{23}^2 - d_{34}^2\right) & d_{23}^2\left(d_{23}^2 - d_{34}^2 - d_{34}^2\right) & 2d_{23}^2 d_{24}^2 d_{34}^2 \end{pmatrix}$$

According to Eq. (3.109), we can get (3.102). This completes the proof.

3.6.3 *NCA-DL Localization*

3.6.3.1 Principles of the Algorithm

Based on the algebraic constraints mentioned in the previous sections, a neighbor constraint assisted distributed localization algorithm (NCA-DL) is hereby presented as a means of improving the localization precision. The main idea behind NCA-DL is to refine the distances to anchors using a neighbor node. In NCA-DL, the regular nodes estimate the initial distances to the anchors using the method similar to DV-Hop. Then according to the algebraic constraints of the distance estimation errors, Lagrangian multiplier method is introduced in order to obtain the optimal errors and refine the distances. The following section will give a full description of the principles and performance analysis of this novel NCA-DL algorithm.

Initially, anchors (set A) are deployed in the sensing field with the regular nodes (set R). We assume each node has the ability of ranging, and for simplicity, the number of the anchors is set to 3. The whole process of NCA-DL is divided into four phases.

1. Distance Estimation. Each regular node is supposed to obtain initial distance estimation to the anchors. So, two times of flooding are required to accomplish

the process of distance estimation. In the first flooding, the anchors start by propagating their location information. Then all nodes receive the location information from every anchor as well as the hop count to these anchors. When an anchor node receives location information from other anchors, it can calculate the average size of a hop based on their locations and the hop count among them. In the second flooding, the average size of a hop is transmitted in a controlled manner into the network as a correction factor. When a regular node receives the correction, it can be able to estimate the distances to the anchors using the correction and the hop count information received in the previous flooding.

2. Neighbor Node Election. The main purpose of this phase is to choose a proper neighbor for each regular node to assist distance refinement in the next phase. For most ranging technology, when sensors are closer, the distance estimations are more accurate. According to the requirement of Theorem 3.4, in order to obtain the constraint equation of the distance estimation errors, it is fundamental to choose an adjacent neighbor because the distance between the regular node and its neighbor can be measured accurately. Meanwhile, the distances among the neighbor and the anchors also should satisfy geometry relationship, that is to say, the distances should be refined by the method in [9]. So, in this phase, the nearest node of each regular node is chosen as an assistant neighbor and its distances to the anchors obtained in the previous phase are supposed to be refined using Theorem 3.3.
   To improve the localization precision of the regular node further, the distances between the neighbor and the anchors need to be estimated accurately. Though the imprecise distances can be refined through Theorem 3.3 which can meet the requirement of Theorem 3.4, the distances are still imprecise. So, in this phase, with the growing research for the mobility of sensors, a mobile anchor also can be used to be a "neighbor" of each regular node. The distances between this "neighbor" and the other static anchors can be accurately calculated by the coordinates of these anchors, which definitely meets the requirement of Theorem 3.3. In this case, the mobile anchor is supposed to move in the sensing field. The aim is just to assist the regular node with localization through the constraint defined in Theorem 3.4 and it does not need to consider the collinear problems of the anchors. So, SCAN could be used for the path planning method, which is the most straightforward one.
3. Distance Refinement. The algebraic equations that define the errors and relate the distances to the anchors for each regular node have now been fully determined from the previous two phases, as described in Theorems 3.3 and 3.4. The next step of the algorithm attempts to quantify these errors in the inaccurate distance estimations between regular nodes and anchors. Let $\varepsilon_i(\forall i \in \{2,3,4\})$ as defined in Eq. (3.96) be the error in the estimated squared distances between a regular node and the anchors. The goal here is to minimize the sum of the squared errors:

$$J = \varepsilon_2^2 + \varepsilon_3^2 + \varepsilon_4^2 \tag{3.110}$$

In Eq. (3.110), J is subjected to a quadratic equality constraint defined in Eq. (3.92) and a linear equality constraint defined in Eq. (3.98). So, to solve the optimization problem with constraints, Lagrangian multiplier method is implemented in the algorithm. It is a mathematical method used to solve the optimization problem, which can convert the constraints to the seeking of extreme values with the help of Lagrangian multipliers. We can get the following Lagrangian multiplier form:

$$H(\varepsilon_2, \varepsilon_3, \varepsilon_4, \lambda_1, \lambda_2) = \sum_{i=2}^{4} \varepsilon_i^2 + \lambda_1 f_1(\varepsilon_2, \varepsilon_3, \varepsilon_4) + \lambda_2(\varepsilon_2, \varepsilon_3, \varepsilon_4) \tag{3.111}$$

where λ_1, λ_2 are the Lagrangian multipliers and f_1, f_2 are the functions of ε_2, ε_3, ε_4, whose coefficients can be obtained in Eqs. (3.92) and (3.98).

By differentiating the Lagrangian H with respect to $\varepsilon_i (i = 2,3,4)$, $\lambda_i (i = 1, 2)$ and equating the result to zero, five equations can be obtained. Solving these five algebraic equations numerically and discarding all the nonoptimal stationary-point solutions, ε_2, ε_3, ε_4 can be solved. Then the distances to the anchors for the regular node are refined. The below example demonstrates the steps in this phase.

The simplest scenario as depicted in Fig. 3.53 is considered, where node **r** is the regular node that needs to be located. Node **r** can measure its distances to three anchors $\mathbf{a}_2$, $\mathbf{a}_3$, $\mathbf{a}_4$ whose coordinates are (10, 10), (90, 10), and (50, 90), respectively. The noisy distance measurements acquired by node **r** are $\overline{d}_{02} = 56.3$, $\overline{d}_{03} = 65.7$ and $\overline{d}_{04} = 41.6$. The distances between the neighbor node **n** and other nodes are estimated in the previous phase as $d_{01} = 5$, $d_{12} = 56.6$, $d_{13} = 56.6$ and $d_{14} = 40$. In this case, the goal is to obtain the estimation errors ε_2, ε_3, ε_4 and refine $\overline{d}_{02}$, $\overline{d}_{03}$, and $\overline{d}_{04}$.

Firstly, two equality constraints as described by Eqs. (3.92) and (3.98) will be determined:

$$\begin{aligned} 0 = f_1(\varepsilon_2, \varepsilon_3, \varepsilon_4) &= \left(-2.5\varepsilon_2^2 - 2.5\varepsilon_3^2 - 2\varepsilon_4^2 + 3\varepsilon_2\varepsilon_3 + 2\varepsilon_2\varepsilon_4 + 2\varepsilon_3\varepsilon_4 \right. \\ &\quad \left. +16597\varepsilon_2 + 7341\varepsilon_3 + 27262\varepsilon_4 + 14159000\right) \times (25600)^{-1} \\ 0 = f_2(\varepsilon_2, \varepsilon_3, \varepsilon_4) &= 0.25\varepsilon_2 + 0.25\varepsilon_3 + 0.5\varepsilon_4 + 310.6 \end{aligned} \tag{3.112}$$

To determine the optimal values for ε_2, ε_3, ε_4, the following problem needs to be solved:

$$\begin{aligned} \min \quad & \varepsilon_2^2 + \varepsilon_3^2 + \varepsilon_4^2 \\ \text{s.t.} \quad & f_1(\varepsilon_2, \varepsilon_3, \varepsilon_4) = 0 \\ & f_2(\varepsilon_2, \varepsilon_3, \varepsilon_4) = 0 \end{aligned} \tag{3.113}$$

By differentiating the Lagrangian H defined in Eq. (3.111), we obtain

$$\begin{aligned}
\frac{\partial H}{\partial \varepsilon_2} &= 2\varepsilon_2 + 0.25\lambda_2 + \lambda_1\left(\frac{3}{25600\varepsilon_3} - \frac{\varepsilon_2}{5120} + \frac{\varepsilon_4}{12800} + 0.65\right) = 0 \\
\frac{\partial H}{\partial \varepsilon_3} &= 2\varepsilon_3 + 0.25\lambda_2 + \lambda_1\left(\frac{3}{25600\varepsilon_2} - \frac{\varepsilon_3}{5120} + \frac{\varepsilon_4}{12800} + 0.28\right) = 0 \\
\frac{\partial H}{\partial \varepsilon_4} &= 2\varepsilon_4 + 0.5\lambda_2 + \lambda_1\left(\frac{\varepsilon_2}{12800} + \frac{\varepsilon_3}{12800} - \frac{\varepsilon_4}{6400} + 1.06\right) = 0 \\
\frac{\partial H}{\partial \lambda_1} &= f_1(\varepsilon_2, \varepsilon_3, \varepsilon_4) = 0 \\
\frac{\partial H}{\partial \lambda_2} &= f_2(\varepsilon_2, \varepsilon_3, \varepsilon_4) = 0
\end{aligned} \tag{3.114}$$

Solving the above five algebraic equations above, we get,

$$\varepsilon_2^* = -67.39, \quad \varepsilon_3^* = -545.76, \quad \varepsilon_4^* = -314.59 \tag{3.115}$$

Correspondingly, the refined distances between regular node **r** and the three anchors are

$$\begin{aligned}
\widehat{d}_{02} &= \sqrt{\overline{d}_{02}^2 + \varepsilon_2^*} = 55.6 \\
\widehat{d}_{03} &= \sqrt{\overline{d}_{03}^2 + \varepsilon_3^*} = 61.4 \\
\widehat{d}_{04} &= \sqrt{\overline{d}_{04}^2 + \varepsilon_4^*} = 37.6
\end{aligned} \tag{3.116}$$

As shown in Fig. 3.54, "." represents the regular node, "□" represents the anchor, and the radius of the circle is the estimated distance between the regular node and the

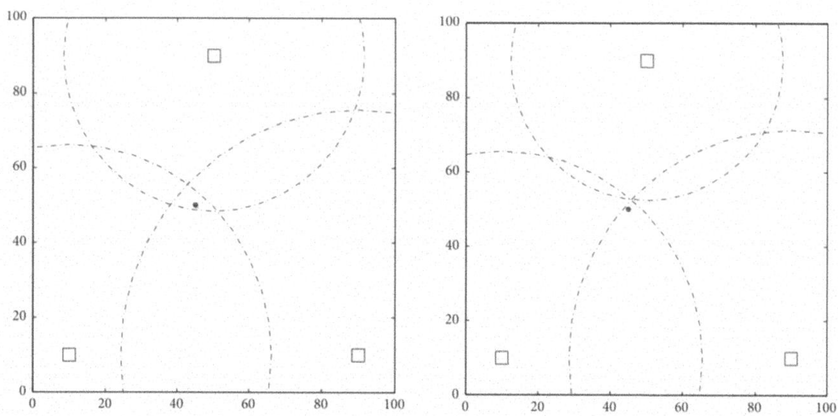

(a) (b)

Fig. 3.54 Locations of anchors and the regular node. (**a**) Performance of noisy distances. (**b**) Performance of refined distances anchor. In Fig. 3.54b, the three circles intersect in one point, which proves that the refined distances satisfy the geometric constraints.

4. Localization. So far, regular nodes have known the refined distances to the anchors according to Eq. (3.116). Based on the above refinement scheme, we know that the refined distances satisfy the geometric constraints. Localization can be carried out by the least square method. $r(x, y)$ represents the regular node, $a(x_i, y_i)(i = 2,3,4)$ represent the locations of the anchors, and $\widehat{d}_{02}$, $\widehat{d}_{03}$, $\widehat{d}_{04}$ represent the refined distances between the regular node and the anchors. The coordinate of the regular node can be estimated by

$$\widehat{X} = \left(\mathbf{A}^T\mathbf{A}\right)^{-1}\mathbf{A}^T b \tag{3.117}$$

where

$$A = \begin{pmatrix} 2(x_2 - x_4) & 2(y_2 - y_4) \\ 2(x_3 - x_4) & 2(y_3 - y_4) \end{pmatrix}$$

$$X = \begin{pmatrix} x \\ y \end{pmatrix}$$

$$b = \begin{pmatrix} x_2^2 - x_4^2 + y_2^2 - y_4^2 + \widehat{d}_{04}^2 - \widehat{d}_{02}^2 \\ x_3^2 - x_4^2 + y_3^2 - y_4^2 + \widehat{d}_{04}^2 - \widehat{d}_{03}^2 \end{pmatrix}$$

Figure 3.55 illustrates the localization of the computational example above with both noisy and refined distances. "$*$"represents the calculated location. The localization errors in Fig. 3.55a and b are 0.67 m and 0.55 m, respectively.

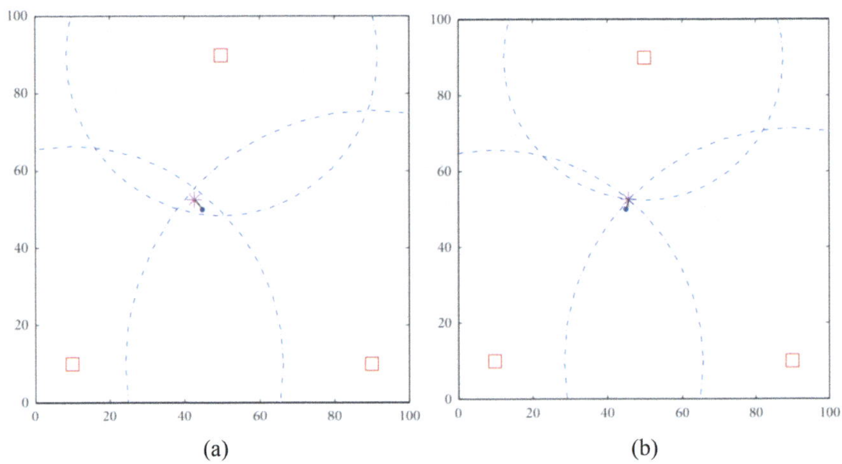

Fig. 3.55 Localization effect of noisy distances and refined distances. (**a**) Localization with noisy distances. (**b**) Localization with refined distances

3.6.3.2 Communication Cost Analysis

For energy cost of NCA-DL, the communication consumption is mainly considered in the distance estimation and refinement phases. n represents the number of the sensors and n_A represents the number of the anchors. Then for the two flooding processes in the distance estimation phase, it gives a bound of $O(2 \times n_A \times n)$ to the communication cost in this process. While in the refinement phase, each node needs to communicate with its neighbors so the communication cost is $O(n)$. The total communication complexity is $O(n \times (2 \times n_A + 1))$. It is known that the communication complexity of DV-Hop is $O(2 \times n_A \times n)$. The cost of NCA-DL is in the same order of magnitude as other algorithms while it can largely improve the localization performance.

3.6.4 Simulation Results

3.6.4.1 Simulation Setup

The basic network setup area is considered to be a 100 m × 100 m square field. The communication radius of the nodes is set to 10 m. In our simulations, sensor nodes are deployed using two models: random placement and perturbed grid. In the random placement model, sensor nodes are randomly deployed in the network by dropping from an airplane or some other methods. In this case, the topology of the network is likely to be irregular. In the second model, nodes are deployed using perturbed grid where the nodes are perturbed with a random shift from grid. In this situation, nodes tend to uniformly occupy the field avoiding large concentration of nodes, which also guarantees the regularity of the topology of the network.

In all cases, regular nodes have the ability of ranging and the results are averaged over 10 trails. The average localization error is defined as follows:

$$\text{error} = \frac{1}{n} \sum_{i=1}^{n} \| \mathbf{x}_i - \overline{\mathbf{x}}_i \| \times \frac{1}{r} \tag{3.118}$$

where n is the number of the regular nodes, $\mathbf{x}_i$ is the actual location of regular node i, $\overline{\mathbf{x}}_i$ is the estimated location of regular node i, and r is the communication radius.

3.6.4.2 NCA-DL Performance

At first, we focus on the distance estimation error between regular nodes and anchors. In this set of simulations, we varied the number of the nodes (avg. degree) from 124 (4.8) to 628 (14.2). The number of the anchors is set to 3. The average distance estimation errors obtained from these simulations are stated in Table 3.3.

Table 3.3 Average distance estimation error

	Avg. distance estimation error (m)			
	Random placement		Perturbed grid	
Number of nodes (avg. degree)	DV-Hop	NCA-DL	DV-Hop	NCA-DL
124 (4.8)	12.32	6.13	6.15	4.51
147 (5.8)	8.28	4.70	4.01	2.38
203 (7.5)	5.43	3.71	3.92	2.17
403 (11.1)	3.39	2.35	2.99	1.55
628 (14.2)	3.06	2.04	2.72	1.35

As indicated in Table 3.3, we can observe that the distance estimation error of Tboth DV-Hop and NCA-DL under perturbed gird deployment is smaller than that of these two algorithms under random placement. With the increase of the number of nodes (avg. degree), both DV-Hop and NCA-DL can improve the ranging effectiveness. Moreover, DV-Hop suffers large ranging error when the average degree is low while NCA-DL has smaller errors for divering network scales.

Now, it has been proved that the NCA-DL algorithm can significantly decrease the distance estimation errors with the help of a collaborating neighbor; both initial deployment model conditions were therefore simulated in order to graphically verify their respective localization performance.

Figure 3.56 exemplifies the distance estimation and localization error of NCA-DL against number of nodes in both deployment models. It is demonstrated that the distance estimation error in NCA-DL can be decreased with the increase of the

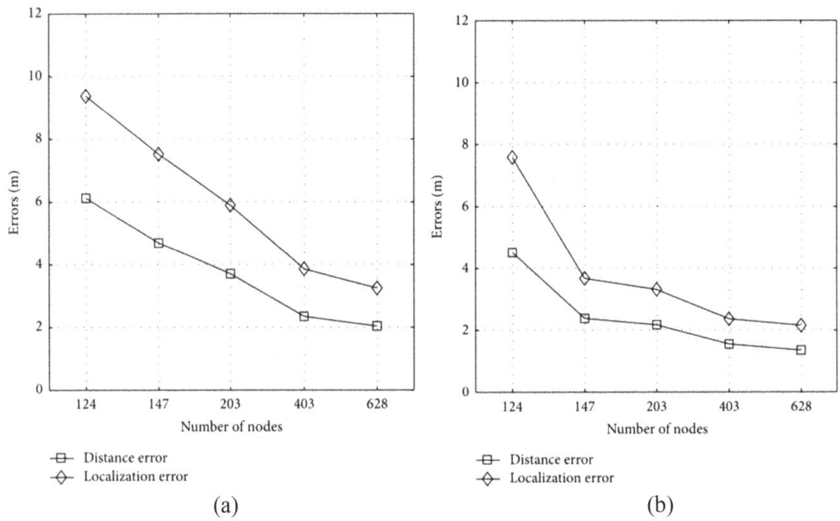

Fig. 3.56 Distance estimation and localization error of NCA-DL. (**a**) Random placement model. (**b**) Perturbed grid model

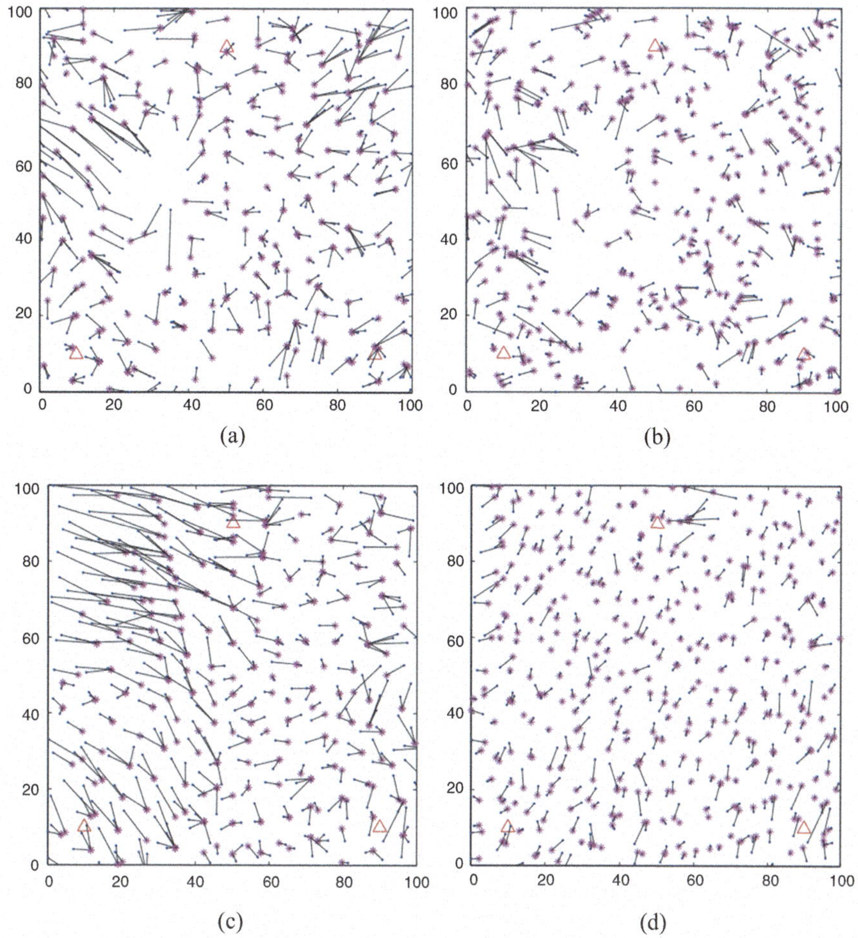

Fig. 3.57 Localization effect in both deployment models. (**a**) DV-Hop in random placement. (**b**) NCA-DL in random placement. (**c**) DV-Hop in perturbed grid. (**d**) NCA-DL in perturbed grid

number of nodes. As shown in Fig. 3.56b, the localization error decreases more obviously when the distance estimation error drops below a critical value (around 2 m).

Figure 3.57 exemplifies the localization performance of DV-Hop and NCA-DL. "Δ" represents the anchors, "." represents the true location, "∗" represents the calculated location, and the line between them represents the localization error. In this set of simulations, the number of deployed nodes (avg. degree) is set to 403 (11.1) and the number of the anchors is set to 3. According to the definition of localization error in (32), in the random placement deployment, the average localization error resulting form DV-Hop is around 66% while the error goes down to around 46% in perturbed grid deployment, as shown in Fig. 3.57a and c. In

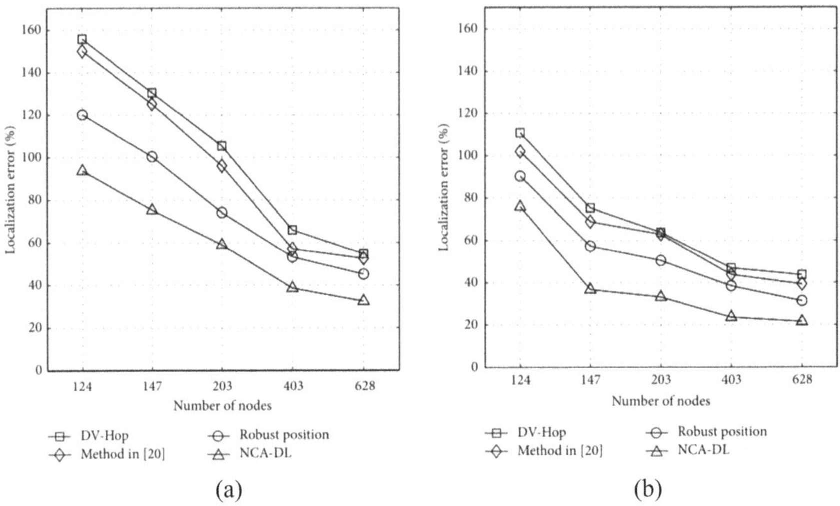

Fig. 3.58 Localization error against number of nodes. (**a**) Random placement model. (**b**) Perturbed grid model

Fig. 3.57b and d, we can see that NCA-DL decreases the localization error to 38% and 23%, respectively, in the two deployment models. In general, NCA-DL can increase the localization precision about 40% compared with DV-Hop under such network environment.

Figure 3.58 illustrates the localization error in both random placement and perturbed grid models. Compared with the other algorithms, NCA-DL has much lower localization error. With the increase of number of nodes, the performance of all the algorithms upgrades. It also demonstrates that the NCA-DL algorithm can achieve high localization precision in the perturbed grid deployment model when the density of sensors is high, as shown by Fig. 3.58b.

3.7 Extended Kalman Filter Multidimensional Scaling Localization

In this section, we present an Extended Kalman Filter Multidimensional Scaling (EKF-MDS) localization algorithm in SSNs with holes [10]. By exploring the virtual node and constructing the shortest paths between nodes, the Euclidean distances between nodes are obtained. The Extended Kalman Filter (EKF) is applied to refine the localized coordinates which are obtained by the MDS-MAP algorithm. Simulation results demonstrate that the algorithm is exceedingly accurate and efficient comparing to state-of-the-art methods in anisotropic networks with holes.

3.7.1 EKF-MDS Localization

3.7.1.1 Mathematical Model

The MDS-MAP algorithm is a precise method for dimensionality reduction. If there are no measured Euclidean distances between every pair of the nodes in the network, they will be achieved using the Dijkstra algorithm. This approximation produces a huge error when the nodes are in the multi-hop communication range. To solve this problem, we need to explore a virtual node and achieve the shortest path as well as to apply the EKF algorithm to improve the accuracy of node localization in anisotropic WSNs with holes.

Consider there are two nodes s and t in an anisotropic network with holes and cannot communicate directly to each other because they cross hole H as shown in Fig. 3.59. The shortest path between s and t intersects with hole H at point o and it is segmented into length of so and ot. As Fig. 3.59 shows, assume that $|so| = d_1$ and $|ot| = d_2$. According to the law of cosines, we can write, $|st|^2 = |so|^2 + |ot|^2 - 2|so||ot| \cdot \cos \angle sot$ and that d_{st} is given by

$$d_{st} = \sqrt{d_1^2 + d_2^2 - 2|d_1||d_2| \cdot \cos \angle sot} \tag{3.119}$$

To find the angle $\angle sot$, we need to explore a virtual node g which is close to the point o. The new shortest path between s and t is augmented. As illustrated in Fig. 3.59, the new shortest path is segmented into the length of sg and gt. Now, assume that $|og| = d_3$, $|sg| = d_4$ and $|gt| = d_5$. The angle value can be derived from Fig. 3.59 as follows:

$$\begin{aligned} \angle \text{sot} &= 2\pi - \angle \text{sog} - \angle got \\ &= 2\pi - \arccos \frac{d_1^2 + d_3^2 - d_4^2}{2d_1 d_3} - \arccos \frac{d_2^2 + d_3^2 - d_5^2}{2d_2 d_3} \end{aligned} \tag{3.120}$$

Considering Eq. (3.120), we are able to achieve the distance d_{st} between node s and t. After all of the Euclidean distance are obtained, we can apply the MDS-MAP algorithm to the distance matrix and obtain the node coordinates.

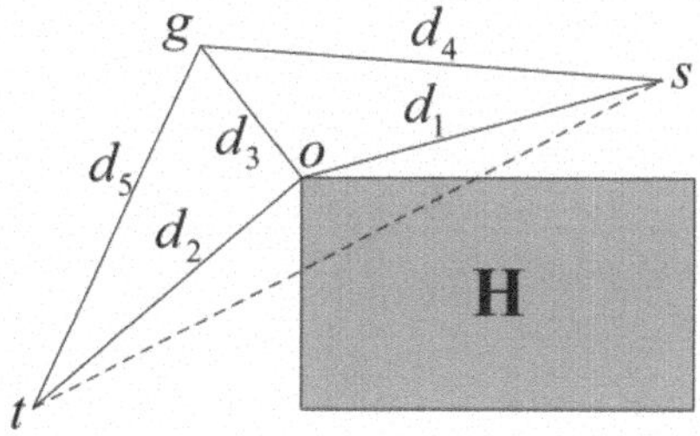

Fig. 3.59 The shortest path

3.7.1.2 EKF Algorithm

The EKF algorithm is applied to the MDS-MAP algorithm in order to improve the accuracy of node localization because the distance between nodes and node coordinates are nonlinear relationship. The node coordinates obtained by the MDS-MAP algorithm are an initial value of the EKF algorithm. The iterative calculation is applied to the information of the entire neighbor of unknown nodes. It will reduce localization error and improve accuracy of node localization. The filter model is given by

$$X(k+1) = f(k,\ X(k)) + V(k) = AgX(k) + V(k) \tag{3.121}$$

where $X(k) = [x_k, y_k]^T$ denotes coordinate vector of unknown node at k time, $V(k)$ denotes process noise and A denotes state transition matrix.

$$A = \begin{bmatrix} 1 & 0 \\ 0 & 1 \end{bmatrix} \tag{3.122}$$

The measured value of the distance between unknown nodes and neighbor nodes is the observed quantity. The measurement equation is given by

$$z(k) = h(k,\ X(k)) + W(k) = HgX(k) + W(k) \tag{3.123}$$

where H denotes Jacobian matrix of $h()$ and $W(k)$ denotes the measurement noise. Figure 3.60 illustrates the EKF refinement.

We assume initial location estimation of an unknown node O is (x_0, y_0) calculated by the MDS-MAP algorithm. There are three unknown nodes, namely A, B and C, and their coordinates are (x_1, y_1), (x_2, y_2) and (x_3, y_3) respectively. The refinement of initial coordinates of node O is applied to the EKF algorithm and it can be close to the real coordinate of node O. In each iteration, it is required to take advantage of all the information of the neighbors of node O. The initial iteration value of EKF is given by

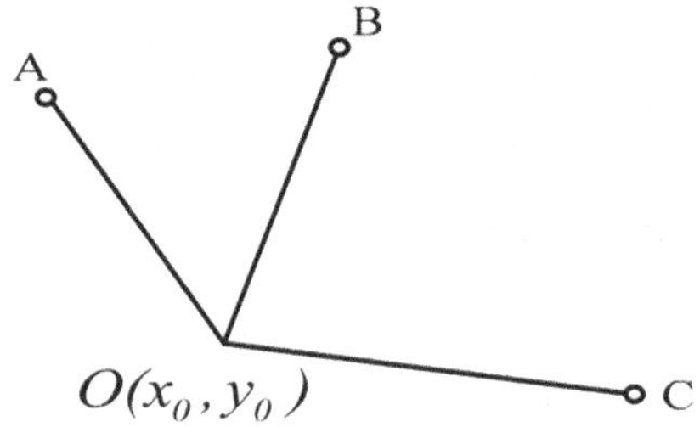

Fig. 3.60 Example of EKF refinement

$$X_0 = (x_0, y_0) \tag{3.124}$$

$$P_0 = \begin{bmatrix} 1 & 0 \\ 0 & 0 \end{bmatrix} \tag{3.125}$$

$$H_0 = \begin{bmatrix} \frac{x_0 - x_1}{\sqrt{(x_0 - x_1)^2 + (y_0 - y_1)^2}} & \frac{y_0 - y_1}{\sqrt{(x_0 - x_1)^2 + (y_0 - y_1)^2}} \\ \frac{x_0 - x_2}{\sqrt{(x_0 - x_2)^2 + (y_0 - y_2)^2}} & \frac{y_0 - y_2}{\sqrt{(x_0 - x_2)^2 + (y_0 - y_2)^2}} \\ \frac{x_0 - x_3}{\sqrt{(x_0 - x_3)^2 + (y_0 - y_3)^2}} & \frac{y_0 - y_3}{\sqrt{(x_0 - x_3)^2 + (y_0 - y_3)^2}} \end{bmatrix} \tag{3.126}$$

$$\widehat{z}_0 = \begin{bmatrix} \sqrt{(x_0 - x_1)^2 + (y_0 - y_1)^2} \\ \sqrt{(x_0 - x_2)^2 + (y_0 - y_2)^2} \\ \sqrt{(x_0 - x_3)^2 + (y_0 - y_3)^2} \end{bmatrix} \tag{3.127}$$

where X_0 is the initial state, P_0 is the initial covariance, H_0 is an observation matrix and $\widehat{z}_0$ is the initial value of matrix which is made up of the predicted value of an unknown node to its neighbor. After completing the initialization of EKF, we start the iterative process. According to the localization accuracy and energy consumption of node in the requirements for practical applications, the iteration termination condition is given by

$$\sqrt{X_k^2 - X_{k-1}^2} \leq \Delta \tag{3.128}$$

where Δ is the fault tolerance value of presupposition, X_k is the localization result of iteration k and $\Delta = 0.01$.

3.7.1.3 EKF-MDS Algorithm Framework

The EKF-MDS algorithm is made up of the MDS-MAP algorithm and the EKF algorithm. The main idea of this algorithm is to amend the preliminary localization result of the MDS-MAP algorithm by means of nonlinear filtering technique and approach the real coordinates of nodes in order to reduce the localization error.

The EKF-MDS algorithm has six steps:

1. The Dijkstra algorithm is applied to calculate the shortest path between every pair of nodes. If two nodes are across the hole, we explore virtual node, recalculate the Euclidean distance by the method described in 3.7.1.1 and construct the shortest path distance matrix D.
2. Calculate the centered matrix J and apply double centered to the square of distance matrix D.
3. Perform singular value decomposition on matrix B, obtain the eigenvalue matrix A and the corresponding vector matrix Q. The two largest eigenvalues are used to formed matrix Ar and the corresponding eigenvectors are used to form matrix Qr. As a result, the relative coordinates of nodes are obtained.
4. The relative coordinates of nodes will be transformed into absolute coordinates of node, and the final localization result of the MDS-MAP algorithm is obtained.
5. The localization result of Step 4 is the initial value of the EKF and it is used to set the initial condition.
6. Start the iterative process. The EKF algorithm is applied to iterate filtering using the ranging information between unknown nodes and all its neighbor nodes, location information of neighbor nodes which is localized until it meets the iteration termination condition and obtains the exact location.

3.7.2 *Simulation Results*

3.7.2.1 Simulation Setup

Our work is concerned with random topology. The basic network setup is a 100 m × 100 m square field with a 40 m × 40 m hole. Other network properties are also investigated. The communication range error: 0, 0.05R, 0.1R, 0.15R, R is communication radius of node. In our simulation study, the anchors are selected randomly.

The system is subjected to both the MDS-MAP and EKF-MDS algorithms to determine the locations of the unknown nodes.

3.7.2.2 Comparison with Other Methods

Simulation results of the MDS-MAP algorithm and the EKF-MDS algorithm with 1200 nodes, 5 anchors and without the communication range error are shown in Figs. 3.61 and 3.62. The blue dots denote the randomly deployed nodes whose locations are to be determined. The red crosses denote the anchors. The pink diamonds denote the localization result of the algorithm and the green lines denote the estimation errors. It can be seen that the estimation error of the EKF-MDS algorithm is much smaller than the MDS-MAP algorithm in this case. The effect of the EKF-MDS algorithm is significantly improved.

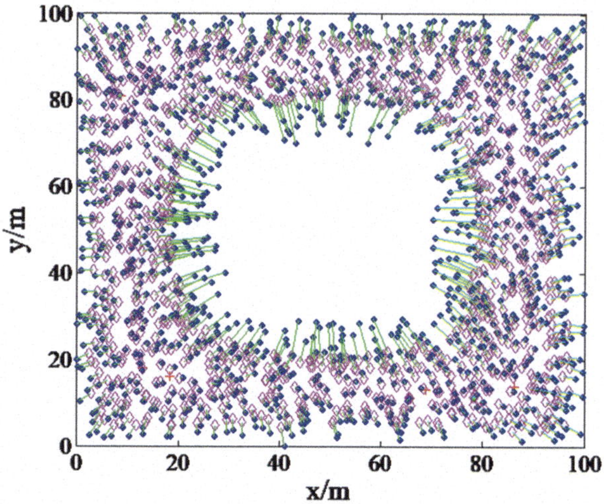

Fig. 3.61 Localization performance of the MDS-MAP algorithm

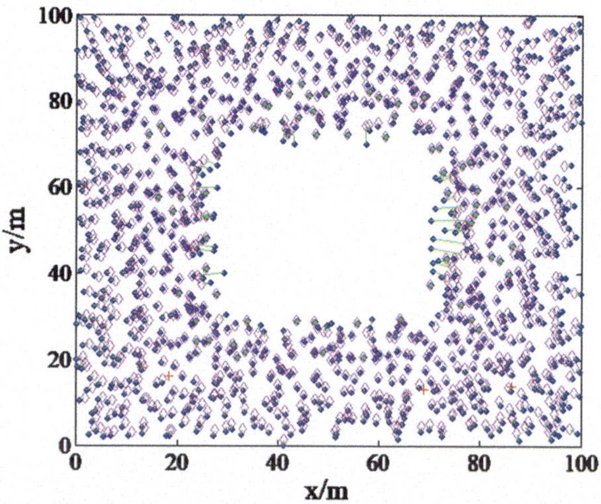

Fig. 3.62 Localization performance of the EKF-MDS algorithm

Figure 3.63 shows the result of the EKF-MDS algorithm, the REP algorithm and the MDS-MAP algorithm for random topology with 700 nodes, 5 anchors and without the communication range error, 5%, 10% and 15% of R of the communication range error respectively. Increasing the communication range error deteriorates the performance of the three algorithms, especially for the MDS-MAP method and the REP method. Simulation results show that the EKF-MDS algorithm can improve the accuracy of node localization.

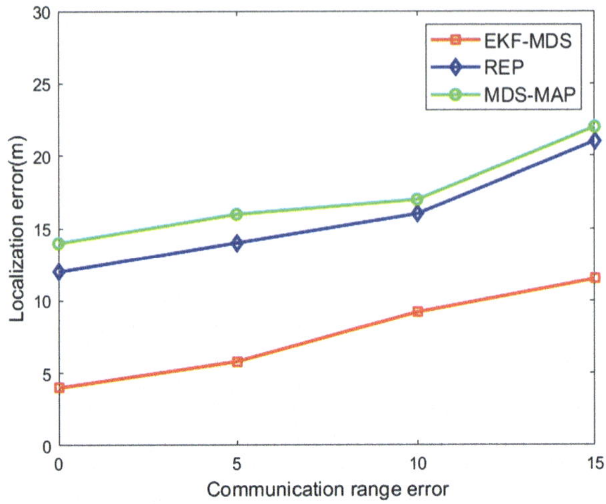

Fig. 3.63 Comparison of the MDS-MAP, the REP and the EKF-MDS algorithm with communication range error

3.8 Conclusion

In this chapter, we present several effective localization methods for WSNs, including HCED, VMCL, VMCL-ORSS, HMDS, ORSS-VBLS, DBLS, NCA-DL and EKF-MDS. The HCED, ORSS-VBLS and DBLS are applied for SSNs without holes. The VMCL and VMCL-ORSS are applied for MSNs. The HMDS and EKF-MDS are applied for anisotropic SSNs with holes. The NCA-DL is effective in refining the distances required for localization.

References

1. Er M J, Zhang S, Zhang B, et al. A novel localization approach towards anchor to node in wireless sensor networks[C]. International Automatic Control Conference, 2016: 143-148.
2. Guan Z, Zhang Y, Zhang B, et al. Voronoi-based localisation algorithm for mobile sensor networks[J]. International Journal of Systems Science, 2016, 47(15): 3688-3695.
3. Zhang S, Er M J, Zhang B, et al. A novel heuristic algorithm for node localization in anisotropic wireless sensor networks with holes[J]. Signal Processing, 2017: 27-34.
4. Wang Y, Gao J, Mitchell J S, et al. Boundary recognition in sensor networks by topological methods[C]. ACM/IEEE International Conference on Mobile Computing and Networking, 2006: 122-133.
5. Li M, Liu Y. Rendered path: range-free localization in anisotropic sensor networks with holes [C]. ACM/IEEE International Conference on Mobile Computing and Networking, 2007: 51-62.
6. Guan Z, Zhang B, Dong L, et al. An optimal region selection strategy for WSNs localization based on Voronoi diagram[C]. The 34th Chinese Control Conference, 2015: 7759-7764.
7. Guan Z, Zhang B, Zhang Y, et al. Delaunay triangulation based localization scheme[C]. The 36th Chinese Control and Decision Conference, 2017: 2627-2631.

8. Wen L, Cui L, Chai S, et al. Neighbor constraint assisted distributed localization for wireless sensor networks[J]. Mathematical Problems in Engineering, 2014: 1-11.
9. Cao M, Anderson B D, Morse A S, et al. Sensor network localization with imprecise distances [J]. Systems & Control Letters, 2006, 55(11): 887-893.
10. Zhang S, Zhang B, Er M J, et al. A novel node localization algorithm for anisotropic wireless sensor networks with holes based on MDS-MAP and EKF[C]. IEEE Region Conference, 2016: 3022-3025.
11. Huang J, Wu Z. Positioning algorithm based on Voronoi diagram[J]. Journal of Qinghai Normal University (Natural Science), 2013(1): 17-21.

Chapter 4
Coverage Control in Wireless Sensor Networks

The coverage control problem is one of the primary issues for Quality of Service (QoS) in WSNs, which effectively improve network connectivity, coverage and self-localization accuracy. In this chapter, we present several effective coverage control algorithms of WSNs. As homogeneity affects coverage rate and algorithm efficiency, we categorize coverage control algorithms into homogeneous algorithms and heterogeneous algorithms. For homogeneous sensor network, we firstly present an autonomous triangular deployment for optimal coverage (ATDOC) algorithm to avoid the nodes stacking problem in traditional virtual force approach. We also present an extended virtual force-based approach (EVFA) that overcome the connectivity maintenance problem in the traditional virtual force algorithm. As the orientation control can eliminate coverage holes and reach reliable connectivity, the Distance and Orientation Control (DOC) algorithm is proposed. For the heterogeneous sensor network, Self-Deployment by Delaunay Triangulation Graph (SDDTG) algorithm is presented and achieves non-redundant coverage. Finally, the Energy Balanced Redeployment Algorithm (EBRA) is proposed to solve the energy-balanced problems in heterogeneous mobile sensor networks.

4.1 Homogeneous Sensor Network Deployment Based on Regular Triangle

In this section, we analyze the influence of the number of neighbor nodes in the coverage control algorithms, and propose a new coverage control algorithm named ATDOC for mobile sensor networks based on regular triangle deployment [1]. The Basic Triangular Deployment Algorithm (BTDA) is firstly introduced which is widely applied in the field of deployment control. To improve BTDA, the extended triangle algorithm is presented to elimate the discontinuous coverage. Then we propose ATDOC algorithm to solve the sparse and stacking problem based on

S. Chai et al., *Wireless Sensor Networks*, Wireless Networks,
https://doi.org/10.1007/978-981-15-5757-6_4

extended triangle algorithm. BTDA and the extended triangular algorithm are finally compared with the proposed algorithm. The simulation results show the proposed algorithm generates better deployment performance than BTDA and the extended triangular algorithm.

4.1.1 *The Basic Triangular Deployment Algorithm (BTDA)*

The purpose of BTDA algorithm is trying to control the movement of nodes so that the center of mass of the triangle formed by node itself and its two neighbors moves toward the circumscribed circle center. By constantly moving the node, the node can finally form a regular triangle with its two neighbor nodes. The BTDA algorithm is described in Algorithm 4.1. It guarantees that any node can be on the vertex of equilateral triangle.

Algorithm 4.1 The Basic Triangular Deployment Algorithm
G_i: the set of neighbors of node i satisfy $d_{ij} \leq R_c$.
Step 1 Find the neighbor p should satisfy $d_{ip} \leq d_{im}$. ($p \in G_i$ while $m \in G_i$ and $m \neq p$)
Step 2 Find the neighbor q should satisfy $d_{iq} + d_{ip} < d_{in} + d_{ip}$. ($q \in G_i$ while $n \in G_i$ and $n \neq p,\ q$)
Step 3 Calculate the coordinates of centroid (x_o, y_0) in triangle Δipq. ($x_o = (x_i + x_p + x_q)/3, y_0 = (y_i + y_p + y_q)/3$)
Step 4 if the orientation of $\overrightarrow{iq}$ is on the clockwise orientation of $\overrightarrow{iq}$
Step 5 θ= the orientation angle of $\vec{p}q$ in Cartesian coordinates;
Step 6 else θ= the orientation angle of $\vec{q}p$ in Cartesian coordinates;
Step 7 end if
Step 8 $x_i' = x_o + D_{th}^*(\cos\ (\theta - \pi\ /2))/\sqrt{3}$;
Step 9 $y_i' = y_o + D_{th}^*(\cos\ (\theta - \pi/2))/\sqrt{3}$;
Step 10 Move node i towards the new position i'.

For BTDA, each node can be on the vertex of an equilateral triangle formed by its two nearest nodes. The sensor node has only two neighbor nodes to control the movement of its position. However, the BTDA algorithm cannot guarantee that the nodes are located on the vertices of six regular triangles, and it may lead to the existence of discontinuous coverage. Furthermore, the deployment structure composed of multiple discontinuous coverages can lead to coverage blind spots.

4.1.2 *The Extended Triangular Algorithm*

The extended triangular algorithm aims to solve the discontinuous coverage and blind spots. As shown in Algorithm 4.1, each sensor node and its two nearest

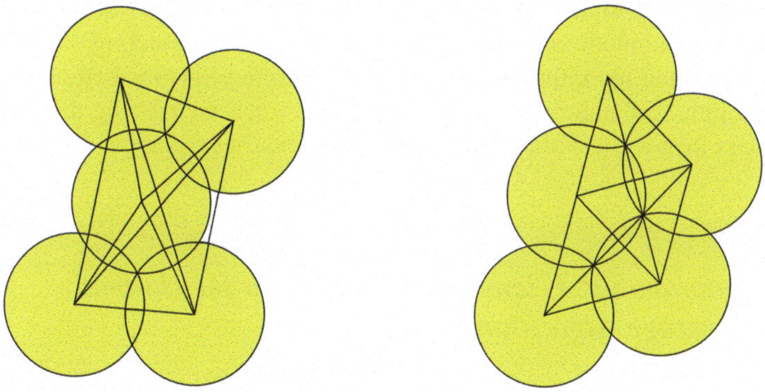

(a) (b)

Fig. 4.1 Continuous equilateral triangles. (**a**) Stable in BTDA, (**b**) Stable in extended triangular algorithm

neighbors are on the vertex of equilateral triangle. Nevertheless, sensor node adjusts the position by information from its only two nearest neighbors that may cause discontinuous coverage. Therefore, an extended triangular algorithm is introduced to elimate the discontinuous coverage in BTDA. The algorithm steps are shown in Algorithm 4.2.

Algorithm 4.2 The Extended Triangular Algorithm

Step 1 Execute BTDA.

Step 2 if $|d_{ip}/D_{th} - 1| < \varepsilon$, $| d_{iq}/D_{th} - 1 | < \varepsilon$, *and* $| d_{pq}/D_{th} - 1 | < \varepsilon$

Step 3 Find the neighbor r should satisfy $d_{ir} + d_{pr} + d_{qr} < d_{im} + d_{pm} + d_{qm}$. ($r \in G_i$ while $m \in G_i$ and $m \neq p,\ q,\ r$)

Step 4 if the orientation of $\overrightarrow{ir}$ is on the clockwise orientation of $\angle pi$

Step 5 Find node s satisfy d_{rs} is minimum. ($s \in (p, q)$)

Step 6 Calculate the coordinates of centroid (x_o, y_0) in triangle Δirs.

Step 7 if the orientation of $\overrightarrow{ir}$ is on the clockwise orientation of $\overrightarrow{is}$

Step 8 θ= the orientation angle of $\overrightarrow{sr}$ in Cartesian coordinates;

Step 9 else θ= the orientation angle of $\overrightarrow{rs}$ in Cartesian coordinates;

Step 10 end if

Step 11 end if

Step 12 end if

Step 13 $x_i' = x_o + D_{th} * (\cos\ (\theta - \pi/2))/\sqrt{3}$;

Step 14 $y_i' = y_o + D_{th} * (\cos\ (\theta - \pi\ /2))/\sqrt{3}$;

Step 15 Move node i towards the new position i'.

To compare the performance of BTDA and extended triangular algorithm, we illustrate a stable deployment in Fig. 4.1. We can see that (b) shows better continuous equilateral triangles ability than (a).

The extended algorithm can make sure sensor nodes at the vertices of rhombus to obtain more continuous and equilateral triangles. The model of the nodes in extended triangular algorithm is similar to the sampling time method in the virtual force algorithm. The algorithm can be set to stop when the moving distance of all nodes is less than a threshold or the running time reaches the upper limit.

4.1.3 Autonomous Triangular Deployment for Optimal Coverage Algorithm

The BTDA and the extended triangular algorithm guarantee any relative distance between node pair cannot be less than D_{th}. Since each sensor node is apart from the ideal distance with its three neighbors, sparse deployment problem cannot be avoided. On the contrary, virtual force algorithm can conceal sparse deployment problem but it may cause stacking problem. The ATDOC algorithm combines virtual force and extended triangular algorithm to simultaneously solve the sparse and stacking problem, it considers the ideal optimal deployment of a sensor network as a regular hexagon deployment structure composed of multiple regular triangles. ATDOC algorithm does not need to think about the truncation of the force field in the highly connected network and avoids the node aggregation problem of the traditional virtual force algorithm (VFA) algorithm. Meanwhile, it makes the deployment of the network to achieve a uniform regular hexagon distribution. ATDOC algorithm completion depends on whether each mobile sensor nodes converges into $\pm 1\ \%\ D_{th}$. Its main steps are described as follows:

Algorithm 4.3 The ATDOC Algorithm
Sum_dis_tri: the sum of distance variations for Algorithm 4.2 in a round;
Sum_dis_VFA: the sum of distance variations for VFA in a round;
ε_2: the threshold of completion criterion for algorithms;
Step 1 Execute Algorithm VFA_GG.
Step 2 if Sum_dis_tri $< \varepsilon_2$ and Sum_dis_VFA $< \varepsilon_2$
Step 3 Sum_dis_tri $= 0$;
Step 4 Execute Algorithm 4.2 and calculate Sum_dis_VFA.
Step 5 if Sum_dis_tri $< \varepsilon_2$
Step 6 Sum_dis_VFA $= 0$;
Step 7 Execute VFA based on force model in Eq. (4.10) and calculate Sum_dis_VFA.
Step 8 end if
Step 9 end if

Traditional VFA may cause stacking problem in mobile sensor networks with non-planar topology graph. To overcome the stacking problem, the first phase of ATDOC algorithm introduces Gabriel Graph (GG). On Gabriel Graph edges, each sensor node gets force information from its logical neighbor nodes. Its virtual force model is as follows:

$$\vec{F}_{ij,j\in G_i(GG)} = \begin{cases} (\omega_A(d_{ij} - D_{th}), \alpha_{ij}) & \text{if } D_{th} < d_{ij} \le R_c \\ 0 & \text{if } d_{ij} = D_{th} \\ (\omega_r(D_{th} - d_{ij}), \alpha_{ij} + \pi) & \text{if } d_{ij} < D_{th} \end{cases} \tag{4.1}$$

where ω_A, ω_R are the virtual force attractive and repulsive coefficients, d_{ij} is the Euclidean distance between sensor node i and node j. α_{ij} is the orientation of the line segment from node i to node j. $G_{i(GG)}$ is the set of neighbors of node i on Gabriel Graph edges. We name the first phrase as VFA_GG.

Little sparse deployment may appear in mobile sensor networks based on the former VFA phase. Then the virtual force model is as follows:

$$\vec{F}_{ij,j\in G_i(GG)} = \begin{cases} (\omega_A(d_{ij} - D_{th}), \alpha_{ij}) & \text{if } D_{th} < d_{ij} \le k_s * R_c \\ 0 & \text{if } d_{ij} < D_{th} \\ (\omega_R(D_{th} - d_{ij}), \alpha_{ij} + \pi) & \text{if } d_{ij} < D_{th} \end{cases} \tag{4.2}$$

In this VFA phase, each sensor node exerts force information to its logical neighbor nodes. k_s approaches to 3. Considering the permissible error, we set $k_s = 2.8$ in this section. Only when the number of node's neighbors is <6 and $d_{ij} \le D_{th}$ satisfies, the attractive force exist.

4.1.4 Simulation Results

In this section, we present simulation results for ATDOC algorithm. Figures 4.2 and 4.3 show the self-deployment results from stochastic initial distribution of 30 mobile sensor nodes in a 120 m × 120 m area under VFA_GG, the extended triangular algorithm and ATDOC algorithm. The communication range is 40 m, while the sensing range is 10 m. Figures 4.2a and 4.3a are the initial concentrated deployment and sparse deployment. Figures 4.2b and 4.3b show results without stacking because GG is a planar graph. In Fig. 4.3a and b, the extended triangular algorithm has obtained a result with sparse deployment though each node is on a vertex of an equilateral triangle. Self-deployment results by ATDOC algorithm are shown in Fig. 4.2d and Fig. 4.3d. It can be seen that regular hexagonal deployment has been attained. We consider the uniform of distance as a performance evaluation for self-deployment. The performance comparison of distance uniformity among algorithm in VFA_GG, the extended triangular algorithm and ATDOC algorithm is shown in Fig. 4.4. It is noticed that after 100 iterations, the distance uniformity are 1.3866, 1.2628 and 0.0551 respectively. Clearly, hexagonal deployment is achieved only in Algorithm 4.3 while others cannot reach precise ideal deployment.

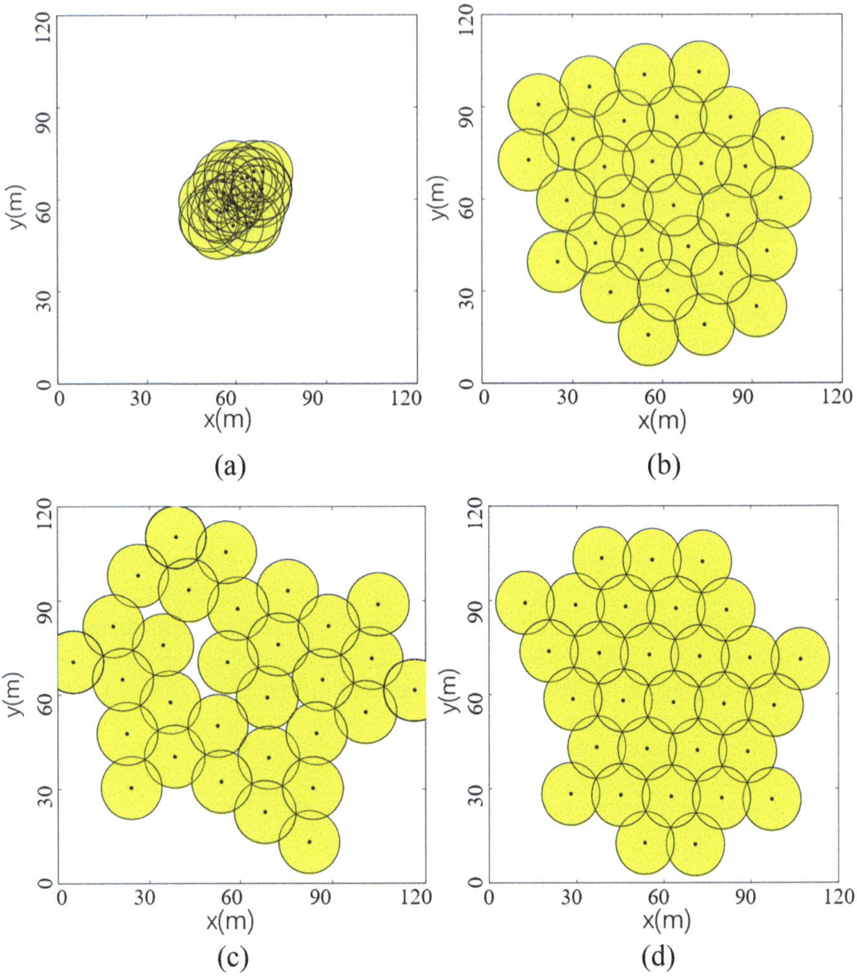

Fig. 4.2 Self-deployment in concentrated deployment. (**a**) initial deployment, (**b**) self-deployment by VFA_GG, (**c**) self-deployment by extended algorithm, (**d**) self-deployment by ATDOC

In order to demonstrate self-deployment of large scale mobile sensor networks, 100 mobile nodes are randomly placed for a concentrated deployment, as shown in Fig. 4.5a. The communication range is 40 m while the sensing range is 10 m. Figure 4.5b shows the self-deployment result by ATDOC algorithm. Figure 4.6 shows the number of nodes' neighbors on equilateral triangle grid after self-deloyment by ATDOC algorithm.

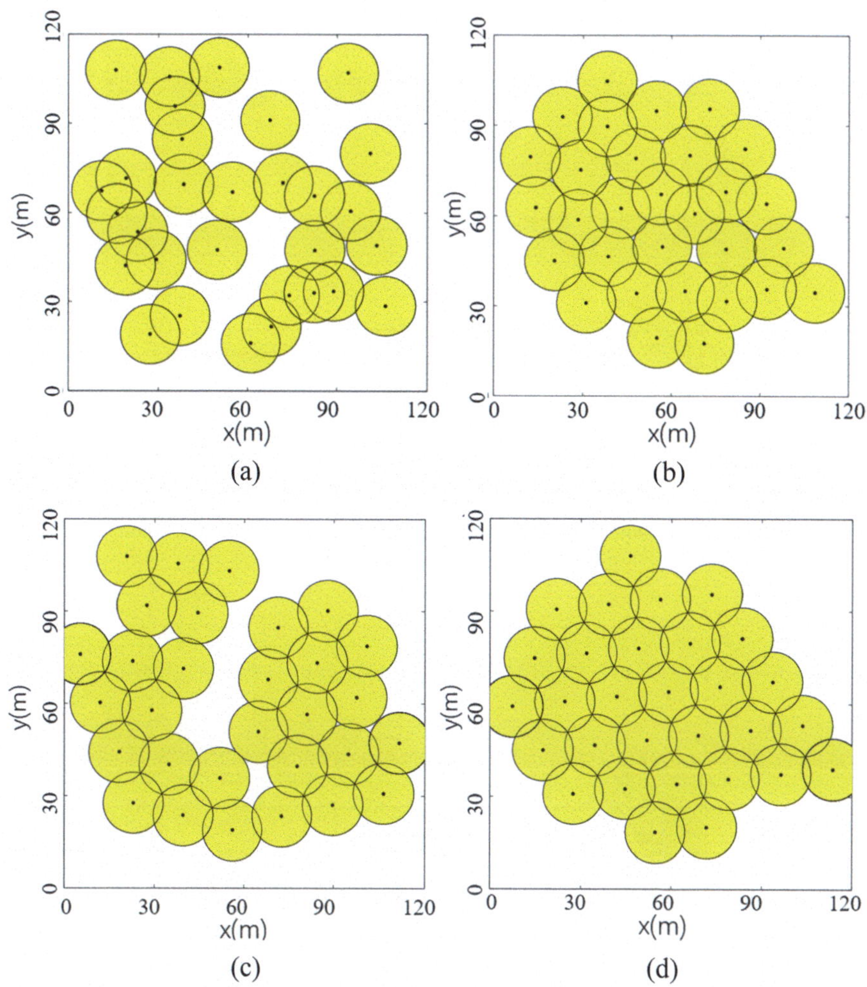

Fig. 4.3 Self-deployment in sparse deployment. (**a**) initial deployment, (**b**) self-deployment by VFA_GG, (**c**) self-deployment by extended algorithm, (**d**) self-deployment by ATDOC

4.2 Homogeneous Sensor Network Deployment Based on Extended Virtual Force

In this section, we propose EVFA coverage control algorithm based on VFA [2]. Traditional virtual force algorithm is one of the major solutions for self-deployment in mobile sensor networks with stochastic distribution of nodes, but it may cause nodes stacking and connectivity maintenance problems. Therefore, the EVFA is investigated to solve these problems. In $low - R_c$ VFA, the orientation force is proposed to guarantee the continuous connectivity. While in $High - R_c$ VFA, the judgment of distance force between node and its faraway nodes is adopted in order to prevent node stacking from nonplanar connectivity.

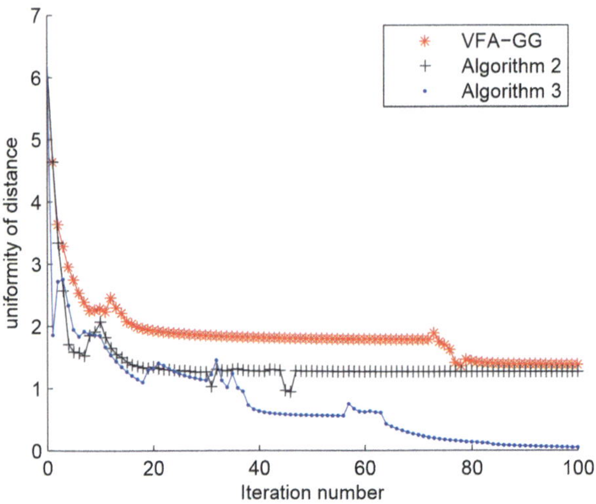

Fig. 4.4 Uniformity comparison among VFA_GG, extended and ATDOC algorithm

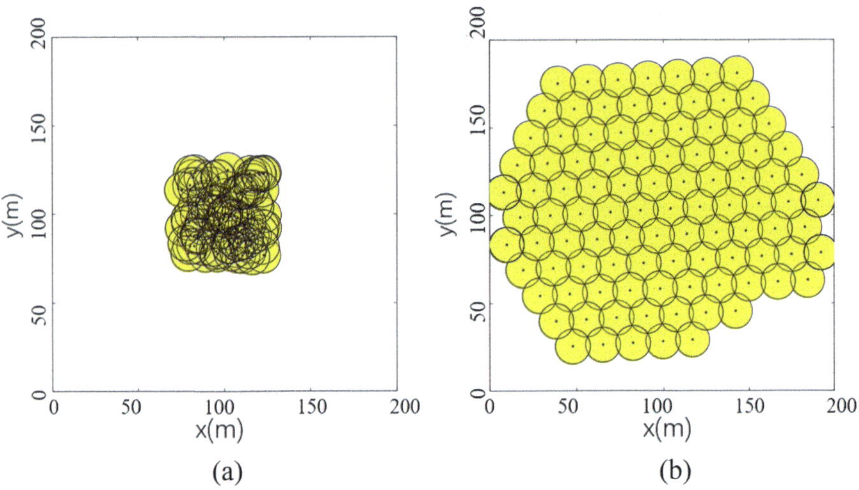

Fig. 4.5 Self-deployment in concentrated deployment. (**a**) Concentrated initial deployment, (**b**) self-deployment by ATDOC algorithm

4.2.1 Virtual Force-Based Approach

This section introduces the traditional virtual force algorithm and focuses on the self-deployment issue in 2-dimensional plane and leverages Euclidian plane R^2 to model the coordinate system. Node's position is represented by its coordinate. The position of node i is described as $S_i(x_i, y_i)$. The distance between node i and node j is defined

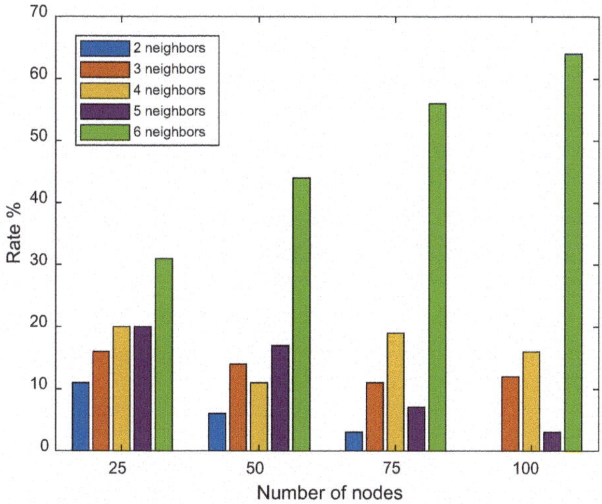

Fig. 4.6 Number of nodes' neighbors

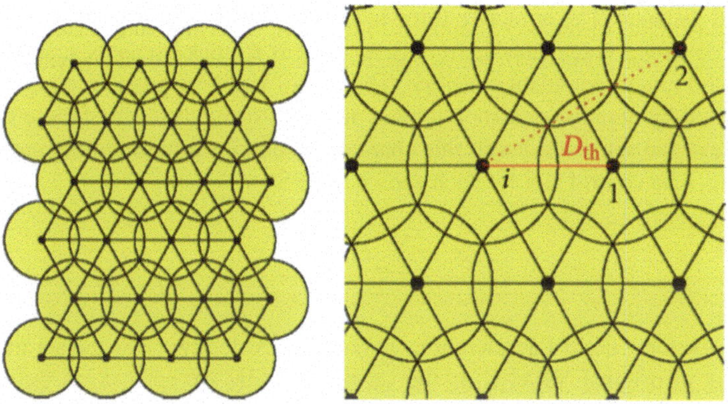

Fig. 4.7 Ideal deployment for full coverage

as Euclidean distance d_{ij}. The initial deployment is a stochastic placement in unknown distribution. It assumes that each sensor node can learn its own exact position via GPS or other localization technology, and sensor node is capable of receiving its neighbors' message without losing data. Besides, the relative distances and orientations between them can also be calculated or measured. The sensor node's communication and sensing models are modeled circular discs. All sensors have the identical communication range R_c and sensing range R_s, respectively. R_c is larger than R_s.

The essential aim of self-deployment is to make the sensor nodes move from their original positions to new positions to form the ideal deployment layout. An ideal deployment grid structure for 1-coverage is shown in Fig. 4.7. Equilateral triangle grid (hexagonal placement structure) has the smallest overlapping area, hence this deployment requires only the least number of sensors for area full coverage. There is no coverage "hole" in an ideal deployment sensor network. The ideal distance D_{th} between sensor node and its nearest neighbor should be $\sqrt{3}R_s$, and the angle formed by one node and its two adjoined neighbors should be $\pi/3$. If $R_c/R_s > 3$, the distance between sensor node and its physical neighbors may be larger than D_{th}.

All virtual force approaches for self-deployment are similar to the framework of virtual force, which combine the ideas of potential field with circle packing [3] by modeling the sensor node to be a particle in the potential field. The potential field exerts forces on the nodes nearby. For nodes i and j, it is useful to write the force as the negative gradient of the potential field. We can construct a potential function V_{ij}. So the control input of a node is the force as

$$F_{ij} = -\nabla V_{ij} \tag{4.3}$$

Sensor nodes move towards the required placement by virtual force. The force can be either attractive force or repulsive force. If two sensor nodes are placed closer than the threshold distance (ideal distance D_{th}), repulsive forces are exerted. Alternatively, attractive forces are exerted if two sensor nodes are farther apart than the threshold distance. The attractive force is to keep a certain density of sensor nodes without coverage holes, and the repulsive force is to make sensors sparse enough without too much redundant coverage. The law of updating position exploits either step method by iteration or sampling time method. Normally, both of these methods can achieve the same results. In this chapter, we choose sampling time method for analyzing. The control law for each sensor node is

$$\ddot{x}_i = \sum_{j \in G_i} F_{ij} - k_d \dot{x}_i \tag{4.4}$$

where $\ddot{x}_i$ denotes node's acceleration, $\dot{x}_i$ is node's velocity, G_i is the set of node's neighbors in original VFA, G_i is the set of all nodes except i, k_d is the positive damping coefficient. According to the traditional VFA, the force law is given as follows:

$$\vec{F}_{ij} = \begin{cases} \omega_A\left(d_{ij} - D_{th}\right), \alpha_{ij}) & \text{if } d_{ij} > D_{th} \\ 0 & \text{if } d_{ij} = D_{th} \\ \left(\omega_R d_{ij}^{-1}, \alpha_{ij} + \pi\right) & \text{if } d_{ij} < D_{th} \end{cases} \tag{4.5}$$

where ω_A and ω_R represent the virtual force attractive and repulsive coefficients, respectively. And d_{ij} is the Euclidean distance between sensor nodes i and j, α_{ij} is the orientation of the line segment from nodes i to j. The sensor moves to its new position under the control law in Eq. (4.4). The total force F_i exerted on node i is decided by the summation of all forces contributed by its all neighbors:

$$F_i = \sum_{j \in G_i} F_{ij} \tag{4.6}$$

However, the existing VFA algorithms may cause connectivity maintenance problem by few neighbors and nodes stacking problem. So we propose a novel extended virtual force algorithm to solve the above problems.

4.2.2 The Extended Virtual Force Algorithm

The extended virtual force algorithm is a self-deployment scheme with some novel features which can overcome the drawbacks of traditional VFA. In mobile sensor networks with small value of R_c/R_s, the stacking problem caused by nonplanar connectivity graph is not serious. Thus, connectivity maintenance should be more important than stacking avoidance. In this section, the algorithm is defined as $Low - R_c$ VFA. On the other hand, in mobile sensor networks with large value of R_c/R_s, the stacking problem caused by nonplanar connectivity graph is quite obvious. The algorithm should eliminate the sensor nodes' stacking. We define the algorithm as $high - R_c$ VFA.

4.2.2.1 $Low - R_c$ VFA

The nonplanar connectivity graph can be formed, when R_c/R_s is larger than 3. So we consider the mobile sensor network as a $Low - R_c$ network, while the value of R_c/R_s is equal to or smaller than 2.5. In $Low - R_c$ VFA, distance force pushes nodes away from neighbors if their distances are less than D_{th} while moving nodes towards to neighbors if their distances are larger than D_{th}, and orientation force is added to improve the force model in order to keep the continuous connectivity. The distance force model is shown in Eq. (4.7) and Fig. 4.8a:

$$\vec{F}_{ij}(d) = \begin{cases} \alpha\left(d_{ij}^{-\beta} - D_{th}^{-\beta}\right)u_{ij} & \text{if } D_{th} < d_{ij} \le R_c \\ 0 & \text{if } d_{ij} = D_{th} \\ \mid \alpha\left(d_{ij}^{-\beta} - D_{th}^{-\beta}\right) \mid u_{ij} & \text{if } 0.5D_{th} < d_{ij}. \le D_{th} \\ 1u_{ij} & \text{if } d_{ij} \le 0.5D_{th} \end{cases} \tag{4.7}$$

where u_{ij} is the unit vector from node i to node j, α and β are constants which can be adjusted according to different situations. Normally, the value of α is equal to $0.25D_{th}^2$, while $\beta = 2$ is equal to 2. As for pair of nodes which distance is less than $0.5D_{th}$, the limit magnitude of virtual force exerted on them is set as 1 in order to avoid the high acceleration and velocity by the control input saturation. The distance force can reduce the connectivity degree to be less than or equal to 6 if the initial

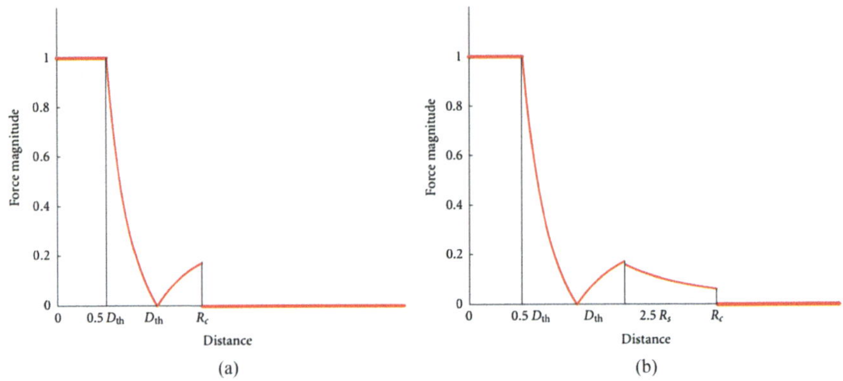

Fig. 4.8 Force models. (**a**) *Low* $- R_c$, (**b**) *High* $- R_c$

connectivity degree is larger than 6. The orientation forces are only exerted on the nodes whose neighbors are less than 6. The aim of orientation force is tried to keep the angle formed by one node and its two adjoined neighbors equal to π/3. Then mobile sensor networks can reach reliable connectivity and eliminate coverage holes caused by continuous connectivity. For node i with connectivity degree m ranging from 2 to 6, we calculate all angles between node i and its two adjoining neighbors as $S(\theta) = \{\theta(i, j, p)\}$. The orientation force model is

$$\vec{F}_{ij(o)} = \begin{cases} \omega_o\left(\theta(i, j, j_c) - \frac{\pi}{3}\right)u_{jjc} \\ \text{if } \theta(i, j, j_{ac}) \text{ is the largest angle} \\ \omega_o\left(\theta(i, j, j_{ac}) - \frac{\pi}{3}\right)u_{jjac} \\ \text{if } \theta(i, j, j_c) \text{ is the largest angle} \\ \omega_o\{\mid \theta(i, j, j_{ac}) - \frac{\pi}{3})u_{jjc} \\ + \mid \theta(i, j, j_{ac}) - \frac{\pi}{3})u_{jjac} \\ \text{otherwise} \end{cases} \tag{4.8}$$

where node j_c is the neighbor of node i on the clockwise orientation under node j, and j_{ac} is the neighbor on the anticlockwise orientation. Then, in *low* $- R_c$ VFA, the force exerted on a node can be calculated by

$$\vec{F}_{ij} = \begin{cases} \vec{F}_{ij(d)} & \text{if } m > 6 \text{ or } m = 1 \\ \vec{F}_{ij(d)} + \sum_{j \in G_i} \vec{F}_{ij(o)} & \text{if } 2 \leq m \leq 6 \end{cases} \tag{4.9}$$

4.2.2.2 *High* − R_c VFA

We consider the mobile sensor network as a $High - R_c$ networks if the R_c/R_s is larger than 2.5. In $High - R_c$ VFA, connectivity maintenance can be guaranteed. Thus, only distance forces are exerted among the nodes and their neighbors. We obtain the virtual force model in $High - R_c$ VFA as follows:

$$\vec{F}_{ij} = \begin{cases} \alpha d_{ij}^{-\beta} u_{ij} & \text{if } D_{th} > 2.5R_s \\ \alpha\left(d_{ij}^{-\beta} - D_{th}^{-\beta}\right) u_{ij} & \text{if } D_{th} < d_{ij} \leq 2.5R_s \\ 0 & \text{if } d_{ij} = D_{th} \text{ or } d_{ij} > R_c \\ \mid \alpha\left(d_{ij}^{-\beta} - D_{th}^{-\beta}\right) \mid u_{ij} & \text{if } D_{th} < d_{ij} \leq 2.5R_s \end{cases} \tag{4.10}$$

In $High - R_c$ VFA, virtual force varies exponentially with distance between each pair of mobile sensor nodes. As shown in Eq. (4.10) and Fig. 4.8b, the virtual forces between node and its faraway neighbors are decreasing with the farther distance. This feature can diminish nodes stacking effectively.

There always exist virtual attractive forces between nodes and its faraway neighbors in the traditional VFA. Therefore, stacking possibly occurs in $High - R_c$ mobile sensor networks. In $High - R_c$ VFA, besides the exponential force model, we consider a judgment of distance force between node and its faraway nodes to prevent node stacking from nonplanar connectivity. The scheme of virtual force between is shown in Algorithm 4.4.

Algorithm 4.4 The Scheme of Virtual Force Between Node *i* and its Neighbor Nodes

G_i: the set of neighbors of node *i* satisfy $d_{ij} \leq R_c$;

S_i: the set of neighbors of node *i* satisfy $d_{ij} \leq D_{th}$;

N_i: the number of nodes belongs to G_i;

n_i: the number of nodes belongs to S_i;

Step 1 For $1 \leq j \leq N_i$

Step 2 If $d_{ij} \leq D_{th}$

Step 3 $\vec{F}_{ij}$ is calculated by Eq. (4.10)

Step 4 End if

Step 5 If $d_{ij} \geq D_{th}$

Step 6 Find the neighbors which belong to both S_i and G_j. *m* is the number of nodes satisfy $d_{jm} < R_c$ and belong toS_i;

Step 7 If $m < 2$

Step 8 $\vec{F}_{ij} = 0$

Step 9 End if

Step 10 If $m < 2$

Step 11 Compute the angle θ formed by node i , j_c, and j_{ac}, where $\overrightarrow{ij}_c$ is on the clockwise side of $\overrightarrow{ij}_c$, while $\overrightarrow{ij}_{ac}$ is on the anti-clockwise side. Node j_c, and j_{ac} belongs to S_i and G_j.

Step 12 If $\theta \leq \pi/3$ and there are not any nodes located in quadrilateral ijj_cj_{ac}

Step 13 $\vec{F}_{ij} = 0$

Step 14 Else

Step 15 $\vec{F}_{ij}$ is calculated by Eq. (4.10).

Step 16 End if

Step 17 End if

Step 18 End if

Step 19 End for

As shown in Fig. 4.9, $d_{14} > D_{th}$. $\angle 213 < \pi/3$ in (a), while $\angle 213 > \pi/3$ in (b). Node 2 and node 3 are the neighbors of node 1, where $d_{1j} < D_{th}$, and they are also the neighbors of node 4, where $d_{4j} < R_c$. And in Fig. 4.9c, node 5 is located in quadrilateral 1234. Both in $low - R_c$ VFA and $High - R_c$ VFA, the total force exerted on node i is decided by adding all forces contributed by its neighbors as Eq. (4.6). Each node is driven for moving by the control law in Eq. (4.4).

4.2.3 Simulation Results

In this section, we present simulation results for the extended virtual force approach including $low - R_c$ VFA and $High - R_c$ VFA. Figure 4.11 shows the self-deployment results from stochastic initial distribution of 20 mobile sensor nodes in a 100 m × 100 m area. In this simulation, the communication range is 25 m, while

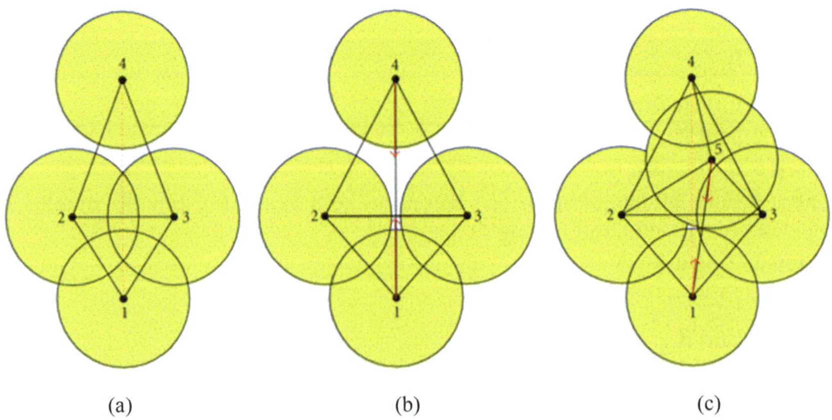

Fig. 4.9 Virtual force between node and its faraway node. (**a**) $\theta < \pi/3$, (**b**) $\theta > \pi/3$, (**c**) $\theta > \pi/3$

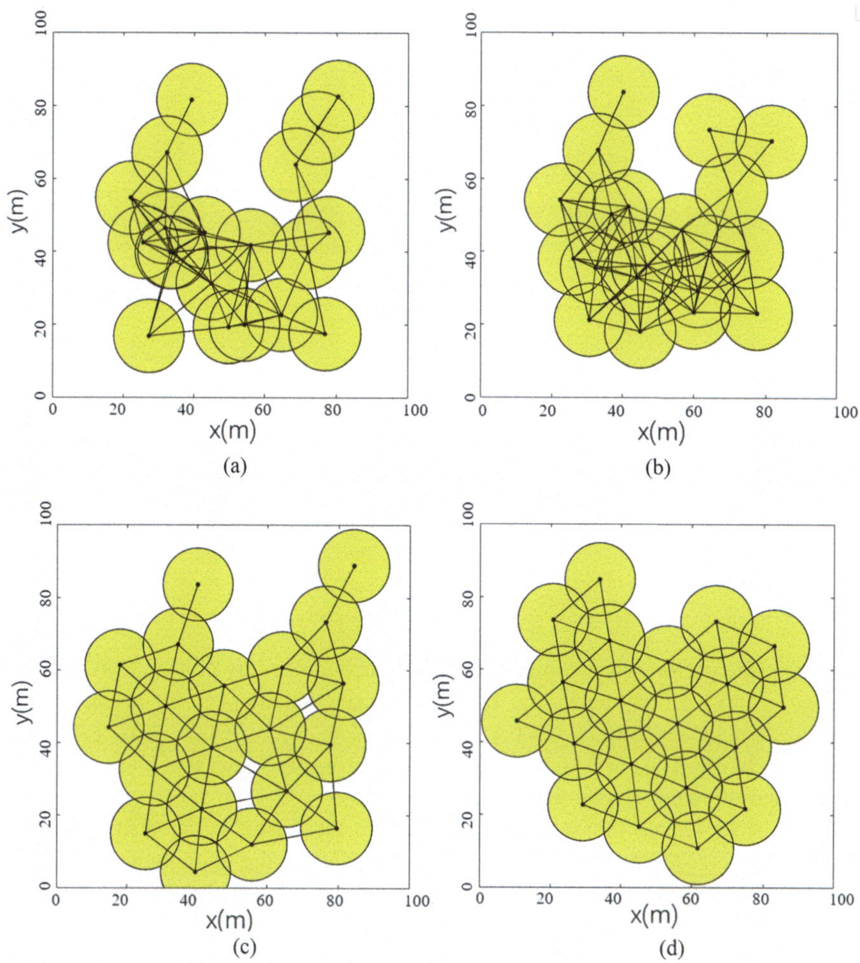

Fig. 4.10 Self-deployment results in $R_c/R_s = 2.5$. (a) initial deployment, (b) self-deployment by VFA, (c) self-deployment by IVFA, (d) self-deployment by $low - R_c$

the sensing range is 10 m. The sampling time is 0.1 s, and the damping coefficient k_d is 0.5. Figure 4.10a is the initial deployment and its unit disk graph (UDG). Figure 4.10b shows the self-deployment result by VFA. The final deployment is compressed because the nonplanar connectivity graph. In Fig. 4.10c, IVFA shows the better result than VFA for avoiding node stacking caused by intersection of connectivity edges. However, the continuous connectivity is not perfect because of significant coverage holes in IVFA. Self-deployment by $low - R_c$ VFA is shown in Fig. 4.10d. We can see that an ideal deployment is achieved. Figure 4.11 shows the performance comparison of coverage rate among VFA, IVFA, and $low - R_c$ VFA. IVFA can reach the highest coverage because the discontinuous connectivity. For the stacking of nodes, VFA cannot guarantee the enhancement of the coverage rate. All three of these algorithms have achieved a convergence state in about 30 s.

Fig. 4.11 Coverage rate comparison among VFA, IVFA, and $low - R_c$ VFA

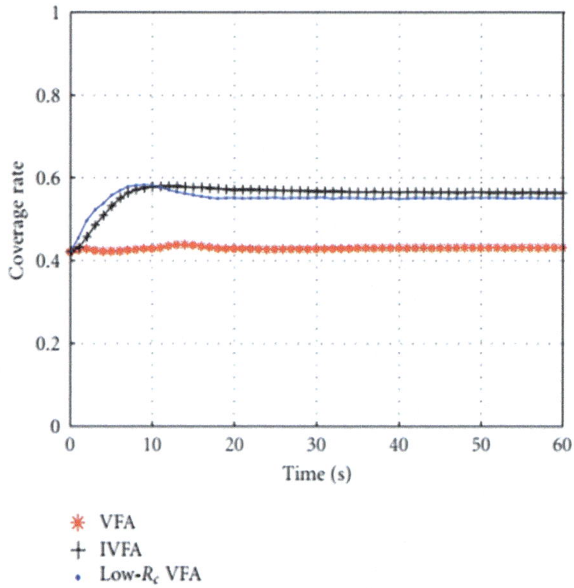

Moreover, the performance comparison of distance uniformity among VFA, IVFA, and $low - R_c$ VFA is shown in Fig. 4.12. It is noticed that after self-deployment with 60 s, the distance uniformity of VFA, IVFA, and $low - R_c$ VFA are 3.01, 1.42, and 0.13 respectively. The performance of distance uniformity shows that $low - R_c$ VFA has reached the most regular ideal deployment.

Figure 4.13 shows the self-deployment from stochastic initial distribution in Fig. 4.13a, when R_c/R_s is equal to 4. In this simulation, the communication range is 40 m, while the sensing range is 10 m. The sampling time is 0.1 s, and the damping coefficient kd is 0.5. Here we investigate self-deployment results under VFA, Gabriel Graph and $High - R_c$ VFA. Figure 4.13a is the initial deployment and its unit disk graph (UDG). It can be seen that more intersections of connectivity edges have been found compared to the network in Figure 4.10a due to the larger R_c. In Fig. 4.13b, a fully connectivity graph and conspicuous nodes stacking are shown in self-deployment results by VFA. Selfdeployment by GG mesh has obtained a result without stacking. Figure 4.13c shows the GG mesh graph after selfdeployment. Although GG is a planar graph, it does not guarantee the hexagonal formation. Self-deployment by $high - R_c$ VFA is shown in Fig. 4.13d. Clearly, it shows that a regular ideal deployment has been attained in this circumstance. In order to demonstrate self-deployment of largescale mobile sensor networks, 100 mobile nodes are randomly placed for a concentrated deployment, as shown in Fig. 4.14a. The communication range is 40 m, while the sensing range is 10 m. The sampling time is 0.1 s, and the damping coefficient kd is 0.5. The simulation result of selfdeployment by $high - R_c$ VFA is shown in Fig. 4.14b. The number of iterations is 6000 (10 min). It is obvious that an asymptotically ideal deployment has been

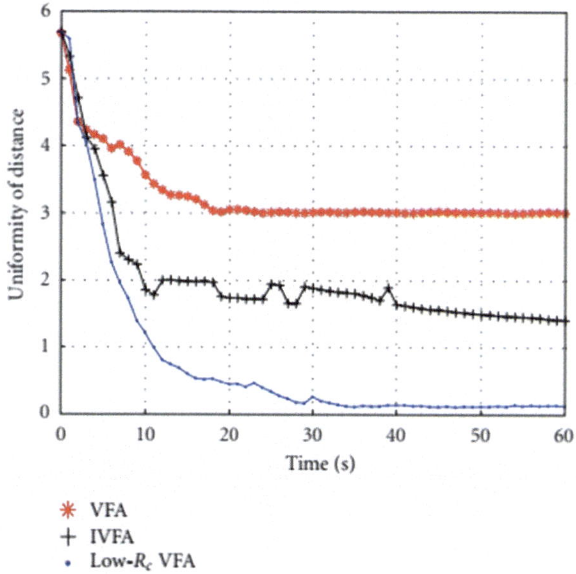

Fig. 4.12 Uniformity comparison among VFA, IVFA, and *low* − R_c VFA

reached. The impact of network size on self-deployment is shown in Fig. 4.15. Figure 4.15a shows the change of distance uniformity with different nodes' numbers. In both IVFA and GG, distance uniformity deteriorates as the number of nodes increased, while *high* − R_c VFA provides almost constant uniformity. And we can also observe that *high* − R_c VFA obtains the best performance uniformity from Fig. 4.15b.

The control coefficients influence the algorithm's performance. In the proposed extended virtual force-based approach, we focus on the sampling time and damping coefficient in this section. Equation (4.4) models the movement of self-deployment as a continuous system. Therefore, more accurate control can be obtained, when the sampling time becomes shorter. Figure 4.16a shows the distance uniformity comparison of different sampling time under *High* − R_c VFA. It can be seen that uniformity oscillation is obvious, when sampling time is 1 s, while convergence state and low uniformity are shown when the sampling time is 0.1 s and 0.01 s. However, a short sampling time needs lots of communication cost, and it also requires the process time should be shorter than sampling time. These requirements may be limited by the hardware of sensor nodes. So, 0.1 s is a reasonable sampling time in this section. The damping coefficient also strongly influences selfdeployment. We conduct simulations with damping coefficient of 0.3, 0.5, and 1 in here. As shown in Fig. 4.16b, the lower damping coefficient takes a faster moving at the beginning of self-deployment. However, it gets a higher value of distance uniformity with oscillation. It means that we should choose a higher damping coefficient to reach the more regular formation. Thus, we consider damping coefficient varying from 0.5 to 1 is reasonable value.

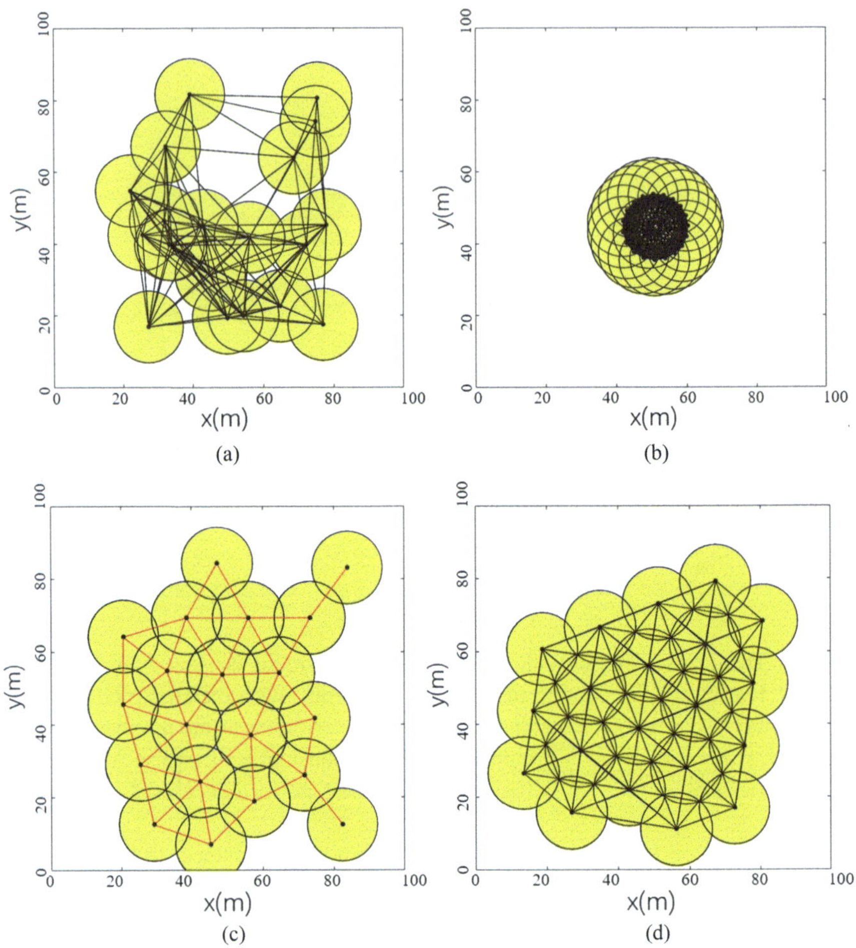

Fig. 4.13 Self-deployment results in $R_c/R_s = 4$. (**a**) initial deployment, (**b**) self-deployment by VFA, (**c**) self-deployment by GG, (**d**) self-deployment by $high - R_c$ VFA

4.3 Homogeneous Sensor Network Deployment Based on Distance and Orientation Deployment

In this section, we propose a self-deployment algorithm for mobile sensor networks based on distance and orientation control (DOC) [4]. The algorithm tries to keep the distance among sensor nodes equal to the idea distance in equilateral triangle grid deployment by distance control while reaching uniform and reliable connectivity by orientation control. Moreover, obstacle avoidance control should be taken into account if there is obstacle lay on the locomotion path. Simulation results of DOC algorithm show that DOC algorithm is effective with good performances in coverage, connectivity and obstacle avoidance.

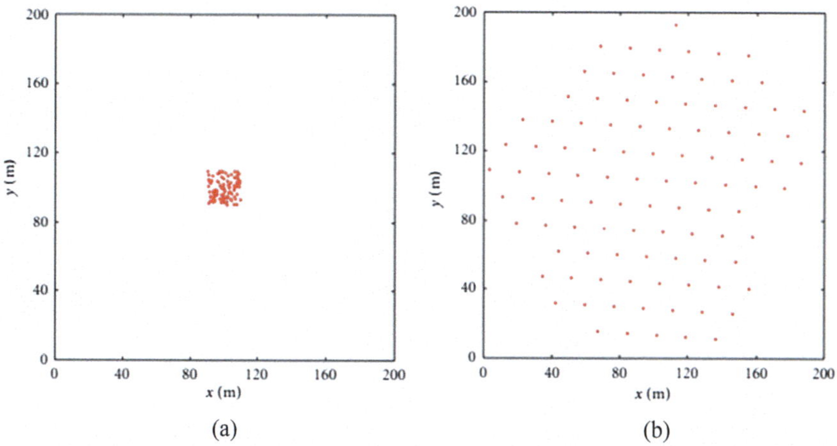

Fig. 4.14 Self-deployment in large-scale concentrated deployment. (**a**) concentrated initial deployment, (**b**) self-deployment by $high - R_c$ VFA

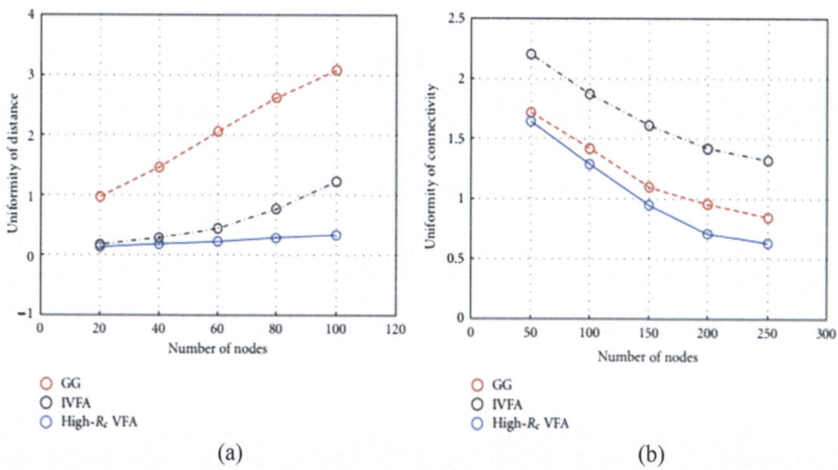

Fig. 4.15 Impact of nodes' number under IVFA, GG, and $high - R_c$ VFA. (**a**) Uniformity of distance, (**b**) Uniformity of connectivity

4.3.1 The DOC Algorithm

The basic idea of DOC is to move the sensor nodes from initial positions to new positions in order to form the ideal deployment, with controlling the distance and angle to be ideal. An ideal deployment grid structure is shown in Fig. 4.17a. Equilateral triangle grid has the smallest overlapping area, hence this deployment requires only the least number of sensors for area whole coverage, and there is no coverage "hole". As shown in Fig. 4.17b, the ideal distance D_s between sensor node

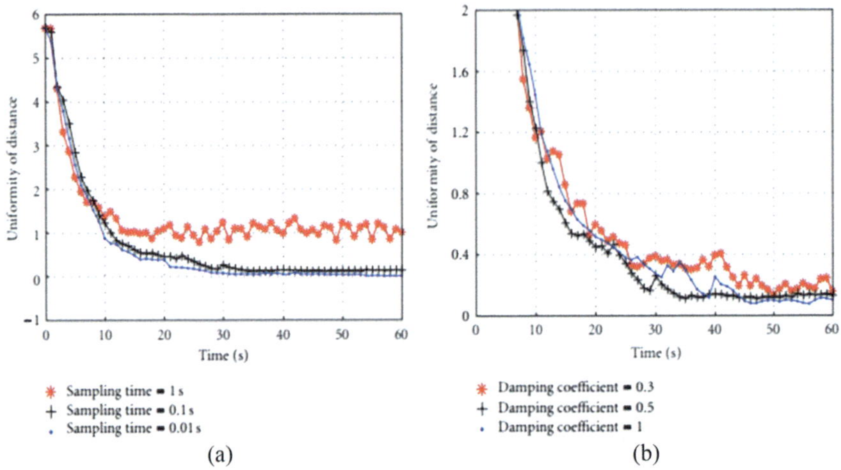

Fig. 4.16 Impact of control coefficients and the nodes' number. (**a**) Uniformity with different sampling time, (**b**) Uniformity with different dampling coeffcient

and its neighbors is $\sqrt{3}Rs$, the relative angle formed by sensor node and its two anti-clockwise continuous neighbors is $\pi/3$.

The DOC algorithm is a distributed self-deployment algorithm. The next position of mobile sensor node is determined by distance and orientation control from its neighbors. Obstacle avoidance should also be taken into account if there are obstacles in deployment area.

4.3.1.1 Distance Control

Similarly to VFA, if the nodes' distance is less than Ds, the distance control in DOC can push nodes away from neighbors. Otherwise it moves nodes towards to neighbors. For node i with connectivity degree k, the distance control strategy of i in DOC is as follows:

$$\begin{cases} \Delta x_{di,j} = k_d^*(d(i,j) - Ds)^*\left(x_j - x_i\right)/d(i,j) \\ \Delta y_{di,j} = k_d^*(d(i,j) - Ds)^*\left(y_j - y_i\right)/d(i,j) \\ \Delta x_{di} = \sum\limits_{j=1}^{k} \Delta x_{di,j}, \Delta y_{di} = \sum\limits_{j=1}^{k} \Delta y_{di,j} \end{cases} \tag{4.11}$$

where k_d is the step size of distance move. It can be easily seen if $d(i,j)$ is larger than Ds, the node i moves towards node j. And in the same way, node i moves away node j when $d(i,j)$ is less than Ds.

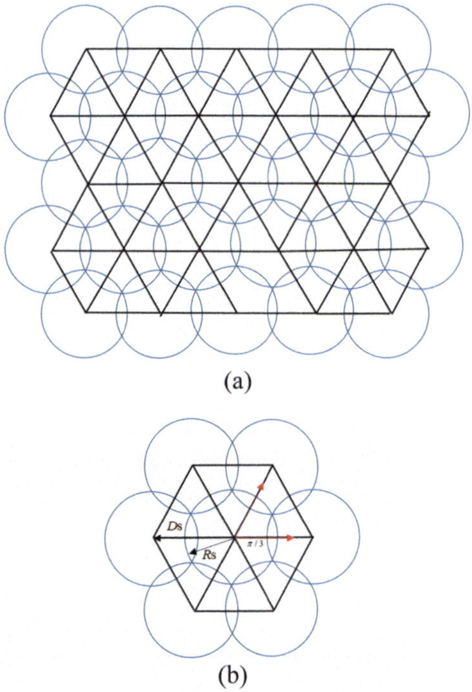

Fig. 4.17 Ideal deployment. (**a**) Ideal deployment: Equilateral triangle grid, (**b**) Ideal distance and angle

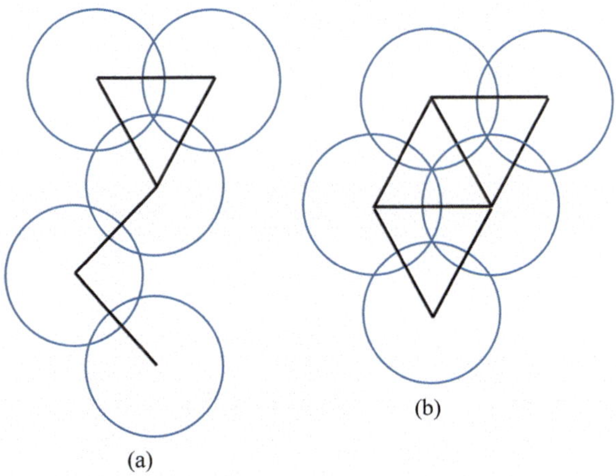

Fig. 4.18 Connectivity performance. (**a**) Discontinuous connectivity, (**b**) Continuous connectivity

4.3.1.2 Orientation Control

The aim of orientation control is to reach reliable connectivity and eliminate coverage holes caused by discontinuous connectivity. As shown in Fig. 4.18, discontinuous connectivity in Fig. 4.18a is the equilibrium state in deployment algorithm based on distance criteria such as VFA. However, it is instable in DOC. The sensor nodes can be moved to form the continuous connectivity shown in Fig. 4.18b by orientation control.

Orientation control tries to make the relative angle form by sensor node and its two contiguous neighbors be $\pi/3$, as shown in Fig. 4.18b. Only the nodes with connectivity degree between 2 and 6 can adjust their neighbors' orientation. Distance control is enough if the connectivity degree is 1 or larger than 6. The distance control can reduce the connectivity degree to less than or equal to 6 if the initial degree is larger than 6.

For a node j with connectivity degree m range from 2 to 6, we define the set of the relative orientations between node j and its neighbors as: $S(\theta) = \{\theta(j, p), 1 \leq p \leq m\}$. Then we sort these angles in an anti-clockwise order between 0 and 2π as: $S'(\theta) = \{\theta'(j, p), 1 \leq p \leq m, \theta'(j, p) \leq \theta'(j, p+1)\}$.

Define a set of all angles between node and its two adjoining neighbors as:

$$S(\theta_1) = \{\theta(j, p), 1 \leq p \leq m\} \tag{4.12}$$

$$\theta_1(j, p) = \begin{cases} \theta'(j, p+1) - \theta'(j, p) & \text{if } q < m \\ 2\pi - \theta'(j, m) + \theta'(j, 1) & \text{if } q = m \end{cases} \tag{4.13}$$

Node j controls its neighbors as follows:

1. Find the largest angle $\theta_1(j, l)$ in $S(\theta_1)$.
2. The require angles' adjustment is calculated as

$$\begin{gathered} \Delta\theta_0(j, l) = k_o * \frac{\theta_1(j, l) - (6 - m + 1) * \frac{\pi}{3}}{2} \\ \text{if } l < m, \Delta\theta_0(j, l+1) = -\Delta\theta_0(j, l) \\ \text{if } l = m, \Delta\theta_0(j, 1) = -\Delta\theta_0(j, l) \end{gathered} \tag{4.14}$$

For other neighbors, nodes' adjoining angles try to be equality, so we can get

$$\Delta\theta_0(j, p) = \begin{cases} k_o * \dfrac{\theta_1(j, p) - \theta_1(j, p-1)}{2}, \text{if } p > 1 \\ k_o * \dfrac{\theta_1(j, p) + 2\pi - \theta_1(j, m)}{2}, \text{if } p = 1 \end{cases} \tag{4.15}$$

where k_o is the step size of angle adjustment.

3. The new position of neighbors is defined as:

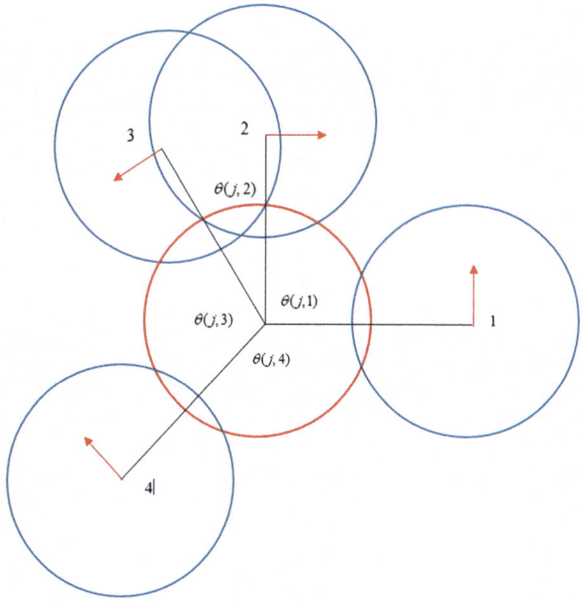

Fig. 4.19 Orientation control of neighbors

$$\begin{aligned} \Delta x_o(j,p) &= d(j,p) * (\cos(\theta'(j,p) + \Delta\theta_o(j,p)) - \cos\theta'(j,p)) \\ \Delta x_o(j,p) &= d(j,p) * (\sin(\theta'(j,p) + \Delta\theta_o(j,p)) - \sin\theta'(j,p)) \end{aligned} \tag{4.16}$$

The orientation control method is used to divide the angles into continuous equal angle $\frac{\pi}{3}$ by margins. As shown in Fig. 4.19, the connectivity degree of node j is 4, four adjoining angles are $\frac{\pi}{2}, \frac{\pi}{6}, \frac{7\pi}{12}$ and $\frac{3\pi}{4}$, respectivly. The largest angle $\theta_1(j,4)$ is less than $(6-4+1) * \frac{\pi}{3}$, so neighbor 4 moves towards clockwise while 1 moves towards the opposite. Node 3 moves towards anti-clockwise and node 2 moves towards clockwise because $\theta_1(j,3) > \theta_1(j,2)$, $\theta_1(j,2) < \theta_1(j,1)$.

After numbers of iteration cycle, the largest angle is equal to π while other angles is equal to $\frac{\pi}{3}$.

The orientation control imposed on each sensor node is determined by all its neighbors. Thus, for node i with connectivity degree k, the position adjustment of i by orientation control strategy in DOC is shown in Fig. 4.20.

4.3.1.3 Obstacle Avoidance Control

After executing distance and orientation control algorithm, the next position for sensor node can be generated. Generally, the movement tracjectory of mobiles sensor node is a straight line. If there is obstacle lay on the locomotion path, obstacle

Fig. 4.20 Position adjustment of i by orientation control

$\Delta x_{oi} = 0, \Delta y_{oi} = 0$
for $j = 1$ to k
    if connectivity degree of j range from 2 to 6
        calculate $\Delta x_o(j,i)$ and $\Delta y_o(j,i)$;
        $\Delta x_{oi} = \Delta x_{oi} + \Delta x_o(j,i)$;
        $\Delta y_{oi} = \Delta y_{oi} + \Delta y_o(j,i)$;
    end
end

avoidance should be considered otherwise mobile sensor node cannot move to ideal position. Define new position $S_i'(x_i', y_i')$ for node i with initial position $S_i(x_i, y_i)$. $S_o(x_o, y_0)$ is the position of locomotion path collide with obstacle. The obstacle avoidance control avoids collision by the strategy as follows:

$$\begin{aligned} \Delta x_{ai} &= x_i' - x_o - \varepsilon \\ \Delta y_{ai} &= y_i' - y_0 - \varepsilon * \frac{y_{i'}' - y_0}{x_i' - x_o} \end{aligned} \tag{4.17}$$

where ε is a positive threshold distance for node and obstacle.

Figure 4.21 shows the flowchart of DOC self-deployment algorithm. Self-deployment terminates when the equilibrium state is reached or the number of iteration equal to the maximum iteration number L we set.

4.3.2 *Simulation Results*

In simulation results, 20 sensors with communication range of 20 m and sensing range of 10 m are randomly placed in 100 m × 100 m area, as shown in Fig. 4.22a. The area coverage ratio in the initial deployment is 0.3116 while the average connectivity degree is 4.0 with mean square deviation 8.72. Figure 4.22b is the self-deployment simulation result for VFA algorithm, and Fig. 4.22c is for DOC algorithm. Figure 4.23 shows the change of coverage during the iterations of DOC and VFA. From the figure, the coverage ration is enhanced from 0.3116 to 0.5524, and in DOC it is almost equal to VFA. However, it's obvious that there are several coverage holes and discontinuous connectivity in VFA. The average connectivity degree is 3.6 and the mean squaredeviation for connectivity is 6.69 in VFA. In DOC,

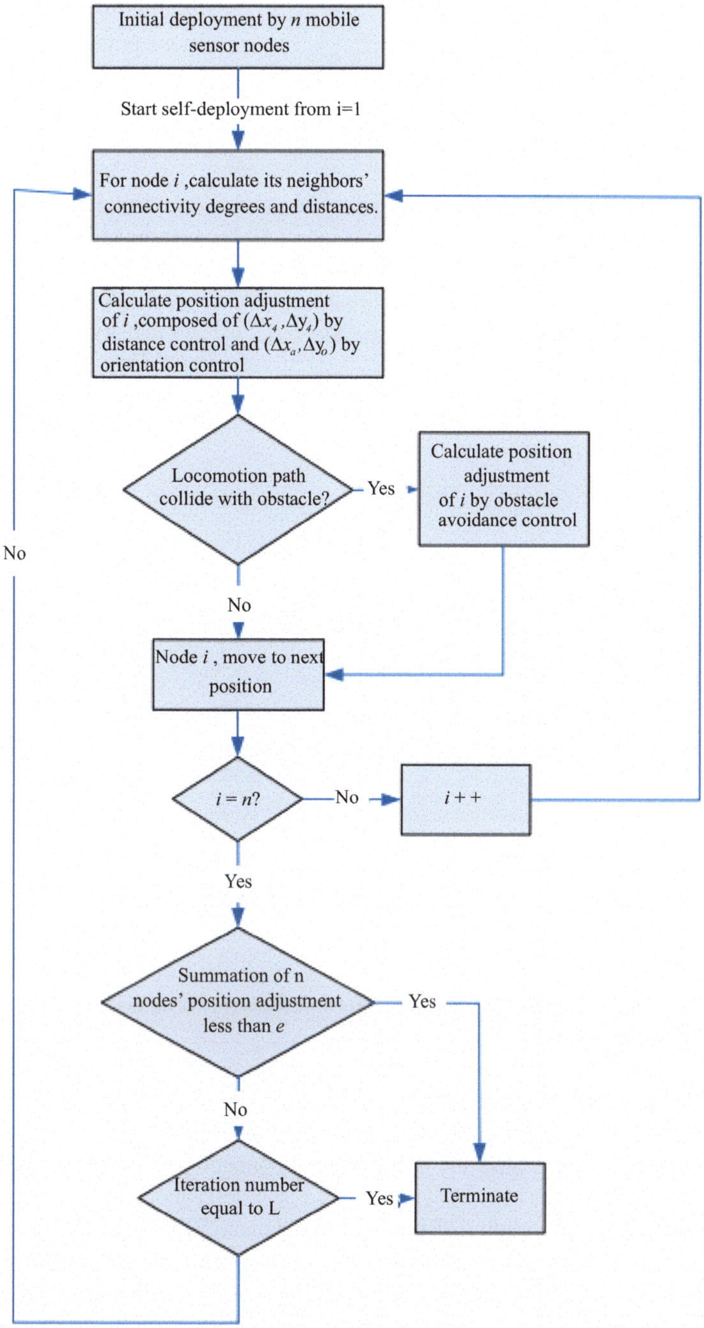

Fig. 4.21 Flowchart of DOC self-deployment algorithm

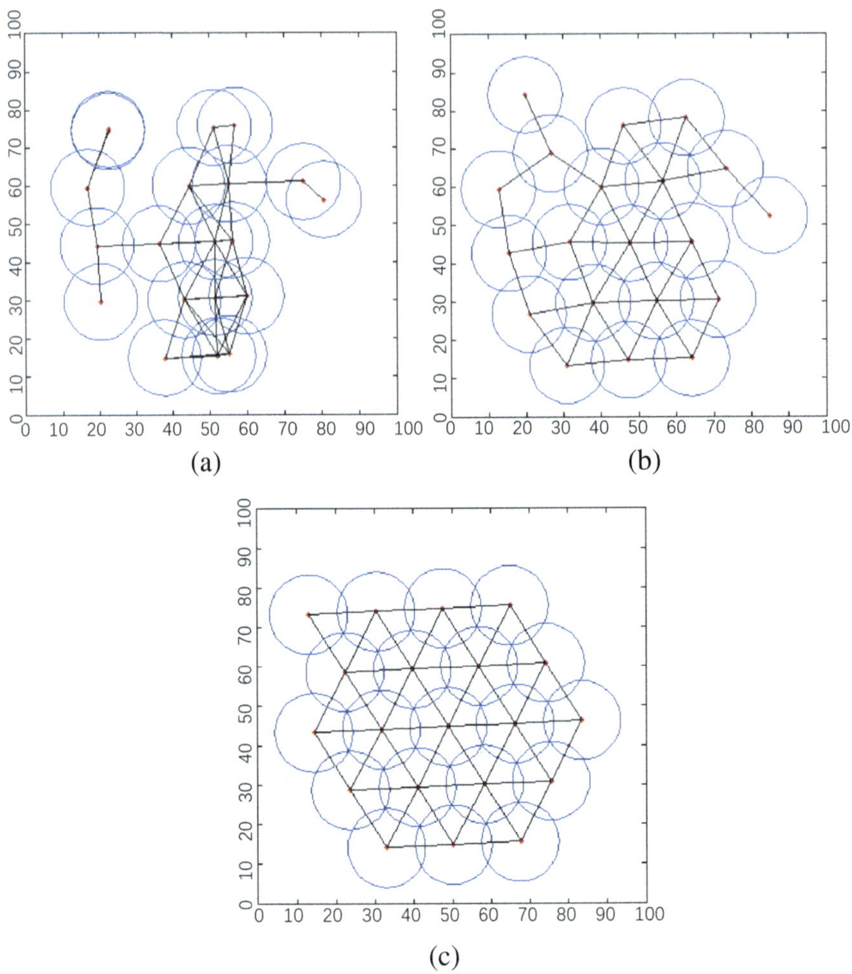

Fig. 4.22 Sef-deployment by VFA and DOC. (**a**) Initial deployment, (**b**) Self-deployment by VFA, (**c**) Self-deployment by DOC

there is no coverage hole, and the average connectivity degree is 4.4 while the mean square deviation is 5.9. The simulation results demonstrate that DOC can reach more reliable connectivity because the uniformity of connectivity in DOC is better than VFA.

For simulating the validity of obstacle avoidance, 50 mobile sensor nodes are placed in 30 m × 30 m area as a concentrated deployment. In addition, there are four obstacles in the field, as shown in Fig. 4.24a. Figure 4.24b is self-deployment result by DOC algorithm. The simulation result shows that mobile sensor nodes can locomote with avoiding obstacle for self-deployment by DOC algorithm.

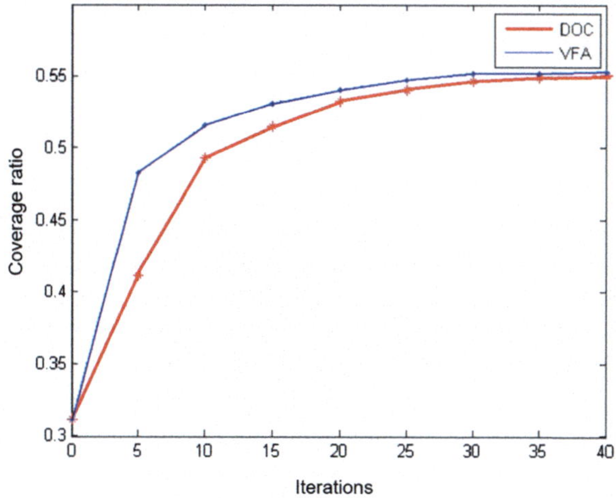

Fig. 4.23 The improvement of coverage ratio by VFA and DOC

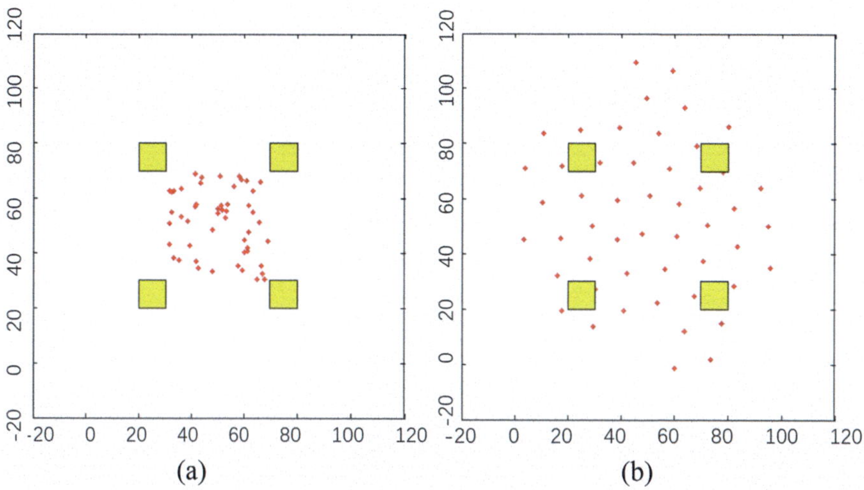

Fig. 4.24 Deployment by DOC. (**a**) Concentrated deployment, (**b**) Obstacle avoidance by DOC

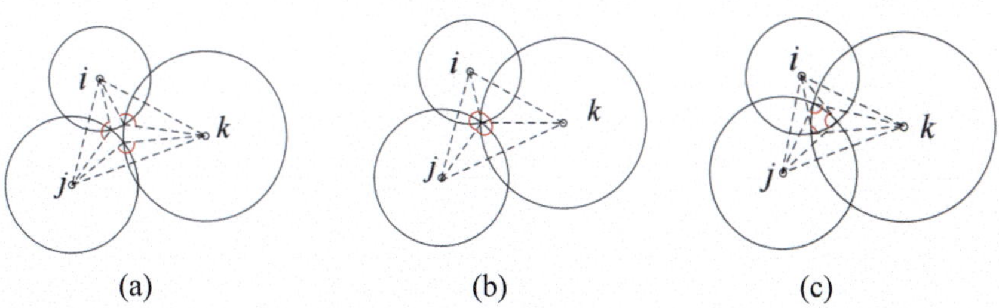

Fig. 4.25 Localization effect in convex and concave holes. (**a**) $\sum\phi_{ijk} = 2\pi$ (**b**) $\sum\phi_{ijk} = 2\pi$ (**c**) $\sum\phi_{ijk} < 2\pi$

4.4 Heterogeneous Sensor Network Deployment Based on Extended Virtual Force

In this section, we propose a heterogeneous sensor network coverage control algorithm based on extended virtual force called Self-Deployment by Delaunay Triangulation Graph (SDDTG) [5]. By analyzing the property of heterogeneous mobile sensor network, a basic algorithm named VFA-HSN, which in view of the fixed distance coefficient is firstly introduced. However, VFA-HSN can only be employed in some specific initial deployment and may not achieve a well coverage result. In order to settle the defects of the VFA-HSN, the Basic Delaunay Triangulation Graph (BDTG) algorithm is introduced to find the logic neighbor of the node. The SDDTG algorithm adds the overlap angle correction to eliminate coverage holes and redundancy of BDTG algorithm. Simulation results show that SDDTG can achieve much higher coverage quality for both gathered initial deployments and scatter initial deployments compared with VFA-HSN and BDTG.

4.4.1 *VFA-HSN Algorithm*

An ideal coverage is achieved when the distance between two nodes is $\sqrt{3}/2 \cdot \left(R_{si} + R_{sj}\right)$, where R_{si} is the sensing range of node *i*, R_{sj} denotes the sensing range of node *j*.

Lemma 4.1 *In a three nodes coverage area as shown in Fig. 4.25, if* $\sum \phi_{ijk} = \phi_{ij} + \phi_{ik} + \phi_{jk} > 2\pi$, *there is a hole in coverage area; if* $\sum \phi_{ijk} = \phi_{ij} + \phi_{ik} + \phi_{jk} < 2\pi$, *there is a redundancy in area; if* $\sum \phi_{ijk} = \phi_{ij} + \phi_{ik} + \phi_{jk} = 2\pi$, *there is neither hole in coverage nor redundancy in area.*

Clearly from Lemma 4.1, there is neither redundant nor hole in coverage when $\sum \phi_{ijk} = 2\pi$.

Theorem 4.1 *Assume that three nodes i, j, k cover an area, and the distance between node i and node j is* $d_{ij} = \lambda(R_{si} + R_{sj})$, *where* λ *is the distance coefficient. Then, when* $\lambda = \sqrt{3}/2$, *the nodes cover the area without any hole.*

The virtual force of the nodes is designed as

$$F_{ij} = \begin{cases} \omega_A\left(d_{ij} - D_{th(i,j)}\right)u_{ij} & \text{if} \quad D_{th(i,j)} < d_{ij} \leq \lambda_2\left(R_{si} + R_{sj}\right) \\ \left| \omega_A\left(d_{ij} - D_{th(i,j)}\right) \right| u_{ij} & \text{if } d_{ij} \leq D_{th(i,j)} \\ 0 & \text{if } d_{ij} > \lambda_2\left(R_{si} + R_{sj}\right) \end{cases} \tag{4.18}$$

where ω_A represents the virtual force attractive and repulsive coefficients, u_{ij} is the orientation of the line segment from nodes *i* to *j*, λ_2 is force field truncation coefficient. When the distance between two nodes is larger than $\lambda_2(R_{si} + R_{sj})$, the virtual force of the nodes is zero. Because long communication radius may cause nodes gathering, we have to set $\lambda_2 = 1$.

The motion dynamic model of the node is given by

$$\begin{cases} \ddot{x}_i = \sum\limits_{j \in G_i} F_{ij} - k_d \dot{x}_i \\ v_i = \dot{x}_i \\ a_i = \ddot{x}_i \end{cases} \tag{4.19}$$

where v_i is the speed of the node, a_i is the acceleration of the node and k_d is damping coefficient. Although this control algorithm is easy to implement, but it provides low-quality deployment. By force field truncation, we can avoid nodes gathering in one place. However, it cannot ensure the connectivity of the networks in the case of scatter initial deployment. Then, we introduce the BDTG algorithm to generate better coverage result.

4.4.2 The Basic Delaunay Triangulation Graph (BDTG) Algorithm

The optimal decision for 1-coverage in homogeneous WSNs is the symmetric deployment of nodes where a hexagonal grid of nodes can be achieved. Unfortunately, symmetric arrangement of WSNs nodes in the real life is uncommon. Comparing with the homogeneous WSNs, the ideal coverage of heterogeneous WSNs is more complicated.

Delaunay Triangulation graph is composed of several disjoint triangles. The vertices of the triangle are sensor nodes. Oelaunay figure is used as the logical topology of the network in the coverage control because it has no intersecting edge topology. The BDTG uses Delaunay graph as network topology where the distance coefficient $\lambda = \sqrt{3}/2$, with virtual force model as in Eq. (4.18) and motion equations as in Eq. (4.19).

To make the algorithm as a distributed algorithm, the Local Delaunay triangulation graph is defined as below.

Definition 4.1 Every node and its neighbor nodes in its communication radius make up the Delaunay graph. This graph is the node's Local Oelaunay Triangulation Graph (Local DTG or LDTG). The LDTGs of all the nodes form Unit Oelaunay Triangulation Graph (Unit DTG or UDTG). Obviously, UDTG is a sub-graph of DTG. The BDTG generation algorithm is shown in Algorithm 4.5. It terminates when the sum of all the nodes' movement distance is smaller than a specified value or the running time ends in time to stop.

Algorithm 4.5 The BDTG Algorithm

// m: the number of the logic neighbors of node *i*// M: the sampling time

Step 1 Find out logic neighbors node *i* in LDTG(i) of node *i*;

Step 2 For $1 \leq j \leq m$

Step 2.1 Set distance coefficient $\lambda = \sqrt{3}/2$

Step 2.2 Calculate F_{ij} the virtual force between node *i* and its neighbor node *j*;

Step 3 Calculate the resultant virtual force;

Step 4 Calculate the accelerated speed a_i and speed v_i of node *I* and then move node *i* to new position;

Step 5 Update $t + \Delta t$;

Step 6 Go back to **Step 1**.

4.4.3 The SDDTG Algorithm

In order to improve the coverage performance of BDTG, SDDTG algorithm is introduced. It can be proved that in Delaunay triangle, the increment of $\angle j_k i j_{k+1}$, causes monotone decrement of R_{si}. Therefore, the node in each Delaunay triangle that meets the distance coefficient must be a determined value. And it's almost impossible to have all nodes meet the value at one time. So another step is used to improve the performance.

From Lemma 4.1, if $\sum\phi_{ijk} = \phi_{ij} + \phi_{ik} + \phi_{jk} = 2\pi$, neither coverage holes nor coverage redundant exist. Then the main aim of SDDTG algorithm is to adjust the distance between the nodes according to the $\sum\phi_{ijk}$ in Delaunay triangle to approach to 2π.

For node *i*:

$$\phi_{th(i,j_k)} = \phi_{ij_k} + \left(2\pi - \sum\phi_{ij_k j_{k+1}}\right)\frac{\phi_{ij_k}}{\sum\phi_{ij_k k_{k+1}}} \tag{4.20}$$

The ideal distance between nodes can be described as:

$$D_{h(i,j_k)} = \left(R_{si}^2 + R_{sj_k}^2 - 2R_{si}R_{sj_k}\cos\phi_{h(i,j_k)}\right) \tag{4.21}$$

Virtual force s between nodes *i* and node *j* are listed as follows:

$$F_{ij} = \begin{cases} \omega_A\left(d_{ij} - D_{th(i,j)}\right)u_{ij} & \text{if } D_{th(i,j)} < d_{ij} \\ \mid \omega_A\left(d_{ij} - D_{th(i,j)}\right) \mid u_{ij} & \text{if } d_{ij} \leq D_{th(i,j)} \\ 0 & \text{if } d_{ij} = D_{th(i,j)} \end{cases} \tag{4.22}$$

Then each node calculates its own resultant virtual force of its logic neighbors. Then it moves to the new position according to the motion dynamic model shown in Eq. (4.19).

SDDTG algorithm is summarized as shown in Algorithm 4.6 below. Algorithm can be terminated when the sum of all nodes' movement distance is smaller than a given value or the running time ends to stop.

Algorithm 4.6 The SDDTG Algorithm
// m: the number of the logic neighbors of node *i*// M: the sampling time

Step 1 Do BDTG when not meet the end condition;

Step 2 Find out logic neighbors node *i* in LDTG(i) of node *i*;

Step 3 For $1 \leq j \leq m$

Step 3.1 Calculate F_{ij} the virtual force between node *i* and its logic neighbor *j*;

Step 4 Calculate the resultant virtual force;

Step 5 Calculate the accelerated speed a_i of node *i* and speed v_i, node *i* move to new position;

Step 6 Update $t + \Delta t$;

Step 7 Go back to **Step 1.**

4.4.4 Simulation Results

In this section, the numerical simulation for VFA-HSN, BDTG and SDDTG are carried out to evaluate their performance. In the simulation environment, 40 nodes are deployed randomly in a 100 m × 100 m square field. Node's communication radius is 30 m and the radius of sensing is in the interval of [2.5 m, 10 m]. Simulations of two different initial deployments are conducted as shown in Fig. 4.26. Figure 4.26a is initialized in the gather deployment manner while Fig. 4.26b is set up in the scatter deployment manner.

The coverage performance by using VFA-HSN under the gather and scatter initial deployment is shown in Fig. 4.27a, b, respectively. According to Fig. 4.27a, it can be noticed that there are several redundant and coverage holes. Based on the analysis in the Sect. 4.4.3, the constant ideal distance coefficient has a great effect on the number of the holes. In Fig. 4.27b, there are several independent groups of nodes. The reason is that, force field truncation is used to avoid nodes gather in one place and the force field truncation long-distance virtual force does not exist. Therefore, the algorithm does not apply to disperse nodes in the scatter initial deployment of heterogeneous WSNs. Since node in WSNs is disconnected in the case of scatter initial deployment, the VFA-HSN cannot be employed for this case.

The simulation results of BDTG are shown in Fig. 4.28. Figure 4.28a shows the deployment of BDTG with the same initial configuration of Fig. 4.26a, while Fig. 4.28b shows the deployment of BDTG with the same initial configuration of Fig. 4.26b. The dotted line shows the Delaunay triangulation of nodes in WSNs.

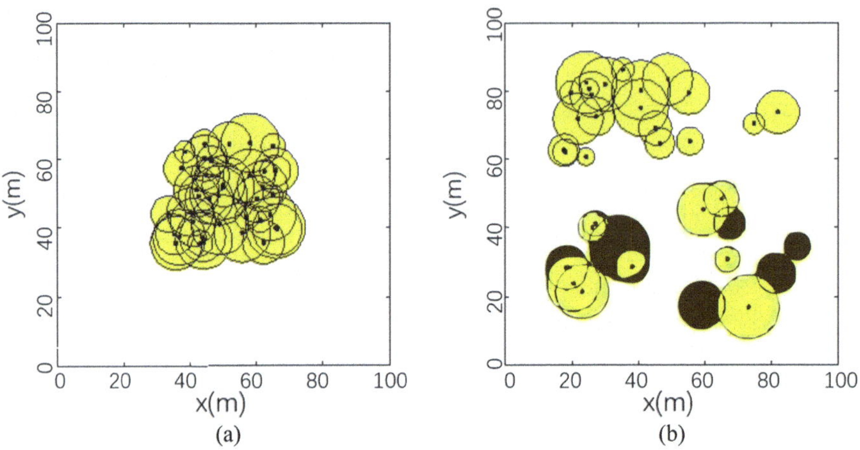

Fig. 4.26 Initial deployment. (**a**) Gather deployment, (**b**) Scatter deployment

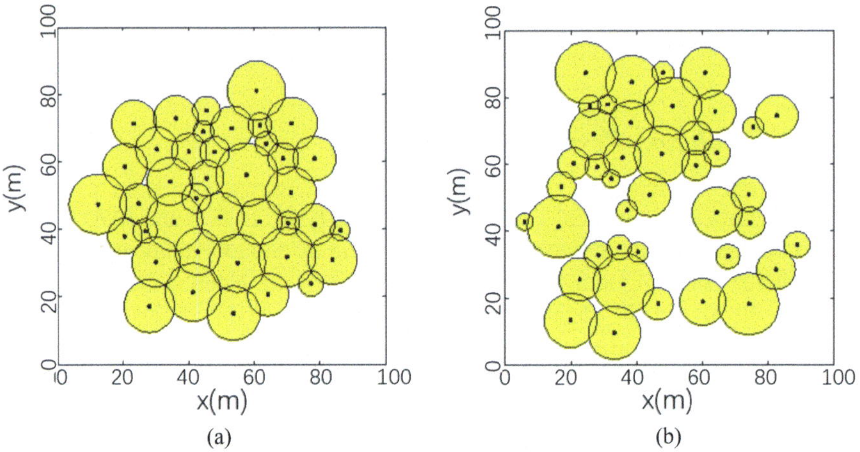

Fig. 4.27 Simulation result of VFA-HSN. (**a**) VFA-HSN in gather deployment, (**b**)VFA-HSN in scatter deployment

Compared with VFA-HSN, BDTG works well either in scatter initial or gather initial deployment. We use logic neighbor nodes instead of real neighbor node to avoid nodes form gathering together. And without force field truncation BDTG can also be employed in scatter initial deployment.

Figure 4.29a shows the deployment of SDDTG with the same initial configuration of Fig. 4.26a. Figure 4.29b shows the deployment of SDDTG with the same initial configuration of Fig. 4.26b. It can be seen clearly from Fig. 4.29a. Compared with Fig. 4.28a, less holes and redundant exist in final deployment. In BDTG, ideal distance coefficient is settled, and neighbor nodes cannot reach the Ideal distance

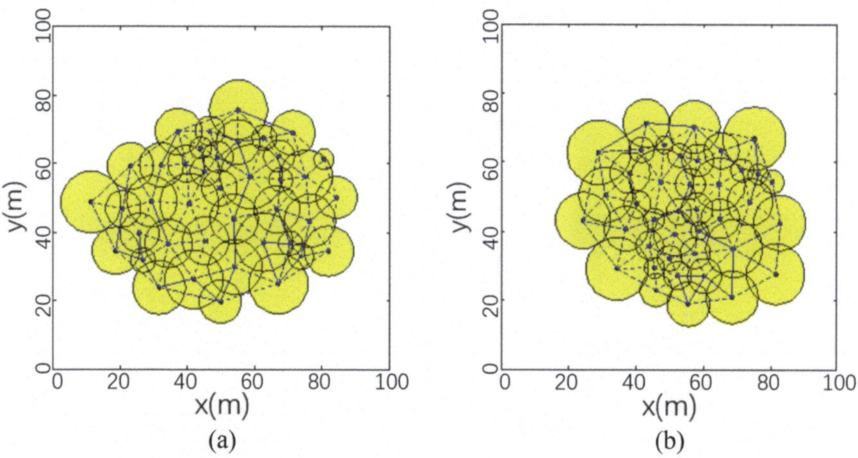

Fig. 4.28 Simulation result of BDTG. (**a**) BDTG in gather deployment, (**b**) BDTG in scatter deployment

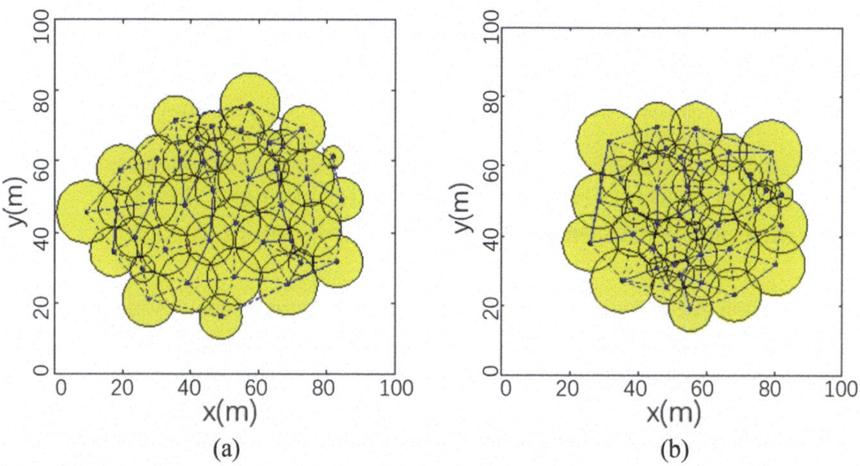

Fig. 4.29 Simulation result of SDDTG. (**a**) SDDTG in gather deployment, (**b**) SDDTG in scatter deployment

coefficient at same time. To improve the performance of the algorithm another round of movement begins with modified ideal distance coefficient. Due to the modified ideal distance coefficient, the change of virtual forces between nodes can cause two scenarios. In area with redundant virtual force makes the nodes move away from each other and in area with holes virtual force makes nodes get closer to each other. The above experiments have been carried out for 100 times, the simulation results are illustrated below. Figure 4.30 shows the k-coverage rates of VFA-HSN, BDTG and SDDTG with the initial deployment as shown in Fig. 4.26b.

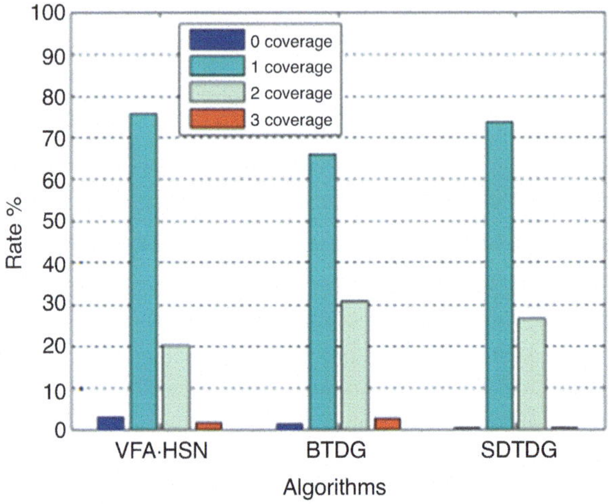

Fig. 4.30 K-coverage rates of the final deployment

In Fig. 4.30 the Y-axis stands for the coverage rate. It can be seen clearly that SDDTG can achieve better deployment than BDTG and VFA-HSN, because it has less redundant and coverage holes.

4.5 Heterogeneous Sensor Network Deployment Based on Energy Balanced Control

In this section, the redeployment problem in mobile heterogeneous WSNs consider in energy balanced is presented [6]. Delaunay triangulation is used in VFA to find the logic neighbors nodes in order to avoid nodes cluster in a small region. Furthermore, according to the research fixed ideal distance is proposed to solve the serious coverage holes and redundant problem in traditional VFA, and the energy model of the nodes is proposed. In order to balance the energy of the whole networks, the energy control function is added to the virtual force model. Simulation results show that EBRA provides better coverage performance (larger coverage area, less redundant, and holes) than traditional VFA.

4.5.1 Energy Balanced Redeployment Algorithm

The extended virtual force algorithm is a redeployment algorithm with some novel features which can overcome the limitations of traditional VFA. In practical networks the sensing range R_s of nodes cannot be always the same due to various reasons; in that case coverage holes may appear which may influence the QoS of the WSNs seriously. Prior algorithm assumed that the value of R_c is two times larger than R_s, but as a matter of fact that R_c is much larger than R_s. $High - R_c$ may induce stacking problem in redeployment.

In order to solve the problem above, Energy Balanced Redeployment Algorithm (EBRA) is proposed.

4.5.1.1 Redeployment Algorithm for Heterogeneous WSNs

As exponential force model can achieve fast convergence, the force model is shown in Eq. (4.23) and Fig. 4.31:

$$F_{ij(d)} = \begin{cases} \alpha\left(d_{ij}^{-\beta} - D_{th}^{-\beta}\right)u_{ij} & \text{if } D_{th} < d_{ij} \leq R_c \\ 0 & \text{if } d_{ij} = D_{th} \\ \mid \alpha\left(d_{ij}^{-\beta} - D_{th}^{-\beta}\right) \mid u_{ij} & \text{if } 0.5D_{th} < d_{ij} \leq D_{th} \\ 1u_{ij} & \text{if } d_{ij} \leq 0.5D_{th} \end{cases} \quad (4.23)$$

where u_{ij} is the unit vector from node i to node j, α and β are constants that may be changed in different situations, and D_{th} is the ideal distance.

In heterogeneous WSNs, there is no ideal deployment to achieve, which is different from homogeneous WSNs, so a new balance point $D_{th(i,j)}$ and the new force field truncation coefficient which keeps the nodes from stacking should be

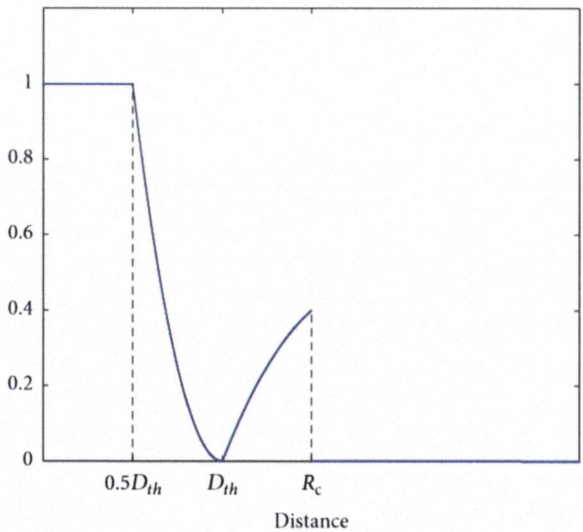

Fig. 4.31 Virtual force model

found. $D_{th(i,j)}$ in homogeneous WSNs is $\sqrt{3}R_s$. However, in heterogeneous WSNs every node has different R_s, so a new ideal distance for heterogeneous WSNs should be found.

As shown in Fig. 4.32, if $\sum\phi_{ijk} = \phi_{ij} + \phi_{ik} + \phi_{jk} = 2\pi$, neither coverage holes nor coverage redundant exist. Then the main aim of the algorithm is to adjust the distance between the nodes according to the $\sum\phi_{ijk}$ in Delaunay triangle to approach 2π. Figure 4.33 shows the Delaunay triangle, and it's definition can be obtained in [6].

For node ŭ56; shown in Fig. 4.32a,

$$\phi_{th(ij_k)} = \phi_{ij_k} + \left(2\pi - \sum \phi_{ij_k j_{k+1}}\right) \frac{\phi_{ij_k}}{\sum \phi_{ij_k k_{k+1}}} \tag{4.24}$$

The ideal distance between nodes can be described as:

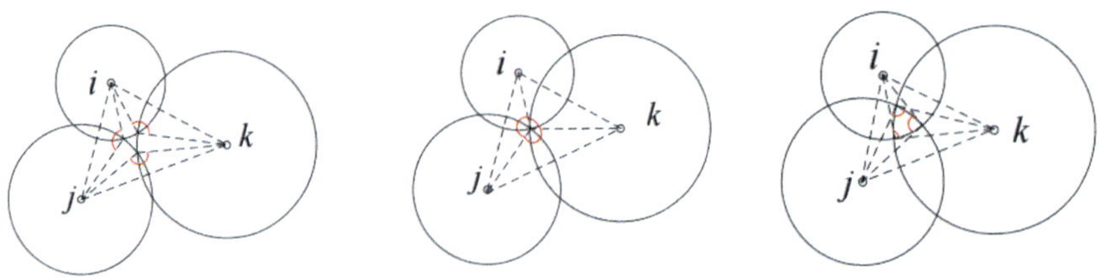

Fig. 4.32 Localization effect in convex and concave holes. (**a**) $\sum\phi_{ijk} = 2\pi$, (**b**) $\sum\phi_{ijk} = 2\pi$, (**c**) $\sum\phi_{ijk} < 2\pi$

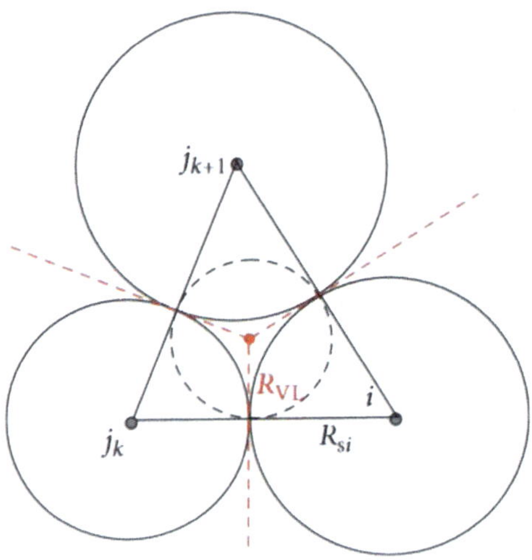

Fig. 4.33 Delaunay triangle

$$D_{h(i,j_k)} = \left(R_{si}^2 + R_{sj_k}^2 - 2R_{si}R_{sj_k}\cos\phi_{h(i,j_k)}\right) \tag{4.25}$$

Virtual force between nodes i and node j are listed as follows:

$$F_{ij}(d) = \begin{cases} \alpha\left(d_{ij}^{-\beta} - D_{th}^{-\beta}\right)u_{ij} & \text{if } D_{th(i,j)} < d_{ij} \le R_c \\ 0 & \text{if } d_{ij} = D_{th(i,j)} \text{ or } E(i,j) \\ \text{is the outside edge} & \\ \mid \alpha\left(d_{ij}^{-\beta} - D_{th}^{-\beta}\right) \mid u_{ij} & \text{if } 0.5D_{th(i,j)} < d_{ij} \le D_{th(i,j)} \\ 1u_{ij} & \text{if } d_{ij} \le 0.5D_{th(i,j)} \end{cases} \tag{4.26}$$

If $E(i,j)$ is the outside edge of the LDTG, the virtual force between nodes ŭ56; and ŭ57; is zero. Then each node calculates its own resultant virtual force of its logic neighbors. Then it moves to the new position according to the motion dynamic model.

Furthermore, the algorithm above can be described in Algorithm 4.7.

Algorithm 4.7

// m: the number of the logic neighbors of node i// Δt: the sampling time

Step 1 Find out logic neighbors node i in LDTG(i) of node i;

Step 2 For $1 \le j \le m$

Step 2.1 Set distance coefficient $\lambda = \sqrt{3}/2$

Step 2.2 Calculate F_{ij} the virtual force between node i and its neighbor node j;

Step 3 Calculate the resultant virtual force;

Step 4 Calculate the accelerated speed a_i and speed v_i of node I and then move node i to new position;

Step 5 Update $t + \Delta t$;

Step 6 Go back to **Step 1.**

Step 7 Find out logic neighbors node i in LDTG(i) of node i;

Step 8 For $1 \le j \le m$

Step 8.1 Calculate F_{ij} the virtual force between node i and its neighbor node j;

Step 9 Calculate the resultant virtual force;

Step 10 Calculate the accelerated speed a_i and speed v_i of node I and then move node i to new position;

Step 11 Update $t + \Delta t$;

Step 12 Go back to **Step 1.**

4.5.1.2 Energy Balanced Revision

Actually, the imbalanced energy cost of the network may cause some nodes to run out of energy in a short time. In this case, the whole structure of the WSNs may change, the QoS of the network cannot be guaranteed, and the lifetime of the network may be greatly shortened. As analyzed above, in WSNs we can prolong the lifetime of the whole network by balancing the energy cost of the nodes during the redeployment process. Minimum energy of node $\overline{E}_{\min}$ is to describe the minimum remain in energy of the nodes in the network.

Definition 4.2 $\overline{L_x}, \overline{L_{-x}}, \overline{L_y}$ and $\overline{L_{-y}}$ are the energy cost of the node when moving standard distance in each direction. Energy cost of node i is shown as follows:

$$\overline{L_i} = \sum_{\forall dir} \mid \overline{L_i} \times \overline{dir_i} \mid \tag{4.27}$$

where $\overline{dir_i}$ is the distance that node i moves in each direction. $\overline{EI}_i$ is the initial energy of node i. $\overline{E}_i$ is the energy of the node i. Thus the lowest energy of the node $\overline{E}_{\min}$ is described as follows:

$$\overline{E}_{\min} = Min\left(\overline{EI}_i - \overline{L_i}\right) \tag{4.28}$$

In order to prolong the lifetime of the network, the algorithm should keep $\overline{E}_{\min}$ as large as possible.

Definition 4.3 Energy differential coefficient φ is used to describe the energy difference between two nodes:

$$\varphi_{i,j} = \frac{\mid \overline{E}_i - \overline{E}_J \mid}{Max\left(\overline{E}_i, \overline{E}_j\right)} \tag{4.29}$$

Obviously, $\varphi \in [0, 1]$, and if φ approaches zero, more effort should be made to protect the nodes which have low energy to save their energy.

Definition 4.4 Node level γ shows the importance of the node. In practical application, some key nodes are more important than other nodes. These key nodes can influence the QoS of the whole network. Therefore, the algorithm should give priority to economize the energy of those key nodes.

To balance the energy cost of the node energy control function is added to Eq. (4.26):

$$e(\varphi_{i,j}) = \begin{cases} \varepsilon \ (rand[0,1] < \varphi_{ij}) \\ 1 \ (rand[0,1] \geq \varphi_{ij}) \end{cases} \tag{4.30}$$

where ε is the attenuation coefficient. Every time node i calculates the virtual force of its logic neighbor j, node i compares its energy with j and gets the energy differential coefficient $\varphi_{i,j}$, and then node i generates a random number $k \in [0, 1]$. If $k < \varphi_{i,j}$, the virtual force would be limited so that the movement of the node i would be reduced and in that case node i may save its energy in this round.

Then the new virtual force between node i and node j is listed as follows:

$$F_{eb}(i) = \sum_{j \in LN} e(\varphi_{i,j}) \cdot F_{ij}(d) \tag{4.31}$$

where LN is the set of logic neighbors of node i.

There are some key nodes in the networks, those nodes should always save energy no matter their neighbor's energy is low or not.

Node level control function is addressed in

$$\delta(\gamma) = (\gamma - \mu)^{-\sigma} \tag{4.32}$$

where μ and σ are constants which can be adjusted according to different situations. And the final resultant virtual force F_i is described in

$$F_i = \delta(\gamma) \sum_{j \in LN} e(\varphi_{i,j}) \cdot F_{ij}(d) \tag{4.33}$$

Then each node calculates its own resultant virtual force of its logic neighbors according to the distance and energy and node level. Then it moves to the new position according to the motion dynamic model.

EBRA can be described in Algorithm 4.8.

Algorithm 4.8 The EBRA Algorithm

// m: the number of the logic neighbors of node i// Δt: the sampling time

Step 1 Find out logic neighbors node i in LDTG(i) of node i;

Step 2 For $1 \leq j \leq m$

Step 2.1 Set distance coefficient $\lambda = \sqrt{3}/2$

Step 2.2 Calculate F_{ij} the virtual force between node i and its neighbor node j;

Step 2.3 Calculate the energy between node i and node j;

Step 2.4 Calculate the energy differential coefficient φ to fix the Virtual force;

Step 3 Calculate the resultant virtual force;

Step 4 Calculate the accelerated speed a_i and speed v_i of node I and then move node i to new position;

Step 5 Update $t + \Delta t$;

Step 6 Go back to **Step 1.**
Step 7 Find out logic neighbors node i in LDTG(i) of node i;
Step 8 For $1 \leq j \leq m$

Step 8.1 Calculate the virtual force F_{ij} between node i and its neighbor node j;

Step 9 Calculate the resultant virtual force considering node levelγ;
Step 10 Calculate the accelerated speed a_i of node i and speed v_i, node i move to new position;
Step 11 Update $t + \Delta t$;
Step 12 Go back to **Step 1.**

4.5.1.3 Performance Evaluation

1. Coverage Rate
   The coverage rate is a measure of the coverage quality for WSNs. In the blanket coverage problem, coverage is defined by the ratio of the union of covered areas of each node to the complete area of interest. In this section, the covered area of each node is defined as the circular area with sensing range R_s. The value of coverage rate C is as follows:

$$C = \frac{\bigcup_{i=1}^{n} A_i}{A} \tag{4.34}$$

where A_i is the area covered by the node i, n is the total number of mobile sensor nodes, and A is the total area.

2. Uniformity of Energy
   In heterogeneous WSNs, the uniformity of energy can be measured by the average of the local standard deviation of the node energy

$$\sigma_E = \left(\frac{1}{n} \sum_{i=1}^{n} \left(E_i - \frac{1}{n} \sum_{j=1}^{n} E_i \right)^2 \right)^{1/2} \tag{4.35}$$

3. Minimum Energy
   $\overline{E}_{\min}$ stands for the minimum energy of all the nodes and is described in (4.28).

4.5.2 Simulation Results

In this section, we present the simulation results of EBRA. In the simulation environment, 40 nodes are deployed randomly in a 100 m × 100 m square field.

Communication range of all nodes is 30 m. The sensing range of 14 nodes is 10 m, and other nodes' sensing range is 8 m.

Firstly, simulation is done in scatter initial deployment. The scatter initial deployment is shown in Fig. 4.34. The redeployment results of VFA and EBRA in scatter initial deployment are shown in Fig. 4.34a, b. It can be seen clearly from Fig. 4.34 that EBRA shows the better result than VFA for guaranteeing the connectivity of the whole network. VFA employed force field truncation to avoid nodes stacking. $\lambda_2 = 1$ makes long-distance virtual force disappear. In EBRA use Delaunay triangulation to seek logic neighbor of nodes instead of real neighbor to settle the defects of VFA.

Secondly, we conducted the simulation in concentrated initial deployment. The concentrated initial deployment is shown in Fig. 4.35. The redeployment results of VFA and EBRA in concentrated initial deployment are shown in Fig. 4.36a, b. It can be seen clearly from Fig. 4.36a. Compared with Fig. 4.36b, less holes and redundant coverage exist in the deployment result. In VFA, ideal distance coefficient is settled, and neighbor nodes cannot reach the ideal distance coefficient at same time. To improve the performance of the algorithm, another round of movement begins with modified ideal distance coefficient. Due to the modified ideal distance coefficient, the virtual forces between nodes change two scenarios. In area with redundant coverage virtual force makes the nodes move away from each other. In area with holes virtual force makes nodes get closer to each other.

The above experiments have been carried out for 100 times, and the simulation results are illustrated below. Figure 4.37 shows the k-coverage rates of VFA and EBRA with the initial deployment as shown in Fig. 4.36. It can be seen clearly that EBRA can achieve better deployment than VFA, because it has less redundant

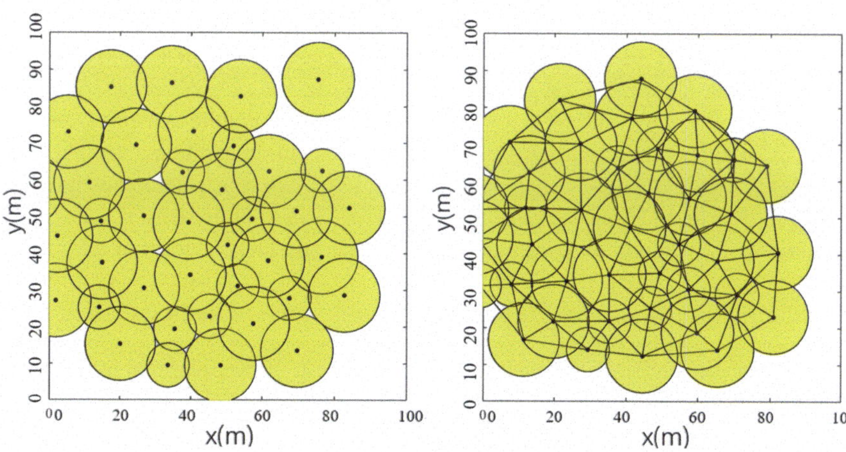

(a) (b)

Fig. 4.34 Simulation result in scatter initial deployment. (**a**) Redeployment by VFA, (**b**) Redeployment by EBRA

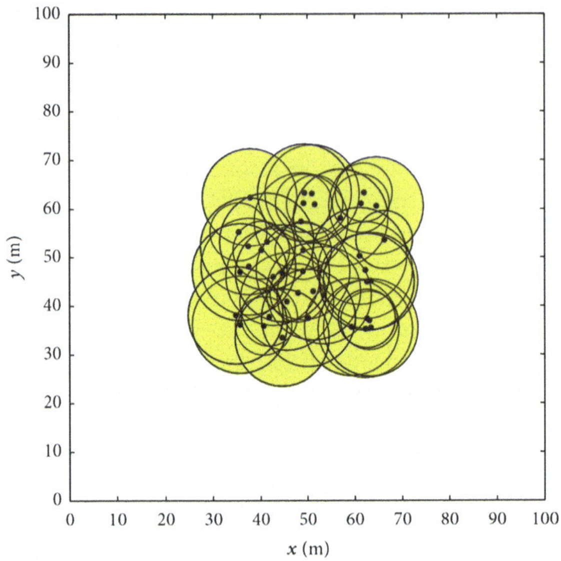

Fig. 4.35 Concentrated initial deployment

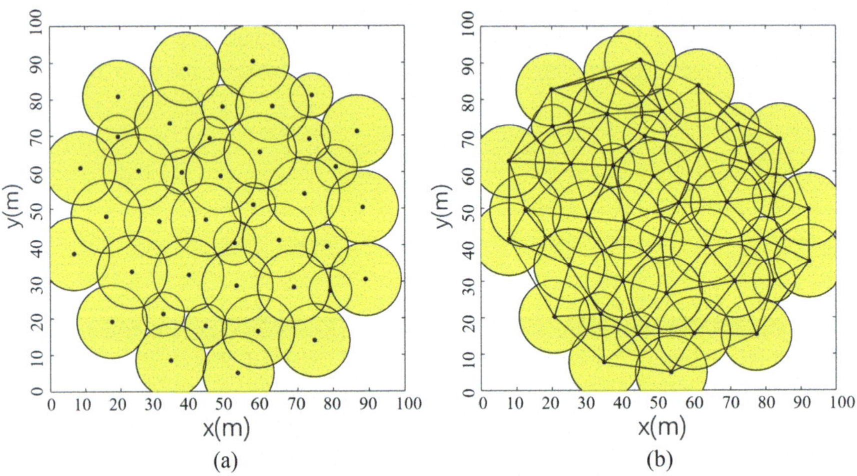

Fig. 4.36 Simulation result in concentrated initial deployment. (**a**) Redeployment by VFA, (**b**) Redeployment by EBRA

coverage, and holes. Figure 4.38 shows the minimum energy of all the nodes in WSNs. Compared with traditional VFA, the minimum energy value of nodes reduces more slowly. At step 100, both algorithms meet the ending condition. EBRA saves more energy for the mobile WSNs which means the mobile WSNs

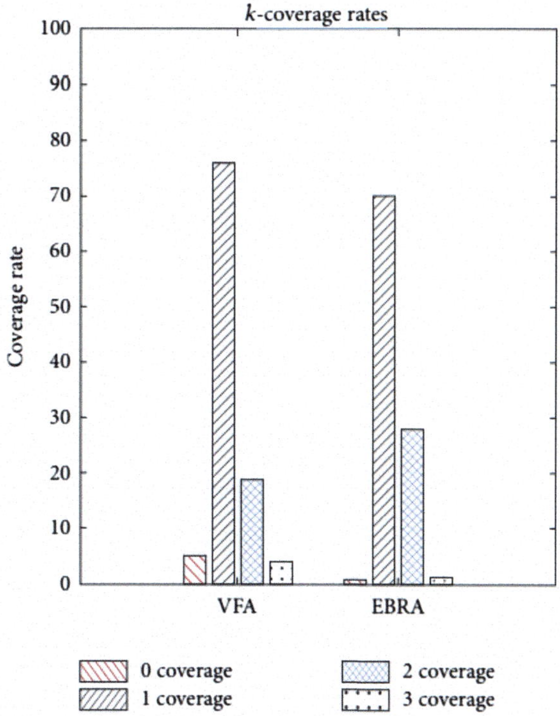

Fig. 4.37 Coverage rates of the final deployment

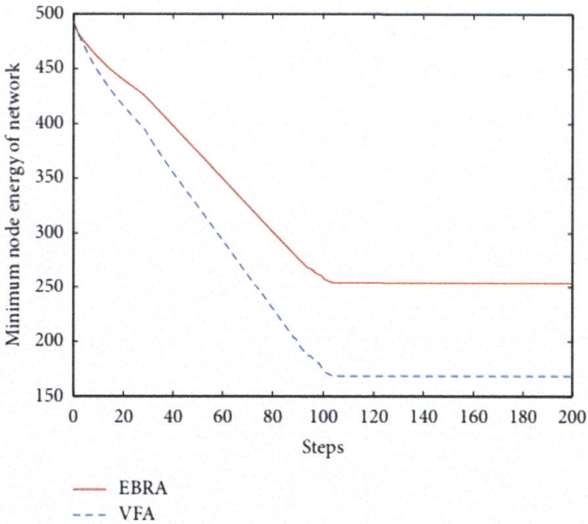

Fig. 4.38 Minimum energy of node

by EBRA can provide longer service time and better QoS. Figure 4.39 shows the uniformity of energy of the whole networks. The uniformity of EBRA energy is lower than traditional VFA which means the energy in EBRA is more uniform distribution which provides better robustness for the whole system. Figure 4.40 shows the energy cost of the key nodes with and without energy protecting. Simulations are also conducted in same initial deployment. Firstly, a random node is selected as a key node whose energy cost is recoded. Secondly, the node which has been selected changes its node level and then recodes the energy cost. It can be seen clearly from Fig. 4.40 that EBRA saves the energy for the key nodes.

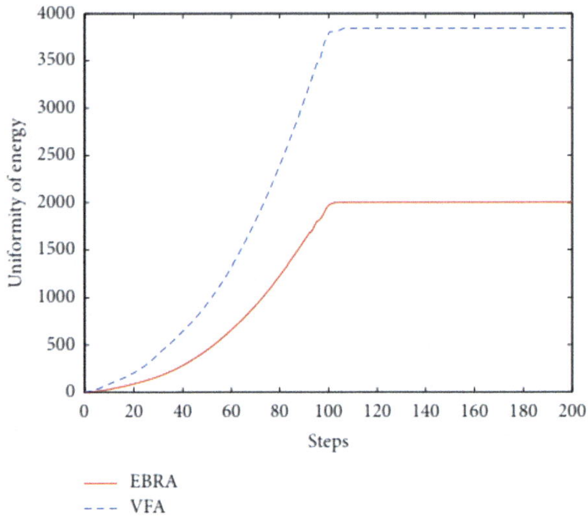

Fig. 4.39 Uniformity of energy

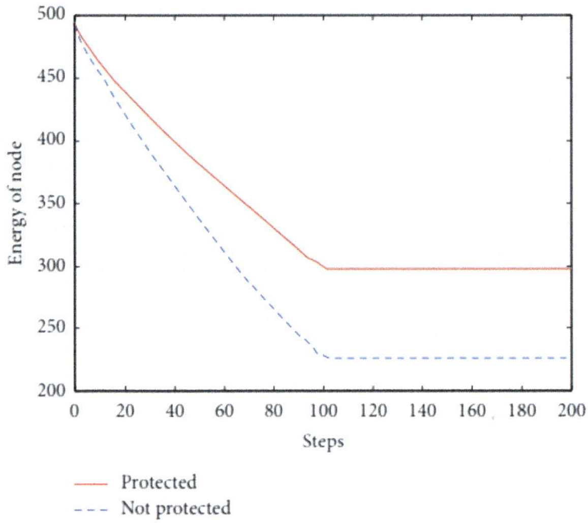

Fig. 4.40 Energy cost of the key nodes with and without energy protectin

4.6 Conclusion

In this chapter, we introduce several algorithms for homogeneous and heterogeneous WSNs, respectively. Considering the problem of discontinuous coverage, the ATDOC algorithm is proposed based on the BTDA algorithm. Aiming at the discontinuous connectivity problem of traditional VFA algorithm and ATDOC algorithm in low connectivity networks, the $low - R_c$ VFA is proposed. On the contrary, the extended virtual force algorithm $high - R_c$ VFA for high-connected networks is also presented to solve the problem of high-connected networks. We also propose DOC algorithm to generate the next position of mobile sensor node and SDDTG to solve the coverage blind area and coverage redundancy problems of VFA-HSN and BDTG algorithms. The EBRA is applied to solve the energy unbalanced, coverage holes, and redundant coverage problems.

References

1. Chai S, Zhang B, Li J, et al. Autonomous hexagonal deployment for optimal coverage in mobile sensor networks[C]. Iet International Conference on Information Science & Control Engineering. IET, 2014.
2. Li J, Zhang B, Cui L, et al. An extended virtual force-based approach to distributed self-deployment in mobile sensor networks[J]. 2012, 2012:184-195.
3. Howard A, Mataric M J, Sukhatme G S. Mobile sensor network deployment using potential fields: A distributed, scalable solution to the area coverage problem[J]. 2002.
4. Li J, Cui L, Zhang B. Self-deployment by distance and orientation control for mobile sensor networks[C]. Proceedings of the IEEE International Conference on Networking, Sensing and Control, ICNSC 2010, Chicago, IL, USA, 10-12 April 2010. IEEE, 2010.
5. Ye G, Zhang B, et al. Extended virtual force-based coverage scheme for heterogeneous wireless sensor networks[J]. Iccas, 2014:1560-1564.
6. Ye G, Zhang B, et al. Energy balanced redeployment algorithm for heterogeneous wireless sensor networks[J]. Mathematical Problems in Engineering, 2015, 2014:1-11.

Chapter 5
Community Detection in Complex Networks

As a ubiquitous feature of complex networks, community characteristic is common in real world networks especially in wireless sensor network. It is vital to detect community structure in complex networks and to take advantage of extracted information. On the basis, community detection has become a popular theme in the field of complex networks. It focuses on uncovering the affiliation of each node through topology analysis. In this chapter, we have presented the essential knowledge and advanced methods of community detection, including proximate support vector clustering (PSVC), deep auto-encoded extreme learning machine (DA-ELM), deep auto-coded clustering (DAC) and local aggregated differential evolution algorithm (LADE). These methods have been applied in tree-based WSNs and have been proved in terms of robustness and fragility.

5.1 Community Detection by Proximate Support Vector Clustering

As one of the most popular interdisciplinary research fields during past decades, complex networks have stimulated many scientific advances in different subjects such as social networks, computer science, epidemiology, biology, and transportation, to name but a few. Tremendous researches focus on exploring different kinds of properties in complex networks, say degree distributions, small world characteristics, scaling features, percolation process, hierarchical structures and community structures. Among these diverse properties, perhaps the most attractive one has been the community structure. Finding out the divisions of nodes in networks, community detection or network clustering, is a fascinating hot spot for investigators, which is a good way to uncover the underlying semantic structure, mechanisms and dynamics of certain networks.

S. Chai et al., *Wireless Sensor Networks*, Wireless Networks,
https://doi.org/10.1007/978-981-15-5757-6_5

In this section, we present a support vector clustering method based on a proximity graph, which surpasses the traditional support vector approach both in accuracy and complexity. Results of extensive experiments undertaken on computer generated networks and real-world data sets illustrate competent performances in comparison with the other counterparts.

5.1.1 Proximate Support Vector Clustering Algorithm

Since being invented by Vapnik and his colleagues in 1995 [1], support vector machine (SVM) has long been dominant as one of the best off the shelf supervised machine learning tools which credits to its precise and deterministic performance. Inspired by the proximity graph modelling method used to accelerate the typical support vector clustering computation, we present a proximate support vector clustering (PSVC) algorithm that implements the SVC on the minimum spanning tree of the underlying graph. Extensive comparison trials on both artificial networks and realistic networks indicate that PSVC outperforms traditional SVC in light of different criterions. Meanwhile, PSVC owns remarkably smaller complexity in comparison with that of traditional SVC [2].

5.1.1.1 Common Network Model of Algorithms

We define undirected network $G(V, \boldsymbol{E})$, where $\boldsymbol{V}$ represents set of vertices with total number of N and $\boldsymbol{E}$ represents set of edges with total number of M. With neglect of self-loop which means an edge starts and ends at the same point, connections among nodes are expressed as a $N \times N$ symmetric adjacency matrix A.

$$A_{ij} = \begin{cases} 1, & \text{if vertex } i \text{ and } j \text{ are connected} \\ 0, & \text{else} \end{cases} \tag{5.1}$$

where $i = 1, \ldots, N$ and $j = 1, \ldots, N$.

According to A, the Laplacian matrix $\boldsymbol{L}$ is defined as

$$L_{ij} = \begin{cases} -1, & \text{if} A_{ij} = 1 \\ k_i, & \text{if } i = j \\ 0, & \text{else} \end{cases} \tag{5.2}$$

In which $k_i = \sum_{j=1}^{N} A_{ij}$ is the degree of vertex i, $i = 1, \ldots, N$ and $j = 1, \ldots, N$.

A community detection algorithm segregates $\boldsymbol{V}$ into mutually exclusive districts by means of attaching labels on each node. Support vector clustering method transforms this label attaching procedure to a minimum enclosing sphere searching process in the corresponding feature space mapped by means of the supposed Gaussian kernel function. However, due to the lack of explicit coordinates there is no way to calculate the Euclidean norm directly. Instead, the diffusion kernel matrix D on graphs has been picked up as our approach to represent the distance between each pair of nodes. Items in D have been defined as the exponential of their negative Laplacian matrix counterparts as

$$D_{ij} = e^{-\alpha L_{ij}} \tag{5.3}$$

where α is the bandwidth parameter, $i = 1, \ldots, N$ and $j = 1, \ldots, N$. After normalization process, D could be used as the delegation of similarities among vertices and $1 - D$, in turn, could indicate distances, which leads to the modification of Gaussian kernel matrix as

$$K_{ij} = e^{-q\left(1-D_{ij}\right)^2} \tag{5.4}$$

In which q is the width of Gaussian kernel, $i = 1, \ldots, N$ and $j = 1, \ldots, N$. Apparently, $\boldsymbol{K}$ is symmetric positive semidefinite that guarantees a Mercer kernel.

5.1.1.2 Support Vector Clustering

Similar to SVC, we suppose each vertex in the graph that could be expressed by a vector $x_i \in \mathbb{R}^d$, where d is the dimension of the vector and $i = 1, \ldots, N$. A nonlinear transformation $\Phi(x_i)$ maps the point into another feature space and endures the property that

$$\Phi(x_i) \cdot \Phi\left(x_j\right) = K\left(x_i, x_j\right) = K_{ij} \tag{5.5}$$

where $i = 1, \ldots, N$ and $j = 1, \ldots, N$.

The particularity of such a transformation is that the minimum enclosing sphere in the new space corresponds to the contours surrounding data sets in the original space. Mathematically speaking, define the distance between x_i and the center of the sphere in feature space

$$R^2(\boldsymbol{x}_i) = \| \Phi(x_i) - \boldsymbol{a}\|^2 \tag{5.6}$$

where a indicates the center of the enclosing sphere, $\| \cdot \|$ is the Euclidean norm and $i = 1, \ldots, N$.

We need to find the smallest radius R that subjects to the constraint:

$$\| \Phi(x_i) - a\|^2 \leq R^2, \ j = 1, \ldots, N \tag{5.7}$$

Taking into account slack variables ξ_i on behalf of soft constraints, the optimization problem is reformulated as follow:

$$\begin{aligned} &\min{}_{R,a,\xi} R^2 + C\sum_{i=1}^{N} \xi_i \\ s.t.\quad &\| \Phi(x_i) - \boldsymbol{a}\|^2 \leq R^2 + \xi_i \\ &\xi_i \geq 0 \end{aligned} \tag{5.8}$$

where C is l_1 regularization constant and $i = 1, \ldots, N$.

Under the Karush-Kuhn-Tucker (KKT) conditions, solving the original optimization problem is equivalent to figuring out the answer to its Lagrangian dual problem

$$\begin{aligned} \min{}_{\beta} \quad &\sum_{i=1}^{N}\sum_{j=1}^{N} \beta_i\beta_j K_{ij} - \sum_{i=1}^{N} \beta_i K_{ij} \\ \text{s.t.} \quad &0 \leq \beta_i \leq C \\ &\sum_{i=1}^{N} \beta_l = 1 \end{aligned} \tag{5.9}$$

where β is the Lagangian multiplier. In the language of SVC, we call a point with $0 < \beta < C$ a support vector and the one with $\beta_i = C$ a bounded support vector. In the feature space, support vectors' images fall on the surface of the minimum enclosing sphere and images of bounded support vectors drop outside that. The other nodes whose $\beta_i = 0$ lie inside the optimal sphere when $C \geq 1$ would be no bounded support vectors because of the contradiction with KKT conditions. Having the optimal solution at hand, Eq. (5.6) could be expressed as

$$R^2(\boldsymbol{x}_i) = K_{ii} - 2\sum_{j=1}^{N} \beta_j K_{ij} + \sum_{j=1}^{N}\sum_{k=1}^{N} \beta_j \beta_k K_{jk} \tag{5.10}$$

where $i = 1, \ldots, N$.

Then the smallest radius R is

$$R = \{R(x_i) | x_i \text{ is a support vector}\}, \quad i = 1, \ldots, \ N \tag{5.11}$$

which implies that all the vertex on the hyper sphere have the same distance to the center practically.

5.1.1.3 Community Assignment

Given any two nodes $i = 1, \ldots, N$ and $j = 1, \ldots, N$ belonging to different communities, a fundamental geometric observation is that if we draw a line connecting these two nodes, then throughout this line there must at least be a point y whose image locates outside the hypersphere. In other word, the image of y satisfies $R(y) > R$.

According to our definition of distance in last subsection, the distance between $i$and this middle point could be written as $\lambda(1 - D_{ij})$, where λ is a fraction, $0 \le \lambda \le 1$. As a result

$$K(x_i, y) = e^{-q\lambda^2\left(1-D_{ij}\right)^2} = \left(e^{-q\left(1-D_{ij}\right)^2}\right)^{\lambda^2} = K_{ij}^{\lambda^2} \tag{5.12}$$

Additionally, we define the distance from y to itself to be 0 and hence $K(y, y) = 1$. Then in the feature space, distance from image y to the center is

$$\begin{aligned} R^2(\boldsymbol{y}) &= K(\boldsymbol{y},\boldsymbol{y}) - 2\sum_{i=1}^{N} \beta_i K(\boldsymbol{x}_i,\boldsymbol{y}) + \sum_{i=1}^{N}\sum_{j=1}^{N} \beta_i\beta_j K_{ij} \\ &= 1 - 2\sum_{i=1}^{N} \beta_i K_{ij}^{\lambda^2} + \sum_{i=1}^{N}\sum_{j=1}^{N} \beta_i\beta_j K_{ij} \end{aligned} \tag{5.13}$$

Based on the above observation, a new matrix $\boldsymbol{C}$ could be generated as

$$C_{ij} = \begin{cases} 1, & \text{if } R(y) > R \text{ for all } \lambda \text{ ranging from 0 to 1} \\ 0, & \text{else} \end{cases} \tag{5.14}$$

where $i = 1, \ldots, N$ and $j = 1, \ldots, N$. In our implementation we uniformly picked up 20 samples from 0 to 1 as λ.

Community labels are now available to be attached on each node in accordance to matrix $\boldsymbol{C}$ by comparing each pair of nodes i and j. Nevertheless, inspired by the proximity graph modelling method, we notice another intuition that we could also identify all the nodes inherently connected in the underlying community structure using a spanning tree rather than using a complete graph. Such intuition encourages us to implement a Minimum Spanning Tree (MST) algorithm on the underlying graph ahead of the generation of $\boldsymbol{C}$. In this way, C_{ij} only shed light on those edges consisting of the spanning tree and leave a bunch of redundant edges behind, which would save a lot of calculation resources [3].

5.1.2 Simulations and Results

In the experiments, we deploy PSVC on computer generated networks as well as four famous real-world networks to compare the performance with the other three competitors, including two divisive algorithms GN and NF, the genetic algorithm GA, and the original support vector clustering algorithm SVC. Following traditional SVCs, the constant parameters of PSVC are set as the bandwidth parameter $\alpha = 0.5$, the width of Gaussian kernel $q = 3.5$, and the regularization constant $C = 2$ implying no bounded support vectors.

Our first criterion, the modularity Q proposed by Newman and Girvan in [4], is defined as the margin between the density of intra community edges on the estimated network and that density on a randomly reorganized network with the same number of nodes and edges.

$$Q = \frac{1}{2M} \sum_{i=1}^{N} \sum_{j=1}^{N} \left(A_{ij} - \frac{k_i k_j}{2M} \right) \delta_{g_i g_j} \tag{5.15}$$

where $k_i = \sum_{j=1}^{N} A_{ij}$ still represents the degree of vertex i as above and g_i indicates the group label of node i. δ_{ij} is the Kronecker delta, leading to

$$\delta_{g_i g_j} = \begin{cases} 1, & \text{if vertex } i \text{ and } j \text{ belong to the same group} \\ 0, & \text{else.} \end{cases} \tag{5.16}$$

A larger Q corresponds to a more apparent community structure in the underlying network, and vice versa.

The normalized mutual information (NMI), another benchmark for community detection, has also been deployed to evaluate the accuracy of algorithms. Based on information theory, the NMI of two division methods A and B to partition N nodes could be defined as

$$I(A,B) = \frac{-2 \sum_{i=1}^{C_A} \sum_{j=1}^{C_B} N_{ij} \log \frac{N_{ij} N}{N_{i \cdot} N_{\cdot j}}}{\sum_{i=1}^{C_A} N_{i \cdot} \log \frac{N_{i \cdot}}{N} + \sum_{j=1}^{C_B} N_{\cdot j} \log \frac{N_{\cdot j}}{N}} \tag{5.17}$$

where C_A and C_B are numbers of communities in A and B, respectively, N_{ij} denotes the number of nodes that both appear in group i of A and in group j of B, $N_{i.} = \sum_{j=1}^{C_B} N_{ij}$ and $N_{.j} = \sum_{i=1}^{C_A} N_{ij}$. The larger $I(A,B)$ is, the more similar A is to B. Let A and B be the candidate algorithm and the predetermined network, $I(A,B) = 1$ means the algorithm has found out all the communities identical to the real structure while $I(A,B) = 0$ indicates a complete fail of the algorithm.

5.1.2.1 Computer Generated Networks

We fabricated networks with community structures in Girvan and Newmans [4] way. 128 vertices are separated into four communities uniformly, meaning 32 nodes in each group, and edges are randomly added between each pair of nodes with probability P_{in} when the two nodes come from the same community and P_{out} for vertices belonging to different groups. In restraint of $P_{in} > P_{out}$ and the average degree of a vertex $z = 16$, through changing P_{out} or the average number of inter-community edges per vertex z_{out} (the two ways are equilibrium) we could achieve different networks with obvious (small z_{out}) or ambiguous (large z_{out}) community characteristics. These computer-generated networks with decided graph knowledge are then fed into our algorithm and its competitors to calculate their NMIs, which is showed in Fig. 5.1.

Each point in Fig. 5.1 represents the average NMI of a certain algorithm on 50 randomly generated scenarios with predefined z_{out} varying from 0 to 8. The error bars accompanying those points have marked out the normalized deviation of NMI in 50 experiments. Lines connecting points are included solely as a guide to read. We can see that our algorithm performs almost as well as GA and better than the GN, FN and traditional SVC, especially for large z_{out}. Albeit slightly falling behind GA in terms of accuracy, PSVC is a deterministic algorithm with identical output all the time while GA based on probabilistic evolution mechanism cannot assure that. Compared with GA and SVC, one outstanding advantage of PSVC is that users do not need to explicitly set the number of communities, which equips the algorithm with stronger scalability. The comparison between PSVC and SVC also shows the improvement on the accuracy of our algorithm, which mainly attributes to the modified method we choose in the calculation of distances and C_{ij}.

Fig. 5.1 NMIs of different algorithms on computer generated networks

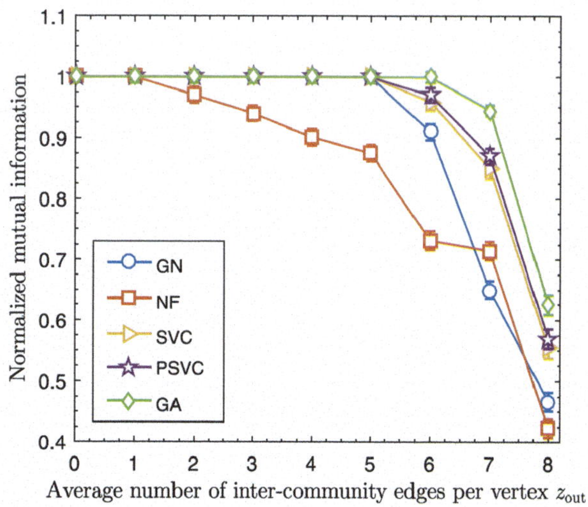

5.1.2.2 Real World Networks

In real world networks experiments, four real world networks with decided number of communities have been selected as test beds for measurement of our algorithms' accuracy. Table 5.1 shows the information of these four datasets in detail. Modularity and NMI results of PSVC and its four counterparts are exhibited in Table 5.2, where each number is the average of 50 trials.

5.1.2.3 Zachary's Karate Club Network

From 1970 to 1972 W. W. Zachary [5] had observed a karate club as an example of social network which captures 34 members in the club and documents 78 pairwise links between these members that interacted outside the club. The decided community structure formed after a series of factional confrontations between the administrator "John" and instructor "Mr. Hi" (pseudonyms), which led to the split of the club into two. Half of the members formed a new club around Mr. Hi, members from the other part found a new instructor or gave up karate. Figure 5.2 shows the partition result of our algorithm in distinct colors and shapes that correctly finds out the separated groups. Furthermore, PSVC is probable to figure out some implicit hierarchical clustering schemes, like in Fig. 5.2 where each group encloses two sub-clusters. The proximity graph created by MST is highlighted in red color.

Table 5.1 Information of datasets

Dataset	Number of nodes	Number of edges	Number of communities
Karate	34	78	2
Dolphins	62	159	2
Polbooks	105	441	3
Football	115	613	12

Table 5.2 The maximum Qs and corresponding NMIs of different algorithms

Dataset	Index	GN	NF	SVC	GA	PSVC
Karate	Modularity	0.4013	0.3807	0.3991	0.3511	0.4198
	NMI	0.7072	0.6925	0.7255	0.8082	0.7286
Dolphins	Modularity	0.5196	0.4955	0.5132	0.3725	0.5241
	NMI	0.5016	0.5727	0.5021	0.8034	0.5124
Polbooks	Modularity	0.5120	0.5020	0.4930	0.4862	0.5269
	NMI	0.4018	0.4396	0.5408	0.5236	0.5737
Football	Modularity	0.5950	0.5682	0.6123	0.6012	0.5984
	NMI	0.4237	0.4436	0.7527	0.8708	0.9095

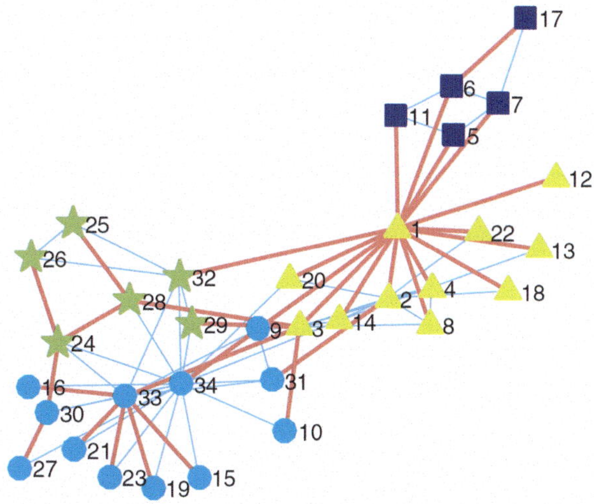

Fig. 5.2 Communities extracted from the Zachary's karate club network using PSVC

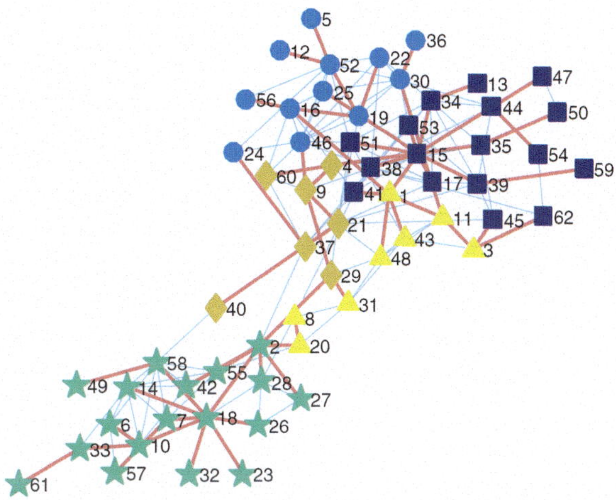

Fig. 5.3 Communities extracted from the dolphin social network using PSVC

5.1.2.4 Dolphin Social Network

The second network is an undirected social network of frequent associations between 62 dolphins in a community living of Doubtful Sound, New Zealand. These dolphins, naturally separated by their sexualities into two categories, had been studied for several years. In Fig. 5.3, we present the snapshot of our algorithm clustering resulting in different colors and shapes along with the highlighted spanning tree in red color. Again, some extra cluster structures have been exposed although they are not marked in the original dataset on behalf of the sexuality.

5.1.2.5 Political Books Network

Our third benchmark is a network of books about US politics published around the time of 2004 presidential election and sold by Amazon.com. Edges between books represent frequent co-purchasing of books by the same buyers. 105 books, denoted as nodes, have been classified into three categories as liberal, neutral and conservative according to their political orientations. The identification of our algorithm in Fig. 5.4 shows that people tend to have relatively stable opinion on politics which is indicated by the obvious segregation of nodes. Yellow triangles and green circles represent the conservative and liberal group, respectively, while the rest light blue pentagons and dark blue squares indicate the neutral part which occupies a small number of nodes. However, it is not very uncommon that some neutral books look like more liberal or conservative in eyes of some critical readers, which explains why our algorithm divides the neutral class into two clusters furthermore. The proximity graph has been highlighted in red as well.

5.1.2.6 US College Football Network

Girvan and Newman introduced this network of American football games between Division I colleges during regular season 2000 that 115 teams (nodes) from 12 conferences play 613 games (edges) with each other in all. On average, there are seven intra-conference games and four inter-conference games waiting for a team to play, which means it has more chance to confront with a competitor in the same conference. This feature gives us a good opportunity to deploy our algorithm to check its accuracy. We display our detection result in Fig. 5.5 that 11 communities corresponding to 11 colored shapes have been recognized with one minority conference being absorbed into big ones. The minimum spanning tree is also

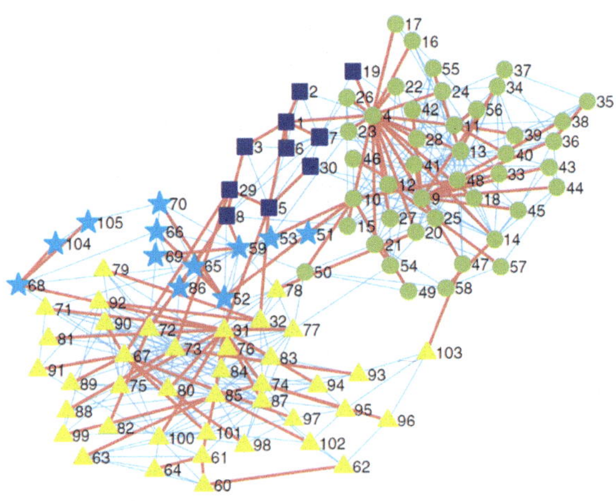

Fig. 5.4 Communities extracted from the political books network using PSVC

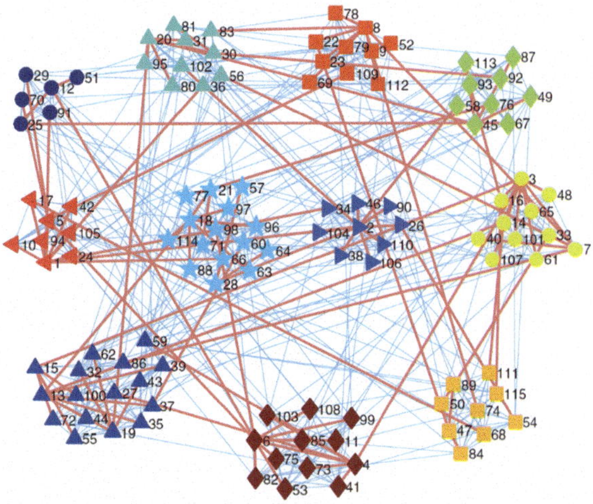

Fig. 5.5 Communities extracted from the US college football network using PSVC

distinguished in highlighted red lines. From Table 5.2, we can see that despite of those misunderstanding, our algorithm still outperforms the others in terms of NMI index.

Time consumption in SVC algorithms is contributed by two parts. The first one comes from the quadratic programming calculation of Eq. (5.9) which, with the assistance of SMO algorithm, could be solved in time of $O(N^2)$ asymptotically. The second part accounts for the worst-case running time of the community labels assignment process. In works of traditional SVCs, attention is only necessarily paid to support vectors and the complexity of labeling procedure could be as low as $O\left((N - N_{bsv})N_{sv}^2\right)$, where N_{sv} and N_{bsv} are numbers of support vectors and bounded support vectors, respectively. While in our scenario, amounts of waste have been avoided thanks to the imposing of the proximity graph that speeds up PSVC a lot in comparison with the traditional SVC. The complexity of MST algorithm we used in PSVC was given out as $O(M \log N)$. In all, the whole complexity of PSVC is approximately $O(N^2)$, which is as good as that of other commonly used community detection algorithms such as GN's $O(N^3)$, NF and GA's $O(N^2)$.

From the algorithms and experiments, we have invented a community detection algorithm based on methods of support vector clustering and proximity graph modelling. Compared with the traditional support vector clustering method, the most evident advantages of the introduced algorithm, named PSVC, fall into the increase in precision and the decrease in complexity. Meanwhile, PSVC have also been tested on Girvan and Newman artificial networks and another four real world networks in contrast with GN, NF, GA, and the traditional SVC in the light of modularity and NMI indexes. Attributing to the modification of the distance definition and the utilization of spanning tree, PSVC demonstrates competitive performance. Our targeted networks in this chapter are unweighted and undirected networks, which leaves weighed and directed networks as open issues worthwhile for studying. We have also noticed that the community detection procedure is a clustering problem in essence, which means our proposed PSVC could be qualified to fulfill much more tasks in broader fields. Therefore PSVC is planned to be restudied to solve problems in networks with weighed and/or directed edges. Moreover, it is implemented and tested in many other scenarios that are suitable for unsupervised learning methods.

5.2 Community Detection by Deep Auto-Encoded Extreme Learning Machine

In this section, we put the cascaded stacked autoencoders and the unsupervised extreme learning machine (ELM) together in a two-level embedding process and propose a novel community detection algorithm named deep auto-encoded ELM (DA-ELM). Extensive comparison experiments in circumstances of both synthetic and real-world networks manifest the advantages of the proposed algorithm [6].

5.2.1 Algorithm Based on Extreme Learning Machines

With the development of optimization algorithms, many methods have been referred to solve network clustering. For example, spectral algorithms explore eigenvalues of Laplacian matrices of graphs using traditional clustering techniques. Similar to spectral clustering, similarity matrix factorization and blocking can also be applied. Label propagation, that is, attaching labels to each node based on the neighbor information, is known as a fast and effective clustering method. *k*-means++ chooses distinct initial seeds far from each other in a probabilistic manner, which leads to more stable clustering results but involves increased complexity.

One of these methods is extreme learning machines (ELM) and its popularity should be largely credited to the simplicity of its implementation. It has been demonstrated that given random input weights and biases of the hidden layer, a single-layer feedforward network (SLFN) could approximate any continuous functions simply by tuning the output weights. As a result, the abstracted task in ELM is equivalent to a regularized least square problem that can be solved in closed form using the Moore-Penrose generalized inverse. Regarding the clustering task, the unsupervised ELM is interpreted as an embedding process that maps the input data into low-dimensional space. As a result, we present an extreme learning machine community detection (ELM-CD) algorithm based on the combination of *k*-means and unsupervised ELM to fulfill the community detection task.

Following the universal semi supervised and unsupervised ELM framework, given N_h neurons in the hidden layer, we define $\boldsymbol{E}_i \in \mathbb{R}$ as the output vector from ELM with respect to input vector $\boldsymbol{a}_i$

$$\boldsymbol{E}_i = \sum_{j=1}^{N_h} \boldsymbol{\beta}_j g\left(\boldsymbol{w}_j^{\mathrm{T}} \boldsymbol{a}_i + b_j\right) \tag{5.18}$$

where $w_j \in \mathbb{R}^n$ and $b_j \in \mathbb{R}^n$ represent input weights and bias of the jth neuron in the hidden layer, respectively; $\boldsymbol{\beta}_j \in \mathbb{R}^n$ is the output weight from the jth neuron to n dimensional output elements; and $g(\cdot)$ is the sigmoid function

$$g(x) = \frac{1}{1 + e^{-x}} \tag{5.19}$$

Assume that the input dataset $\boldsymbol{A}$ can be classified as unlabeled set $\{\boldsymbol{a}_i\}$, $i = 1, \ldots,$ u, and u represents the number of unlabeled nodes and labeled set $\{\boldsymbol{a}_i, \boldsymbol{y}_i\}$, $i = 1, \ldots,$ l where $\mathbf{y}_i \in \mathbb{R}^n$ is the corresponding community label and l is the number of labeled nodes so that $u + l = N$. The calculation of ELM would fall into two parts. In the first part, we randomly decide the input weights and biases according to a uniform distribution. In the second, let (5.18) be expressed as the inner product

$$\boldsymbol{E} = \boldsymbol{H\beta} \tag{5.20}$$

where

$$\boldsymbol{H} = \begin{bmatrix} g\left(\boldsymbol{w}_1^{\mathrm{T}} \boldsymbol{a}_1 + b_1\right) & \cdots & g\left(\boldsymbol{w}_{N_h}^{\mathrm{T}} \boldsymbol{a}_1 + b_{N_h}\right) \\ \vdots & \cdots & \vdots \\ g\left(\boldsymbol{w}_1^{\mathrm{T}} \boldsymbol{a}_N + b_1\right) & \cdots & g\left(\boldsymbol{w}_{N_h}^{\mathrm{T}} \boldsymbol{a}_N + b_{N_h}\right) \end{bmatrix} \in \mathbb{R}^{N \times N_h} \tag{5.21}$$

$$\boldsymbol{\beta} = \begin{bmatrix} \boldsymbol{\beta}_1^{\mathrm{T}} \\ \vdots \\ \boldsymbol{\beta}_{N_h}^{\mathrm{T}} \end{bmatrix} \in \mathbb{R}^{N_h \times N} \tag{5.22}$$

We define each line of $\boldsymbol{H}$ as $\boldsymbol{h}_i \in \mathbb{R}^{1 \times N_h}$, indicating the output with respect to $\boldsymbol{a}_i$. The target of the second stage could be interpreted as a manifold-regularized optimization problem

$$\begin{aligned} \min_{\boldsymbol{\beta}} \quad & \frac{1}{2} \sum_{i=1}^{l} C_i \parallel \boldsymbol{e}_i \|^2 + \frac{1}{2} \parallel \boldsymbol{\beta} \|^2 + \frac{\lambda}{2} \operatorname{Tr}\left(\boldsymbol{E}^{\mathrm{T}} \boldsymbol{L} \boldsymbol{E}\right) \\ \text{s.t.} \quad & \boldsymbol{e}_i^{\mathrm{T}} = \boldsymbol{y}_i^{\mathrm{T}} - \boldsymbol{h}_i \boldsymbol{\beta}, \; i = 1, \ldots, l \end{aligned} \tag{5.23}$$

In which $\|\cdot\|$ is the Euclidean distance, $\mathrm{Tr}(\cdot)$ calculates the trace of a matrix, and λ is a trade-off parameter. The first term of Eq. (5.23) is the loss function taking into account all labeled nodes with diverse penalty coefficients C_i. The second term refers to the classical anti-overfitting regularization item that constrains the output weight to be as small as possible. The third term is the manifold regularization where the unlabeled data comes into play. Concretely speaking, the manifold regularization framework believes in the assumption that if two points on the manifold sphere are close to each other, then they would also result in similar predicted outcomes. Consider the adjacency matrix as a measure of distance that two nodes are close to (connected to) each other if $A_{ij} = 1$. Otherwise they are far away (disconnected) from each other. The approximation of the manifold regularization is provided as

$$R_{\mathrm{m}} = \frac{1}{2}\sum_{i=1}^{N}\sum_{j=1}^{N} A_{ij} \parallel \boldsymbol{E}_i - \boldsymbol{E}_j\|^2 = \ \mathrm{Tr}\ \left(\boldsymbol{E}^{\mathrm{T}}\boldsymbol{L}\boldsymbol{E}\right) \tag{5.24}$$

This regularization penalizes the large difference in prediction of two connected nodes.

When there exist no labeled nodes in the input dataset and substituting Eq. (5.19) into Eq. (5.23), the optimization formulae could be reformulated as

$$\begin{aligned} &\min_{\beta} \quad \parallel \boldsymbol{\beta}\|^2 + \lambda \mathrm{Tr}\left(\boldsymbol{\beta}^{\mathrm{T}} H^{\mathrm{T}} L H \boldsymbol{\beta}\right) \\ &\mathrm{s.t.} \quad (\mathrm{H}\boldsymbol{\beta})^{\mathrm{T}}\mathrm{H}\boldsymbol{\beta} = \boldsymbol{I}_n \end{aligned} \tag{5.25}$$

where $\boldsymbol{I}_n$ is an n dimensional identity matrix.

Let γ and $\boldsymbol{v}$ denote the eigenvalue and the corresponding eigenvector. When $N \geq N_h$ and $\boldsymbol{H}$ have full column rank, it has been proven in that solving Eq. (5.25) is equivalent to selecting n generalized eigenvectors corresponding to the n smallest eigenvalues of the problem

$$\left(\boldsymbol{I}_{N_h} + \lambda \boldsymbol{H}^{\mathrm{T}}\boldsymbol{L}\boldsymbol{H}\right)\boldsymbol{v} = \gamma \boldsymbol{H}^{\mathrm{T}}\boldsymbol{H}\boldsymbol{v} \tag{5.26}$$

To organize the matrix $\boldsymbol{\beta}$, we abandon the first eigenvector because it corresponds to eigenvalue 0 and contributes little in the embedding process. Thus, given the first $n + 1$ smallest eigenvalues $\gamma_1, \gamma_2, \ldots, \gamma_{n+1}$, sorted by ascending order and their corresponding eigenvectors $\boldsymbol{v}_1, \boldsymbol{v}_2, \ldots, \boldsymbol{v}_{n+1}$, the output weights $\boldsymbol{\beta}$ are

$$\boldsymbol{\beta} = \left[\overline{\boldsymbol{v}}_2, \overline{\boldsymbol{v}}_3, \ldots, \overline{\boldsymbol{v}}_{n+1}\right] \tag{5.27}$$

In which $\overline{\boldsymbol{v}}_i = \overline{\boldsymbol{v}}_i / \|\boldsymbol{H}\boldsymbol{v}_i\|$, $i = 2, \ldots, n + 1$, indicates the normalized eigenvectors.

If $N < N_h$, meaning that the dimension of $\boldsymbol{A}$ is smaller than the number of neurons in the hidden layer, the proven formulation of (5.26) is underdetermined and has the following alternative formulation:

$$\left(\boldsymbol{I}_n + \lambda \boldsymbol{L}\boldsymbol{H}\boldsymbol{H}^{\mathrm{T}}\right)\boldsymbol{v} = \gamma \boldsymbol{H}\boldsymbol{H}^{\mathrm{T}}\boldsymbol{v} \tag{5.28}$$

In turn, the solution is

$$\boldsymbol{\beta} = \boldsymbol{H}^{\mathrm{T}}[\overline{\boldsymbol{v}}_2, \overline{\boldsymbol{v}}_3, \ldots, \overline{\boldsymbol{v}}_{n+1}] \tag{5.29}$$

where the normalized eigenvectors are given by $\overline{\boldsymbol{v}}_i = \overline{\boldsymbol{v}}_i / \left\| \boldsymbol{H}\boldsymbol{H}^{\mathrm{T}}\boldsymbol{v}_i \right\|, i = 2, \ldots, n + 1$.

Finally, we substitute $\boldsymbol{\beta}$ into Eq. (5.19) to obtain the embedding matrix ***EE*** that would be fed into a *k*-means clustering algorithm to determine the community labels **t**.

The implementation of the original *k*-means clustering, owing to its low computational complexity, has been integrated into ELM-CD. First, ELM-CD randomly selects *k* rows in $\boldsymbol{E}$ as the initial centroids of *k* clusters. Second, taking the Euclidean distance as the standard, each row in $\boldsymbol{E}$, which is represented by $\boldsymbol{e}_i$, $i = 1, \ldots, N$, is assigned to a cluster whose centroid is closest to a certain row. Then, for each cluster, we calculate the mean value of all its members and designate this mean vector as the new centroid. We then iterate the cluster assignment and the centroid update processes until no row in $\boldsymbol{E}$ changes its community label. We show the entire procedure of the ELM-CD algorithm in Algorithm 5.1.

Algorithm 5.1 Algorithm of the ELM-CD

Step 1 ELM-CD($\boldsymbol{A}$)

Step 2 initialize constant parameters, including λ,N_h,n

Step 3 generate $\boldsymbol{L}$ from $\boldsymbol{A}$

Step 4 normalize $\boldsymbol{A}$

Step 5 randomly generate input weights and biases of neurons in hidden layer

Step 6 calculate $\boldsymbol{H}$ according to Eq. (5.21)

Step 7 if $N < N_h$

Step 8 select the first $n + 1$ smallest eigenvalues of Eq. (5.26) and assemble $\boldsymbol{\beta}$ using the corresponding eigenvectors as in Eq. (5.27)

Step 9 else

Step 10 select the first $n + 1$ smallest eigenvalues of Eq. (5.28) and assemble $\boldsymbol{\beta}$ using the corresponding eigenvectors as in Eq. (5.29)

Step 11 archive the embedding matrix $\boldsymbol{E} = \boldsymbol{H\beta}$

Step 12 randomly select k rows from $\boldsymbol{E}$ as centroids, say $\boldsymbol{c}_1, \boldsymbol{c}_2, \ldots, \boldsymbol{c}_k$

Step 13 while $\boldsymbol{t}$ changed

Step 14 for $i = 1$ **to** N

Step 15 for $j = 1$ **to** k

Step 16 $dist_j = \| \boldsymbol{e}_i - \boldsymbol{c}_j \|$

Step 17 find the smallest element in ***dist*** and its corresponding index s

Step 18 $\boldsymbol{t}_i = s$

Step 19 for $i = 1$ **to** k

Step 20 update $\boldsymbol{c}_i$ to be the mean value of all members in cluster i

Step 21 return $\boldsymbol{t}$

5.2.2 *Simulations and Results*

Based on the algorithm, we deploy ELM-CD on computer-generated networks as well as four well-known real-world networks to compare its performance with that of three competitors, including two divisive algorithms: GN and NF, and conventional k-means. The constructed ELM network has $N_h = 2000$ neurons and randomly initialized input weights between -1 and 1 with all biases being ignored (set to 0). Following the empirical configuration in, we set the embedding feature space dimension $n = 3$ and the trade-off parameter $\lambda = 0.1$.

5.2.2.1 Computer Generated Networks

The fabricate networks with community structures and NMI are same as the Sect. 5.1.2. These computer-generated networks with decided graph knowledge are then fed into our algorithm and its competitors to calculate their NMI, and the results are presented in Fig. 5.6.

Each point in Fig. 5.6 represents the average NMI of a certain algorithm on 50 randomly generated scenarios with predefined z_{out} varying from 0 to 8. The error bars accompanying those points mark the normalized deviation of NMI in 1000 experiments. Lines connecting points are included solely as a guide. The community detection resolution of all methods drops with increasing z_{out}; however, our algorithm outperforms all competitors, especially when z_{out} is approaching 8, which means the community structure is becoming extremely ambiguous. Even when $z_{out} = 8$, ELM-CD could still correctly classify half the number of nodes. Another

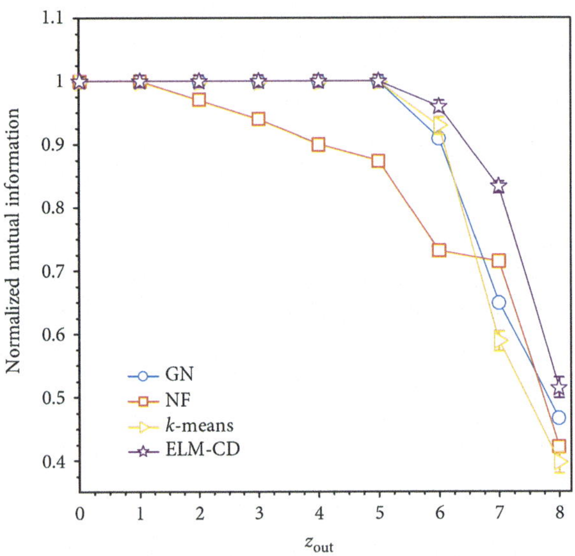

Fig. 5.6 NMI of different algorithms on computer-generated networks

outstanding feature of our algorithm is its comparative robustness. Different from what happens to ELM-CD, both GN and *k*-means run competitively in terms of accuracy until $z_{out} = 6$, but performance noticeably deteriorates after that.

5.2.2.2 Real World Networks

Based on the four real-world networks have been introduced in Sect. 5.1.2 in Table 5.1. For each scenario, we took 1000 trials for each of the three probabilistic algorithms NF, ELM-CD, and *k*-means. For the decided algorithm GN, only one trial is sufficient.

First, we check each algorithm's understanding of community structures based on the definition of modularity. For GN, ELM-CD, and *k*-means, we need to predefine the number of communities as one of the input parameters. Practically, we varied this parameter from 1 to 10 for the first three networks (Karate, Dolphins, and Polbooks) and from 1 to 20 for the last (football). Among all partition results, the clustering scheme with the highest Q is chosen along with its corresponding NMI scores. Table 5.3 lists these experimental results and shows that ELM-CD exhibits obvious but not dominant superiority compared with other algorithms. Specifically, compared with *k*-means, ELM-CD demonstrates a stronger capability to determine more modular structures, simultaneously with higher NMI in most instances. We cannot ignore the higher modularity scores of GN and NF obtained in the dolphin network, but we should bear in mind that in this test, GN and NF determined five and four clusters, respectively, compared to ELM-CD's three clusters.

Second, we fixed the predefined number of communities as the inherent number for each network and calculate the NMI of each algorithm to evaluate how close the difference between the discovered community structure and the given underlying structure could be. Meanwhile, the consumed time was recorded to compare the complexity of the algorithms. We have displayed the maximum NMI accompanied by the corresponding modularity and the calculation time in Table 5.4. From this table, in spite of the minor improvement in precision, GN is much slower in comparison with the other three algorithms. Regarding NF, although it runs fast

Table 5.3 Maximum Q values and corresponding NMI of different algorithms

Dataset	Index	GN	NF	*k*-means	ELM-CD
Karate	Modularity	0.4013	0.3807	0.3715	0.4198
	NMI	0.7072	0.6925	1.0	0.6873
Dolphins	Modularity	0.5196	0.4955	0.4216	0.4617
	NMI	0.5016	0.5727	0.4964	0.6548
Polbooks	Modularity	0.5120	0.5020	0.4509	0.5221
	NMI	0.4018	0.4396	0.5525	0.5538
Football	Modularity	0.5950	0.5682	0.5779	0.5797
	NMI	0.4237	0.4436	0.8490	0.8207

Table 5.4 Compared results for maximum NMI, corresponding Q values, and time elapsed of different algorithms

Dataset	Index	GN	NF	k-means	ELM-CD
Karate	Modularity	0.3600	0.3870	0.3715	0.3715
	NMI	0.8365	0.6925	1.0	1.0
	Running time	5.5980	7.6e-4	8.6e-4	0.0098
Dolphins	Modularity	0.3787	0.4923	0.3555	0.3787
	NMI	0.8888	0.6208	0.8084	0.8888
	Running time	10.2244	0.0020	0.0012	0.0171
Polbooks	Modularity	0.4831	0.5020	0.4248	0.4530
	NMI	0.5754	0.5308	0.4144	0.6650
	Running time	199.2434	0.0064	0.0033	0.0293
Football	Modularity	0.5973	0.5773	0.5523	0.5585
	NMI	0.9214	0.7624	0.8954	0.8944
	Running time	808.1076	0.0120	0.0021	0.0308

Table 5.5 Average NMI and Q values of the three fastest algorithms

Dataset	Index	NF	k-means	ELM-CD
Karate	Modularity	0.3807	0.1071	0.2374
	NMI	0.6925	0.3960	0.4340
Dolphins	Modularity	0.4941	0.2818	0.2788
	NMI	0.5943	0.3336	0.5728
Polbooks	Modularity	0.5019	0.3360	0.4861
	NMI	0.5282	0.3739	0.5461
Football	Modularity	0.5680	0.5126	0.5086
	NMI	0.7424	0.8396	0.8262

enough, the low modularity and NMI results represent a shortfall in the overall performance.

ELM-CD no doubt consumes more time than k-means, whose computational complexity is $O(N \cdot k)$ where k is the predefined number of communities, because ELM-CD requires the ELM calculation expenditure, which costs $O(N_h)$ time. Theoretically, once k is determined and when $N > > N_h$ the complexity of ELM-CD is asymptotically $O(N)$, of the same as k-means. On the other hand, owing to the application of ELM, the stability of ELM-CD is much better than that of k-means. We demonstrate this by listing the average modularity and NMI results of the three fast algorithms in Table 5.5.

In Table 5.5, k-means illustrates very poor stability (even worse than that of NF) in light of the large differences between the average results and the maximum results shown in Table 5.4. In contrast, ELM-CD overcomes this inherited shortcoming due to the inclusion of the manifold regularization term in the equivalent optimization process as depicted in Eq. (5.23). Another trend exposed is that the robustness of k-means and ELM-CD is stronger for large networks with higher numbers of communities. The graphical clustering results of ELM-CD in Tables 5.3 and 5.4 have been illustrated in the left and right parts of Figs. 5.7, 5.8, 5.9, and 5.10, respectively.

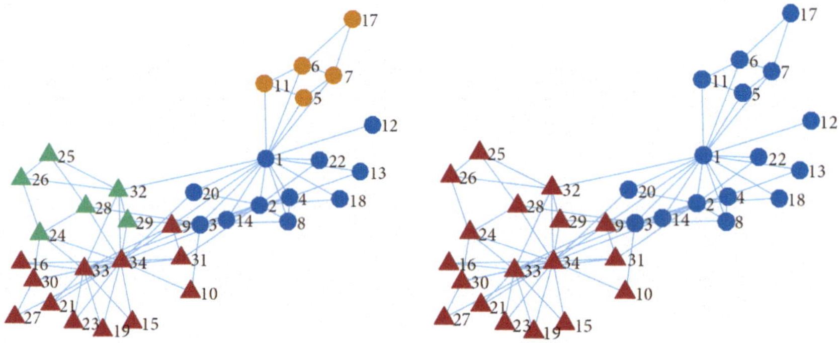

(a) (b)

Fig. 5.7 Communities extracted from Zachary's karate club network using ELM-CD. (**a**) Clustering result with the largest Q. (**b**) Given $k = 2$ the scenario with the best NMI

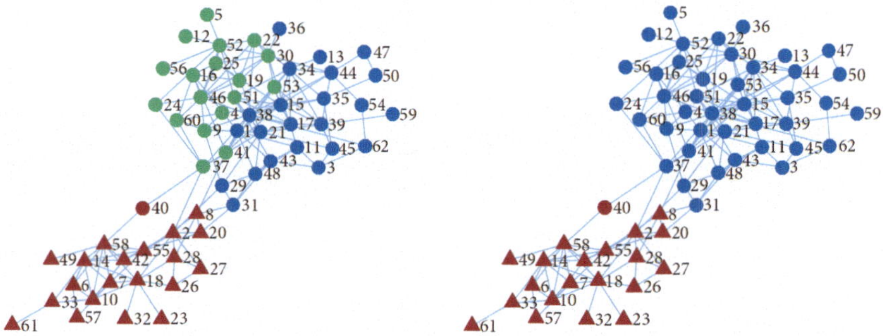

(a) (b)

Fig. 5.8 Communities extracted from the dolphin social network using ELM-CD. (**a**) Clustering result with the largest Q. (**b**) Given $k = 2$ the scenario with the maximum NMI

5.2.2.3 Zachary's Karate Club Network

Figure 5.7 shows the partition results of our algorithm corresponding to the scenarios in Tables 5.3 and 5.4. We use triangles to mark the new group belonging to the administrator, and circles represent the instructors. From Fig. 5.7a, ELM-CD converges to four groups to reach the largest modularity, although it redundantly separates each new club into two smaller ones. When the number of communities is decided, ELM-CD was able to detect the difference after the confrontation occurred in the club, which is demonstrated in Fig. 5.7b.

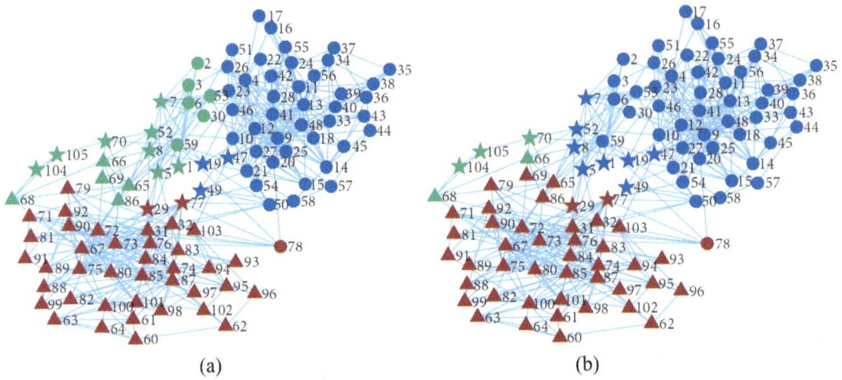

Fig. 5.9 Communities extracted from the political book network using ELM-CD. (**a**) Clustering result with the largest Q. (**b**) Given $k = 3$ the scenario with the maximum NMI

5.2.2.4 Dolphin Social Network

In Fig. 5.8, we present snapshots of ELM-CD clustering results for the largest Q and the maximum NMI. Again, triangle and circle markers have been used to represent the two identical genders known ahead of time. For the best effort segregation in light of the modularity criteria, ELM-CD splits the female community into two subgroups, as shown in Fig. 5.8a. Node 40 appears to be the key factor that has the greatest effect on NMI results. In both images of Fig. 5.8 node 40 is the only one that has been incorrectly classified. As a matter of fact, in Figure 5.8b, the maximum NMI would be 1 rather than 0.8888 if node 40 could be assigned to the correct group.

5.2.2.5 Political Books Network

In Fig. 5.9, these political tendencies have been identified by triangle (conservative), circle (liberal), and pentagon (neutral) markers, respectively. The identifications made by our algorithm, on the other hand, are shown in three different colors. Red and blue colors represent the conservative and liberal groups, respectively, while green indicates the neutral part, which occupies a small number of nodes. It is clear from Fig. 5.9 that people tend to have relatively stable opinions on politics as indicated by the obvious segregation of these books. However, it is quite common that some neutral books appear more liberal or conservative oriented in the eyes of some critical readers, while the political orientations of some liberal or conservative books are obscure and may be considered as neutral. This explains why our algorithm can make mistakes on the boundaries between different classes. Moreover, from these two images, some books could have been classified more reasonably. For example, node 78 has four connections to conservative books while only sharing two

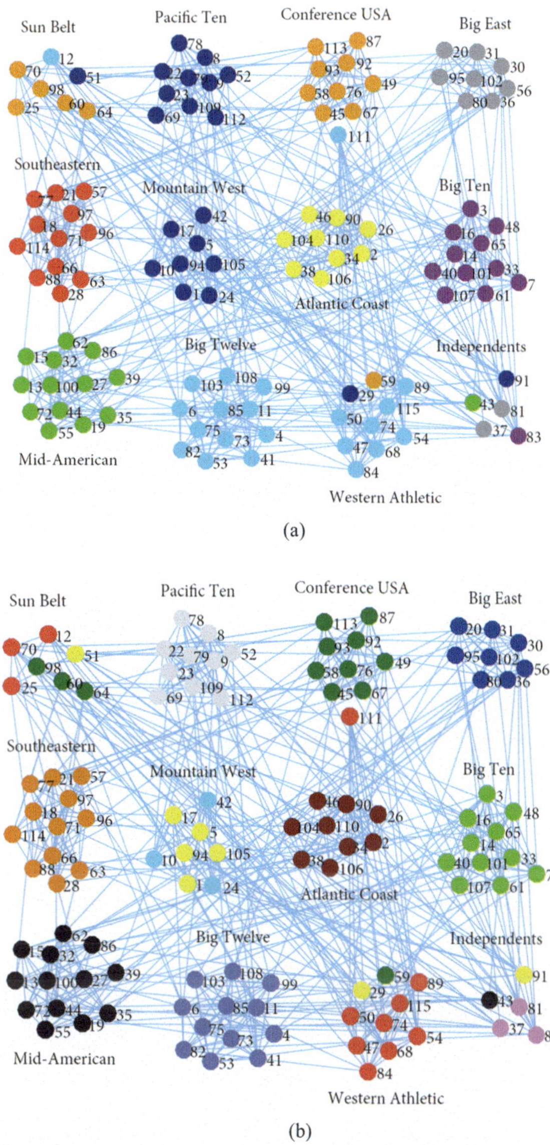

Fig. 5.10 Communities extracted from the US college football network using ELM-CD. (**a**) Clustering result with the largest Q. (**b**) Given $k = 12$ the scenario with the maximum NMI

connections with liberal ones; nevertheless, it is defined as a liberal book. Another example is the neutral node 77 that has no link to any other neutral books. Both the best efforts in Figure 5.9a, b have successfully corrected these two unreasonable classifications.

5.2.2.6 US College Football Network

In Fig. 5.10, each conference has been located in a cluster relatively far from the other conferences and the community detection results have been recolonized using different colors. Aiming at the highest score in modularity, ELM separates these teams into eight classes. From Fig. 5.10a, except for the Southeastern and the Atlantic Coast conferences, all others have been misunderstood to some extent. Some of them, say the Pacific Ten and the Mountain West in dark blue, have been merged into a larger group. Some have been disassembled and displaced into other groups, such as the Independents and the Sun Belt. Such categorization could be rational because conferences in a larger group do compete with each other more frequently than smaller conferences.

In contrast, teams in the independent conference, except for nodes 81 and 83, share no game with each other; hence, it is no surprise to see them in other categories. When the number of communities is restrained to 12, as shown in Fig. 5.10b, ELM-CD performs well, with NMI = 0.8944 according to Table 5.4. This time, most conferences have retained the majority of nodes to organize independent communities, except for the Sun Belt, which has been fragmented, and the Mountain West, which contains two subclasses.

However, we also have to note some unusual clustering behavior in this procedure, such as node 37 being placed in the same group with nodes 81 and 83, having nothing to do with 37 at all. Both of the snapshots in Fig. 5.10 on some level expose the irrationality of the original conference definitions. For instance, node 111 belongs to Conference USA, but this team has always played its games with opponents from other conferences, which is contrary to the definition of a community.

5.3 Deep Auto-Coded Clustering Algorithm

Based on the widely-used DA-ELM algorithm, we present a novel community detection algorithm called the deep auto-encoded clustering algorithm (DAC), in which unsupervised and sparse single autoencoders are trained and piled up one after another to embed key community information in a lower-dimensional representation, such that it can be handled easier by clustering strategies. Extensive comparison tests undertaken on synthetic and real-world networks reveal two advantages of the algorithm: on the one hand, DAC shows higher precision than the k-means community detection method benefiting from the integration of sparsity constraints; on the other hand, DAC runs much faster than the spectral community detection algorithm based on the circumvention of the time-consuming eigenvalue decomposition procedure [7].

5.3.1 *Deep Auto-Encoded Clustering Algorithm for Community Detection*

In recent years, deep learning, one of the most popular topics in artificial intelligence, has been constantly stimulating new revolutions in both the academic and industrial communities. Among these popular methods, deep neural network provides another way of providing the embedded features in lower-dimensional representations. In particular, the unsupervised learning scheme called autoencoder attempts to reconstruct the input information by means of a series of nonlinear transformations. In fact, it demonstrates that for embedding representations, spectral clustering and stacked autoencoder share the same objective but autoencoding has a smaller complexity than spectral clustering. Recently, an optimization model with a loss function that considers both the deep-learning representation and the clustering process has been developed to solve traditional clustering problems. It inspires us to utilize the autoencoder in community detection.

Aiming at community detection, we present a deep auto-encoded clustering (DAC) algorithm that encapsulates stacked autoencoders as the embedding producer and a k-means clustering as the classification strategy. Additionally, in consideration of the sparsity characteristics of underlying networks in many cases, we also integrate sparsity regularization in autoencoder layers, which is a difficult task for spectral clustering methods. Extensive comparative trials in both synthetic and real-world networks show that the algorithm is almost as accurate as the sparsity community detection algorithm while running much faster than the latter one [8].

5.3.1.1 Structure of Auto-Encoder with Deep Learning

The model of network is same as the Sect. 5.1. A community detection algorithm segregates $\boldsymbol{V}$ into mutually exclusive districts by means of attaching labels on each node to indicate its group. As $\boldsymbol{A}$ is symmetrical, each row or column of $\boldsymbol{A}$ could be considered as an N-dimensional input sample. Instead of directly assigning each node with a community label, we first learn the community information as features embedded from a deep-stacked sparse autoencoder. The elementary building block in our deep-learning networks is an autoencoder as depicted in Fig. 5.11, which is a single, hidden layer, and fully connected neural network with identical number of neurons in both end layers.

$$\begin{aligned} a_j^{(i)} &= f\left[z_j^{(i)}\right] = f\left[\left(\boldsymbol{W}_j^{(i-1)}\right)^{\mathrm{T}} \boldsymbol{a}^{(i-1)} + b_j^{(i-1)}\right] \\ &= f\left[\sum_{k=1}^{N_{i-1}} W_{jk}^{(i-1)} a_k^{(i-1)} + b_j^{(i-1)}\right] \end{aligned} \tag{5.30}$$

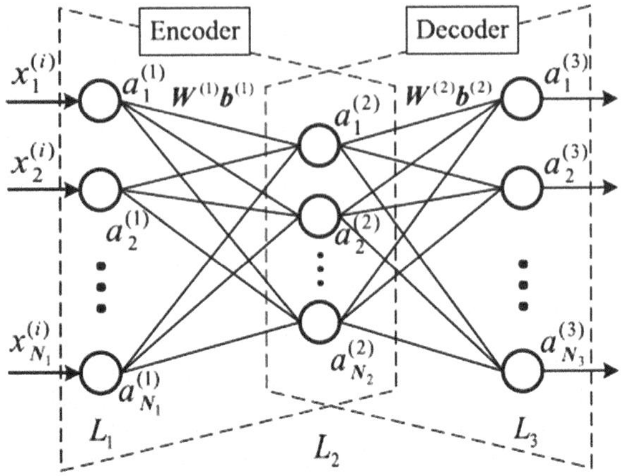

Fig. 5.11 Structure diagram of single autoencoder

We use L_i, $i = 1, \ldots, n$ to represent the ith layer of a network with n layers, considering the input layer as L_1. Subsequently, the number of neurons in L_i is given by N_i. a_i denotes the output of node j in layer L_j. In Fig. 5.11, specifically, except for neurons in L_1 whose input and output are the same, all the other neurons have an output in the form of to the input summation pointing to neuron j in L_i. $W_{jk}^{(i-1)} \in \mathbb{R}^{N_{i-1}\times 1}$ refers to a parameter vector with each member as $W_{jk}^{(i-1)}$ being the weight value from the kth neuron in L_{i-1} to the jth neuron in L_i. Another parameter is the scalar bias $b_j^{(i-1)}$ that differs from neuron to neuron. The activation function $f(\cdot)$ defines the nonlinear transformation, which has been chosen as the sigmoid function $f(x) = 1/(1 + e^{-x})$whose derivative is $f'(x) = f(x)(1 - f(x))$. Furthermore, the transformation from vector $a^{(i-1)} \in \mathbb{R}^{N_{i-1}\times 1}$ to vector $a^{(i)} \in \mathbb{R}^{N_i\times 1}$ could be expressed in the matrix form as

$$\boldsymbol{a}^{(i)} = f\left[\left(\mathbf{W}^{(i-1)}\right)^{\mathrm{T}}\boldsymbol{a}^{(i-1)} + \boldsymbol{b}^{(i-1)}\right] \tag{5.31}$$

where $i \in \{2,3\}$, $W^{(i-1)} \in \mathbb{R}^{N_{i-1}\times N_i}$ and $b^{(i)} \in \mathbb{R}^{N_i\times 1}$. For the convenience of the following narration, we choose $x^{(i)} \in \mathbb{R}^{N_i\times 1}$ as the symbol to represent the ith input training sample, of which the item is $x_j^{(i)}$, $i = 1, \ldots, N$ and $i = 1, \ldots, N_1$. The whole set $\boldsymbol{X} = \{x^{(1)}, x^{(2)}, \ldots, x^{(N)}\}$. Moreover, $a^{(3)}$ for each $x^{(i)}$ would be described as a hypothesis function $h(x^{(i)})$.

5.3.1.2 Learning Process under Sparsity Restriction

As marked in the dashed lines in Fig. 5.11, an autoencoder can be represented as a combination of an encoder and a decoder. The encoder embeds the original input $a^{(1)}$ into a new dimensional representation $a^{(2)}$ that to the utmost extent maintains the useful information while filtering useless noise. If the subsequent decoder is capable of reconstructing $a^{(1)}$ from $a^{(2)}$, *i.e.*, if $a^{(3)} \approx a^{(1)}$, then we could use $a^{(2)}$ as an approximation of $a^{(1)}$ for further study. Hence, the loss function for all the training examples could be represented as

$$L(\boldsymbol{W}, \boldsymbol{b}) = \frac{1}{N} \sum_{i=1}^{N} \frac{1}{2} \| h\left(\boldsymbol{x}^{(i)}\right) - \boldsymbol{x}^{(i)} \|^2 + \frac{\lambda}{2} \sum_{i=1}^{2} \sum_{j=1}^{N_{i+1}} \sum_{k=1}^{N_i} \left(W_{jk}^{(i)}\right)^2 \tag{5.32}$$

where $\boldsymbol{W}$ and b covers all the weight and bias parameters; λ is a constant that maintains the balance between the two terms. In Eq. (5.32), the first term, in which $\| \cdot \|$is the Euclidean norm, exhibits the difference between the original and the recovered training data. The second term shows the regularization rule used to prohibit overfitting.

Sparsity characteristics are ubiquitous in complex networks, especially in networks with community structures owing to the relatively sparse intergroup connections. Consequently, we incorporate the sparsity information into the autoencoder as another constraint term in the loss function

$$L_s(\boldsymbol{W}, \boldsymbol{b}) = L(\boldsymbol{W}, \boldsymbol{b}) + \beta \sum_{i=1}^{N_2} \mathrm{KL}(\rho \| \widehat{\rho}_{\mathrm{i}}) \tag{5.33}$$

The sparsity regularization term expressed in a relative entropy form penalizes the large variance between $\widehat{\rho}_i = \frac{1}{N} \sum_{j=1}^{N} a_i^{(2)} x^{(j)}$, which is the average output values of L_2, and the typically small sparsity constant ρ. β is a weight constant.

$$\mathrm{KL}(\rho \| \widehat{\rho}_{\mathrm{i}}) = \rho \log \frac{\rho}{\widehat{\rho}_{\mathrm{i}}} + (1 - \rho) \log \frac{1 - \rho}{1 - \widehat{\rho}_{\mathrm{i}}} \tag{5.34}$$

The unfolded relative entropy is also known as the Kullback–Leibler (KL) divergence. It is noteworthy that the inclusion of the sparsity regularization enables the hidden lay to have more nodes than the other two layers. Once the feed forward process has finished, the Batch gradient descent (BGD) method would be used in the back-propagation stage to update the parameter sets. For each weight and bias value,

$$W_{jk}^{(i)} = W_{jk}^{(i)} - \alpha \frac{\partial L_s(\boldsymbol{W}, \boldsymbol{b})}{\partial W_{jk}^{(i)}} \tag{5.35}$$

$$b_j^{(i)} = b_j^{(i)} - \alpha \frac{\partial L_s(\boldsymbol{W}, \boldsymbol{b})}{\partial b_j^{(i)}} \tag{5.36}$$

where $i \in \{1, 2\}, j = 1, \ldots, N_{i+1}$, and $k = 1, \ldots, N_i$. The partial derivatives could be calculated as

$$\frac{\partial L_s(\boldsymbol{W},\boldsymbol{b})}{\partial W_{jk}^{(i)}}=\frac{1}{N}\sum_{l=1}^{N}\frac{\partial L_s\left(\boldsymbol{W},\boldsymbol{b};\boldsymbol{x}^{(l)}\right)}{\partial W_{jk}^{(i)}}+\lambda W_{jk}^{(i)} \tag{5.37}$$

$$\frac{\partial L_s(\boldsymbol{W},\boldsymbol{b})}{\partial b_j^{(i)}}=\frac{1}{N}\sum_{l=1}^{N}\frac{\partial L_s\left(\boldsymbol{W},\boldsymbol{b};x^{(l)}\right)}{\partial b_j^{(i)}} \tag{5.38}$$

In which $L_s(\boldsymbol{W}, b, x^{(l)})$ means the sparse loss function corresponding to the lth input vector. Furthermore, we introduce another notation

$$\delta_j^{(i)}=\frac{\partial L_s\left(\boldsymbol{W},\boldsymbol{b};\boldsymbol{x}^{(l)}\right)}{\partial z_j^{(i)}} \tag{5.39}$$

where $i \in \{2, 3\}$ and $j = 1, \ldots, N_i$, indicating the contribution of the jth neuron in L_i to the final output offset

$$\begin{aligned}\delta_i^{(3)}&=\frac{\partial}{\partial z_i^{(3)}}\frac{1}{2}\parallel h\left(\boldsymbol{x}^{(l)}\right)-\boldsymbol{x}^{(l)}\parallel^2\\&=\left(a_i^{(3)}-x_i^{(l)}\right)f'\left(z_i^{(3)}\right)\end{aligned} \tag{5.40}$$

$$\delta_i^{(2)}=\left[\sum_{j=1}^{N_3}W_{ji}^{(2)}\delta_j^{(3)}+\beta\left(-\frac{\rho}{\widehat{\rho}_i}+\frac{1-\rho}{1-\widehat{\rho}_i}\right)\right]f'\left(z_i^{(2)}\right) \tag{5.41}$$

It is noteworthy that in Eq. (5.42) $i = 1, \ldots, N_3$ and $l = 1, \ldots, N$, while in Eq. (5.43) $i = 1, \ldots, N_2$ The partial derivatives for each training sample could now be derived as

$$\frac{\partial L_s\left(\boldsymbol{W},\boldsymbol{b};\boldsymbol{x}^{(l)}\right)}{\partial W_{jk}^{(i)}}=a_k^{(i)}\delta_j^{(i+1)} \tag{5.42}$$

$$\frac{\partial L_s\left(W,\boldsymbol{b};\boldsymbol{x}^{(l)}\right)}{\partial b_j^{(i)}}=\delta_j^{(i+1)} \tag{5.43}$$

where i, j, k, l are the same as in Eqs. (5.39) and (5.40).

One single sparse autoencoder may not be able to reproduce all the source data precisely. Stacked autoencoders, alternatively, have been demonstrated to be powerful in accomplishing this task. Figure 5.12 illustrates an abbreviated architecture of our proposed community detector, with each layer being abstracted as a rectangle whose length is in proportion to the number of neurons in it. With respect to the output layer of the encoder, the network has a symmetrical layout of layers in that $N_4 = N_2$ and $N_5 = N_1$. As mentioned earlier, A could be viewed as N vectors, each of which is a training example. At the start, we extract two individual sparse

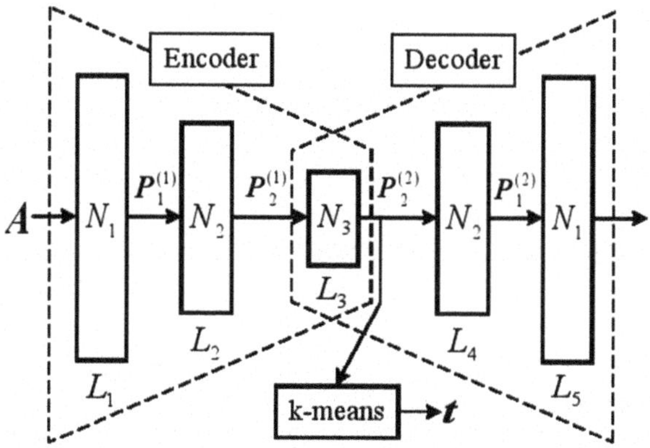

Fig. 5.12 Structure diagram of a stacked autoencoder

autoencoders of structures $N1 \to N2 \to N1$ and $N2 \to N3 \to N2$, respectively. Subsequently, we train them one after another, meaning the output of L_2 in the first autoencoder would be fed into the second autoencoder as input data and so on. For a general stacked autoencoder with n layers, $(n - 1)/2$ single autoencoders would be extracted. A new symbol $\boldsymbol{P}_j^{(i)}$ is defined herein as the parameter set$\{\boldsymbol{W}^{(i)}, b^{(i)}\}$ in the jth-extracted autoencoder where $i \in \{1, 2\}, j = 1, \ldots, (n - 1)/2$. With the obtained trained weight and bias parameters, we could assign the initial values of the parameters in the stacked autoencoder in the manner as shown in Fig. 5.12.

Subsequently, the whole network would be trained again to fine tune the parameters. Further, the stacked encoder is selected as our desired embedding processor from which the dimension-reduced representation originates. Finally, a k-means clustering model deployed immediately after the end of the encoder proceeds to output a vector $t \in \mathbb{N}_+^N$, whose items are community labels of each node. A summary of the whole algorithm could be found in the following Algorithm 5.2.

Algorithm 5.2 Algorithm of DAC

Step 1 Normalize $\boldsymbol{A}$

Step 2 According to size of, properly choose the scale of each layer and create a stack autoencoder SA with n layers (n is odd).

Step 3 Build $(n - 1)/2$ single autoencoders as *block*, $i = 1, \ldots, (n - 1)/2$

Step 4 $block^{(i)}.\ \boldsymbol{X} = \boldsymbol{A}$

Step 5 **for** $i = 1$ **to** $(n - 1)/2$

Step 6 Train *block*

Step 7 if $i < (n - 1)/2$

Step 8 $block^{(i+1)}.\ \boldsymbol{X} = block^{(i)}.\ \boldsymbol{a}^{(2)}$

Step 9 Initialize SA with block weights and bias as in Fig. 5.12

Step 10 Train SA with $\boldsymbol{A}$ again and obtain output from $L_{(n+1)/2}$ as the embedded representation $\boldsymbol{E}_{(n+1)/2}$

Step 11 Feed $\boldsymbol{E}$ into a clustering algorithm and obtain $\boldsymbol{t}$

Step 12 return $\boldsymbol{t}$

5.3.2 Simulations and Results

Based on the algorithm, we deploy the algorithm using DAC on computer-generated networks and four famous real-world networks. Two baseline algorithms chosen for comparison are the k-means and spectral clustering method for community detection. In practice, the dimension of the embedded output $\boldsymbol{E}$ is five and we use a multiplier to enlarge the scale of the layers one after another until it reaches the scale of $\boldsymbol{A}$. For example, if the multiplier is two, the stacked encoder for a network with 34 nodes has a structure:$2 \rightarrow 10 \rightarrow 20 \rightarrow 34$. In this work, the constants in DAC are set as $\alpha = 0.85$,$\rho = 0.2$,$\lambda = 0$,$\beta = 1.8$, with the weights and bias of single autoencoders randomly initialized between -0.5 to 0.5. Additionally, only the first single autoencoder has been assigned with sparsity regularization, which means $\beta = 0$ for the others.

5.3.2.1 Synthetic Networks

The fabricate networks with community structures and NMI are defined as same as the Sect. 5.1.2. These computer-generated networks with decided graph knowledge are then fed into our algorithm and its competitors to calculate their NMIs and Qsm which are presented in Figs. 5.13 and 5.14, respectively. The corresponding time spent in these scenarios has been displayed in Fig. 5.15.

In Figs. 5.13, 5.14, and 5.15, each point represents an average value of a certain algorithm on 50 randomly generated scenarios with predefined z_{out} varying from 0 to 12. The error bars accompanying those points show the normalized deviation of values in 1000 experiments. The lines connecting the points are included solely as a

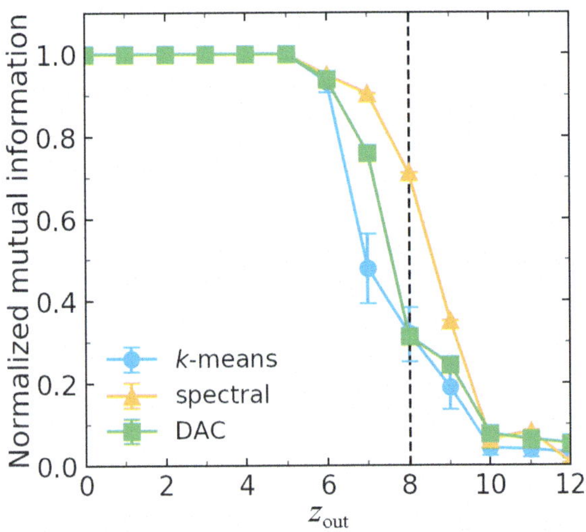

Fig. 5.13 NMI values of different algorithms on computer-generated networks

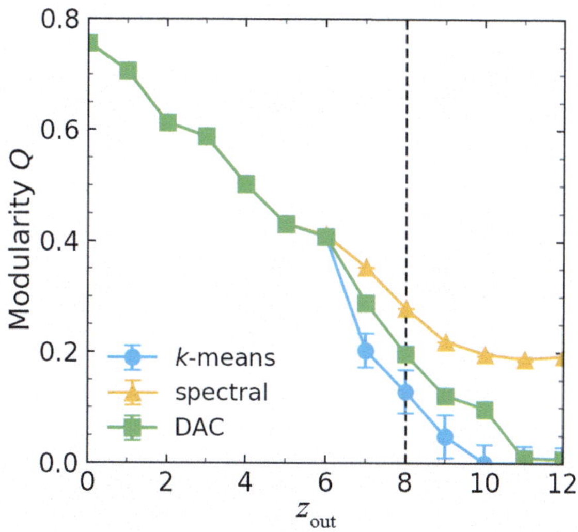

Fig. 5.14 Q values of different algorithms on computer-generated networks

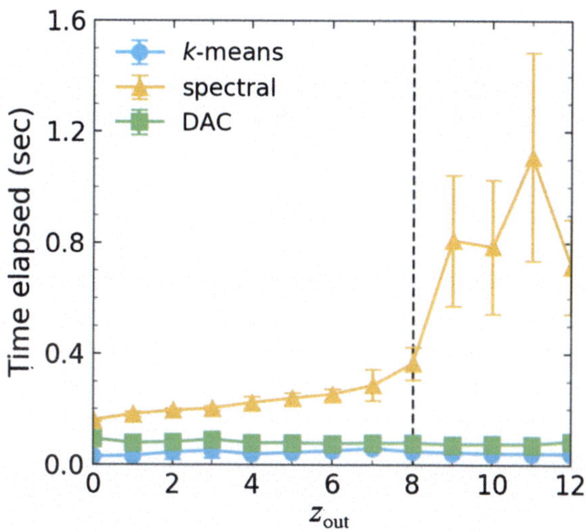

Fig. 5.15 Time consumed of different algorithms on computer-generated networks

visual guide. The black dashed line indicates the boundary whether community structures would exist in the networks.

We find that the community detection resolution of all methods decreases with the increase in z_{out}, and the resolution of our algorithm falls in between its two counterparts. We have to admit the advantage of the spectral clustering method in precision especially when z_{out} is above eight, which means the community structure is becoming increasingly ambiguous; however, we also noticed a tradeoff in that scenario.

Figure 5.15 shows a sharp-climbing and fluctuating time curve of the spectral method, which is clearly different compared with DAC. Inevitably, as a result of the time consuming eigenvalue decomposition, the computational complexity of spectral clustering cannot effectively decrease from $O(N^{2.367})$, in contrast with the approximate linear $O(N\bar{d}N_2)$ where $\bar{d}$ is the average degree and N_2 is the scale of the first hidden layer in the encoder. Theoretically, for a given network topology, once the deep-learning network has been constructed, the complexity of DAC could be asymptotically as small as k-means' $O(N)$. It is noteworthy that spectral clustering shows higher Q values than DAC when z_{out} is approaching 12 because of its modularity-oriented design; however, the corresponding NMI values are not as good.

5.3.2.2 Real World Networks

We chose three real-world networks (Zachary's karate club network, Political books network and US college football network) in Table 5.1 with the decided number of communities have been selected as test beds for the measurement of our algorithms' accuracy as well as its running times. For each scenario, we use 1000 trials of each algorithm.

Given the number of communities as the inherent number of the networks, we calculated the NMIs of each algorithm to see how close the difference between the discovered community structure and the given underlying structure could be. Meanwhile, the consumed time had been recorded to compare the complexity of the algorithms. We present the maximum and mean values of NMI accompanied by the corresponding modularity and the calculation time in Tables 5.6 and 5.7, respectively.

An obvious trend in general is the deteriorating accuracy as the network scale increases. DAC shows the best performance in terms of the maximal NMI values in the first three networks and wins in stability for the last one. DAC possesses higher accuracy and stability than k-means and executes much faster than spectral clustering. In the remainder of this section, we explore the rationality of the proposed algorithm on each real-world network.

Table 5.6 Comparison of the maximum NMI with the corresponding Q and time consumed (seconds)

Dataset	Index	k-means	spectral	DAC
Karate	Q	0.3715	0.3715	0.3715
	NMI	1.0	1.0	1.0
	Time	0.1033	0.1794	0.0157
Polbooks	Q	0.4248	0.5112	0.4543
	NMI	0.4144	0.5915	0.6222
	Time	0.0404	0.2322	0.0571
Football	Q	0.5523	0.5999	0.5597
	NMI	0.8954	0.9362	0.9213
	Time	0.0865	1.0145	0.1329

Table 5.7 Comparison of the mean NMI with the corresponding Q and time consumed (seconds)

Dataset	Index	k-means	spectral	DAC
Karate	Q	0.1071	0.3715	0.2555
	NMI	0.3960	1.0	0.8022
	Time	0.0141	0.0282	0.0159
Polbooks	Q	0.3411	0.5112	0.3425
	NMI	0.3430	0.5915	0.5390
	Time	0.0457	0.2462	0.0622
Football	Q	0.5398	0.5982	0.5914
	NMI	0.8606	0.9078	0.9186
	Time	0.0781	1.4987	0.1421

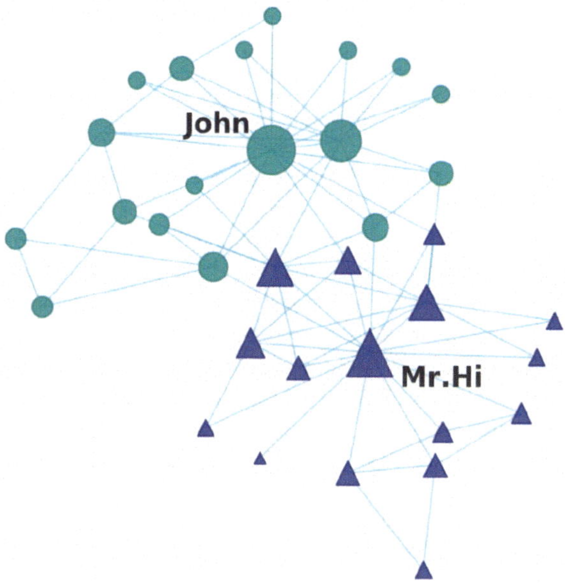

Fig. 5.16 Communities extracted from Zachary karate club network using DAC

5.3.2.3 Zachary's Karate Club Network

Figure 5.16 shows the NMI best-shot partition results of our algorithm. The size of the nodes illustrates their degree. We use circles and triangles to represent the inherent classification. The different colors mark the detection results of DAC. We show that DAC can clearly tell the differences after a confrontation happened in the club.

5.3.2.4 Political Books Network

In Fig. 5.17, these political tendencies have been identified as circle (conservative), pentagon (liberal), and triangle (neutral) markers, respectively. The identifications

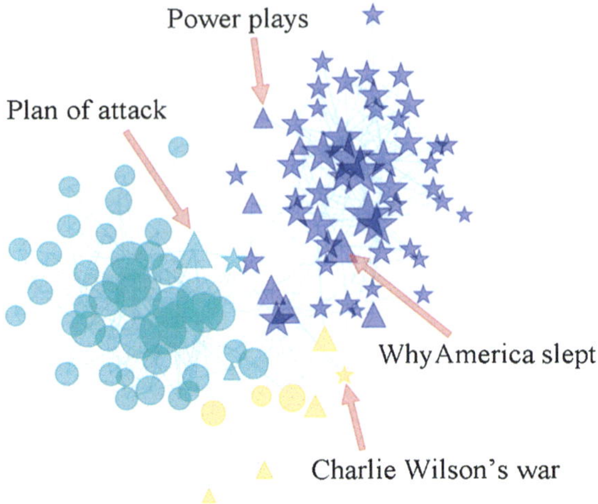

Fig. 5.17 Communities extracted from the political books network using DAC

from our algorithm, however, show the differences in three different colors. Green and blue represent the conservative and liberal groups, respectively. Orange indicates the neutral part, which occupies a small number of nodes. Figure 5.17 shows that people tend to have a relatively stable opinion on politics, as indicated by the obvious segregation of these books. However, some neutral books may look more liberal or conservative according to some critical readers, while the political orientations of some liberal or conservative books are obscure that one may think they are the neutral ones. This explains why our algorithm can easily misjudge the boundaries between different classes. For example, with names as indicated in Fig. 5.17, all the four "incorrectly" labeled books have more fans from other orientation groups than from the community of their own.

5.3.2.5 US College Football Network

Figure 5.18 shows that each conference has been located in a cluster relatively far from the other conferences, and the community detection results have been recolonized in different-colored shapes. As shown in Fig. 5.18, seven groups have been detected correctly according to the predefined partition; for the other imperfectly found communities, some rational explanations can be given. At the bottom of the Conference USA, one eye-catching pink hexagon node that plays no game with any team in the same conference has been assigned with the Western Atlantic conference, from which most of its opponents originate. Teams in the independents indeed shared a few games with each other, hence it appears reasonable for DAC to split these four teams and assign them to two distinct communities.

To solve community detection problems, we design a deep embedding preprocessor using stacked autoencoders in this section. Compared with other dimensionality-reduction-based algorithms as a spectral community detection

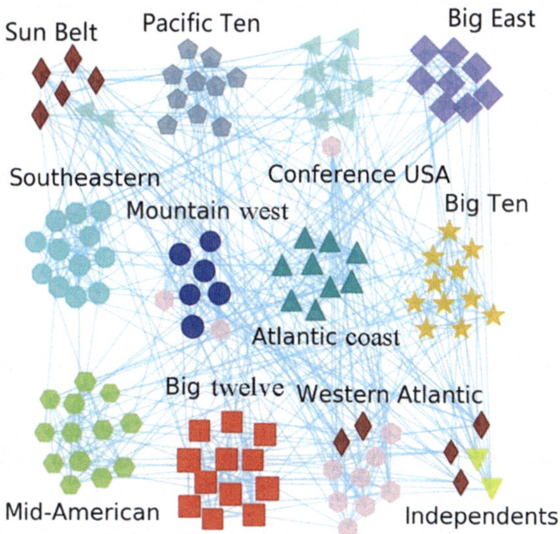

Fig. 5.18 Communities extracted from the US college football network using DAC

method, owing to the avoidance of implementing the notoriously time-consuming eigenvalue decomposition, the DAC undoubtedly executes faster. Experiments perform on artificial and real-world networks have demonstrated that DAC could be a competitive choice in the comprehensive consideration of both precision and efficiency. Despite of all its benefits, DAC also has some drawbacks that may serve as future discussions. The first is the hyperparameters that appear in the construction and training process of DAC. As in other neural networks, tuning them decisively by hand consumes a lot of time.

5.4 Local Aggregated Differential Evolution Algorithm

In this section, we present a differential evolution-based community detection algorithm with local aggregation ability called local aggregated differential evolution (LADE), which is lightweight and highly efficient both in design and implementation. Inspired by local search ideas we integrate a local aggregation operator into the original differential evolution algorithm to speed up the optimizing process converging to the maximal fitness, which is the modularity in our scenario.

5.4.1 Process of Local Aggregated Differential Evolution

Backing to 1995 an artificial intelligent optimization algorithm, named differential Evolution (DE) was published, which in pursuit of an individual evolved from a population pool resulting in a global optimum on behalf of the fitness function. During the past years, DE has evolved itself to a bunch of siblings and descendants that have prevailed in a great variety of application scenarios. Here we present a lightweight edition of DE, which we call it Local Aggregated Differential Evolution (LADE), to solve the community detection problem [9].

5.4.1.1 Initialization

We define the population pool as a matrix $\boldsymbol{X} \in \mathbb{R}^{N_P \times N}$. Each row is a candidate solution, called individual, to the community detection problem. Each item in an individual labels out which community that node belongs to, *i.e.* X_{ij} represents the group number that node j is put into by the ith individual. To start with, all the elements in $\boldsymbol{X}$ is initialized as a uniformly randomized integer within the searching space, which in our context ranges between 1 and N. So

$$X_{ij} = [1 + \text{rand}(0, 1) \cdot (N - 1)] \tag{5.44}$$

where $i = 1, 2, \ldots, N_{P,}$ $j = 1, 2, \ldots, N$, and rand(0,1) is a uniformly distributed random variable in range of [0, 1].

5.4.1.2 Local Aggregation

It is quite common to find that two connected nodes are coming from the same community. Taking this prior information into account, we manipulate the population in accordance with the adjacency matrix via an operation, what we call local aggregation, that for each individual $\boldsymbol{X}_i$ we randomly pick up a node and switch all its neighbors community labels to the same as that certain nodes. By the end of this reorganization procedure, the survived competitor is the individual who has the maximal modularity.

Algorithm 5.3 Algorithm of Local Aggregation
Step 1 for $i = 1$ **to** N_{P}
  Step 2 $\overline{\boldsymbol{\tau}} = \boldsymbol{X}_i$,$\boldsymbol{\tau}' = \boldsymbol{X}_i$
  Step 3 for $k = 1$ **to** N_{L}
  Step 4 let j be a randon integer from 1 to N
  Step 5 for $q = 1$ **to** N
  Step 6 if $A_{jq} = 1$
  Step 7 $\overline{\tau}_q = X_{ij}$
  Step 8 if $Q(\overline{\boldsymbol{\tau}}) \geq Q(\boldsymbol{\tau}')$
  Step 9 $\boldsymbol{X}_i = \overline{\boldsymbol{\tau}}$
  Step 10 if $Q(\overline{\boldsymbol{\tau}}) \leq Q(\boldsymbol{\tau}')$
  Step 11 $\boldsymbol{X}_i = \boldsymbol{\tau}'$
  Step 12 return $\boldsymbol{X}$

In Algorithm 5.3 the detail of local aggregation implementation has been given out, where two vectors $\overline{\tau}$ and τ' are used to save the temporary best resolutions. A constant parameter $N_L < N$ is defined as the iteration time for each individual's aggregation. Q is the modularity function corresponding to Eq. (5.44).

5.4.1.3 Mutation

Mutation diversifies the population through a way of combining individuals in the current generation pool $\boldsymbol{X}$ to create a new vector for each individual. The new generated vector is called the mutant vector $\boldsymbol{M}_i$, $i = 1, 2, \ldots, N_P$, and the corresponding new matrix is named mutant matrix $\boldsymbol{M}$. An implementation of mutation in LADE has been listed in Algorithm 5.4, in which the mutation strategy randomly picks up three individuals and fuses them into a new one. F is a constant positive scaling factor.

Algorithm 5.4 Algorithm of Mutation
Step 1 for $i = 1$ **to** N_{P}
  Step 2 let r_1, r_2, r_3 be mutually exclusive random integers from 1 to N_{P}
  Step 3 $M_i = X_{r1} + F \cdot (X_{r2} - X_{r3})$
  Step 4 return M

5.4.1.4 Crossover

After mutation, crossover operation comes out to make mixtures between original candidates in X and their mutated counterparts in M. What is left after the crossover process is a new trial matrix $\boldsymbol{T} = \{T_{ij}\}, i = 1, 2, \ldots, N_P, j = 1, 2, \ldots, N$, generated according to the binomial crossover strategy as follows:

$$T_{ij} = \begin{cases} M_{ij} & \text{if and } [0,1) \leq C \text{ or j } = j_{\text{rand}} \\ X_{ij} & \text{else.} \end{cases} \quad (5.45)$$

In Eq. (5.45), Crefers to a constant parameter named crossover rate that constrains the fraction of community labels inherited from M, j_{rand} is a random integer between 1 to N for each individual. Algorithm 5.5 shows an implementation of the binomial crossover strategy.

Algorithm 5.5 Algorithm of Crossover
Step 1 for $i = 1$ **to** N_{P}
  Step 2 let j_{rand} be a random integer from 1 to N
  Step 3 for $j = 1$ **to** N
  Step 4 let k be a random integer from 0 to 1
  Step 5 if $k \leq C$ **or** $j = j_{rand}$
  Step 6 $T_{ij} = M_{ij}$

Step 7 else
Step 8 $T_{ij} = X_{ij}$
Step 9 return $\boldsymbol{T}$

5.4.1.5 Selection

The selection operator has two jobs to deal with. For the first one, it reevaluates the modularity of each candidate in $\boldsymbol{T}$ and $\boldsymbol{X}$ in order to reassemble the next generation of population using the ones with larger fitness function outputs. After that, for the second purpose of the selection operator, the best candidate in the population is found and saved as the global optimal resolution of the community detection problem by the end of all the iterations of our algorithm.

Algorithm 5.6 Algorithm of Selection
Let $\boldsymbol{\tau}$ be the candidate in $\boldsymbol{X}$ that $Q_{\max} = Q(\boldsymbol{\tau})$ outweighs all the other vectors modularity
Step 1 for $i = 1$ **to** N_{P}
Step 2 if $Q(\boldsymbol{T}_i) \geq Q(\boldsymbol{X}_i)$
Step 3 $\boldsymbol{X}_i = \boldsymbol{T}_i$
Step 4 if $Q(\boldsymbol{T}_i) \geq Q_{\max}$
Step 5 $\boldsymbol{\tau} = \boldsymbol{T}_i, Q_{\max} = Q(\boldsymbol{T}_i)$
Step 6 return $\boldsymbol{X}$

In Algorithm 5.6, and $Q_{\max}$ should be defined as global variables outside the selection function.

The LADE algorithm terminates after a user specified N_{I} generations and gives out the best solution to the community detection problem. We provide a whole description of the LADE algorithm in Algorithm 5.7.

Algorithm 5.7 Algorithm of LADE
Step 1 initialize constant parameters including $N_{\mathrm{P}}, N_{\mathrm{L}}, N_{\mathrm{I}}, F$ and C
Step 2 let matrix $\boldsymbol{X} = X_{ij}, i = 1, 2, \ldots, N, X_{ij}$ are random integers from 1 to N
Step 3 for 1 **to** N_{I}
Step 4 $\boldsymbol{X} = \mathrm{LocalAggregate}\,(\boldsymbol{X}, \boldsymbol{A})$
Step 5 $\boldsymbol{M} = \mathrm{Mutation}(X)$
Step 6 $\boldsymbol{T} = \mathrm{Crossover}(\boldsymbol{X}, \boldsymbol{M})$
Step 7 let $\boldsymbol{\tau}$ be the candidate in $\boldsymbol{X}$ that $Q_{\max} = Q(\boldsymbol{\tau})$ outweighs all the other vectors modularity
Step 8 $Xs = \mathrm{Selection}(\boldsymbol{X}, \boldsymbol{T})$
Step 9 return $\boldsymbol{\tau}$

At the end of this section, let shed our light on the asymptotic complexity of our algorithm. Given a complex network with N vertices, it is not that difficult to calculate the worst-case time consumed by each part of the algorithm, which has been listed in Table 5.8.

Table 5.8 Complexity of LADE

Operations	Worst-case running time
Initialization	$O(N_P \cdot N) \sim O(N)$
Local aggregation	$O(N_P \cdot N_L[O(N) + O(N^2)]) \sim O(N^2)$
Mutation	$O(N_P \cdot O(N_P)) \sim O(1)$
Crossover	$O(N_P \cdot O(N)) \sim O(N)$
Selection	$O(N_P \cdot O(N^2)) \sim O(N^2)$
In all	$O(N^2)$

It is obvious that the complexity of modularity evaluation accounts for $O(N^2)$. The total complexity of LADE, which asymptotically is $O(N^2)$, consists of time wasted by initialization and N_I iterations of local aggregation, mutation, crossover and selection.

5.4.2 *Simulations and Results*

We deploy our algorithm LADE on computer generated networks and compare the performance with that of the other three famous competitors, including two fast greedy algorithms GN and NF, the modified genetic algorithm GAS, and the original differential evolution algorithm DE. Meanwhile, our algorithm has also been tested on four real world networks with certain number of community structures for the demonstration of the accuracy as well as on the other two real world networks whose community structures are undetermined to inspect the scalability of the algorithm. The constant parameters of LADE are set as $N_P = 50,\ N_L = 4,\ N_I = 1000,\ F = 0.5$, and $C = 0.3$ according to experiences.

5.4.2.1 Computer Generated Networks

The fabricate networks with community structures and NMI are same as the Sect. 5.1.2. And these computer-generated networks with decided graph knowledge are then fed into our algorithm and its competitors to calculate their NMIs, which is presented in Fig. 5.19.

In Fig. 5.19, for DE, GAS and LADE, each point represents the mean NMI value taken from 500 repeated experiments in circumstances of the computer-generated scenarios with predefined z_{out} varying from 0 to 8. Since GN and NF have decided outputs, there is no necessary to test them redundantly. Error bars accompanying those points mark out the normalized deviation of NMI in 500 experiments. Lines connecting points are included solely as a guide to read. We could see that our algorithm performs as well as GAS and much better than the GN, FN and DE, especially for large z_{out}. Notice that even when $z_{out} = 8$ LADE is still capable of identifying nearly 70% community labels of nodes correctly, which is slightly better

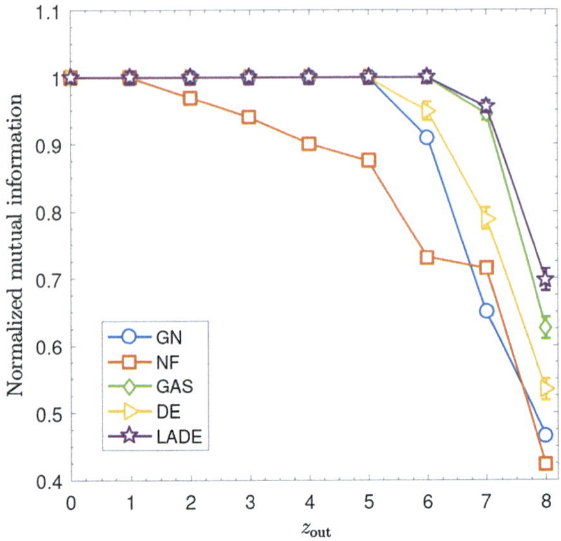

Fig. 5.19 NMIs of different algorithms on computer generated networks

Table 5.9 Comparison of NMI and Modularity (Q)

Dataset	Index	GN	NF	DE	GAS	LADE
Karate	Q	0.4013	0.3807	0.3624	0.3517	0.3723
	NMI	0.7072	0.6925	0.7692	0.8062	0.9837
Dolphins	Q	0.5196	0.4955	0.3658	0.3611	0.3804
	NMI	0.5016	0.5727	0.7629	0.8618	0.8891
Football	Q	0.5950	0.5682	0.6012	0.6026	0.6054
	NMI	0.4237	0.4436	0.7743	0.8832	0.8903
Polbooks	Q	0.5120	0.5020	0.4760	0.4873	0.4992
	NMI	0.4018	0.4396	0.5020	0.5366	0.7194

than GAS. The comparison LADE and DE also shows the improvement on the accuracy of our algorithm, which mainly attributes to the local aggregation operator.

5.4.2.2 Real World Networks

In this part, four real world networks as before have been selected as test beds for measurement of our algorithm's accuracy, and the other one popular web data set with obscure community partition information have been used to check out the detection scalability of the algorithm.

For GA, DE and LADE, we make 500 repeated tests of each algorithm in each real world network and record their best Q value achievements along with the corresponding NMI values, as illustrated in Table 5.9. For GN and NF, we test them with different predefined number of communities which ranges from 1 to 10 in networks of Karate, Dolphins and Polbooks, and from 5 to 15 in the Football network. The maximum Q value and the corresponding NMI values of GN and NF could also be found in Table 5.9.

5.4.2.3 Zachary's Karate Club Network

Figure 5.20 shows one partition result indicated by two distinct colors using our algorithm that totally corresponds to the real groups in which node 1 and 34 represent the instructor and the administrator respectively. As a probabilistic algorithm, the LADE gives out the 100% correct result accounting for 90% out of all the 50 experiments accompanied by the other five times with only one node misunderstood. Note that in Table 5.9 GN and NF have higher modularity than LADE because they separate the graph into more than two communities, which is not in accordance with the real situation.

5.4.2.4 Dolphin Social Network

In Fig. 5.21, we present one snapshot of our algorithm performance that, except for node 31, the two communities have been labeled out completely by two different colors. The average NMI of LADE in Table 5.9 also demonstrates the accuracy of our algorithm and the advancement compared with others.

5.4.2.5 US College Football Network

We display our detection result in Fig. 5.22 that ten communities corresponding to ten colors have been recognized with some minority conferences being absorbed

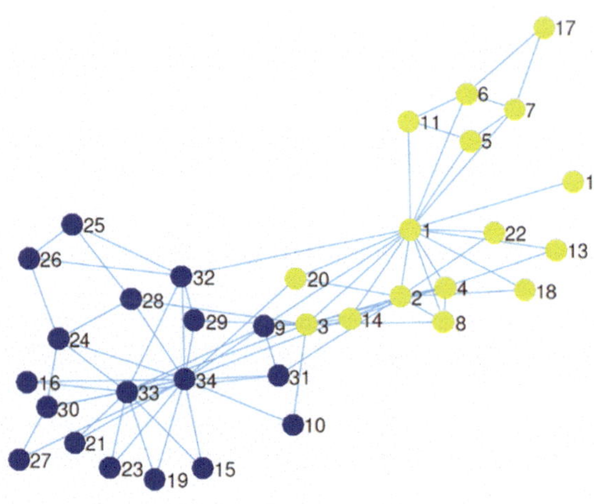

Fig. 5.20 Communities extracted from the Zacharys karate club network using LADE

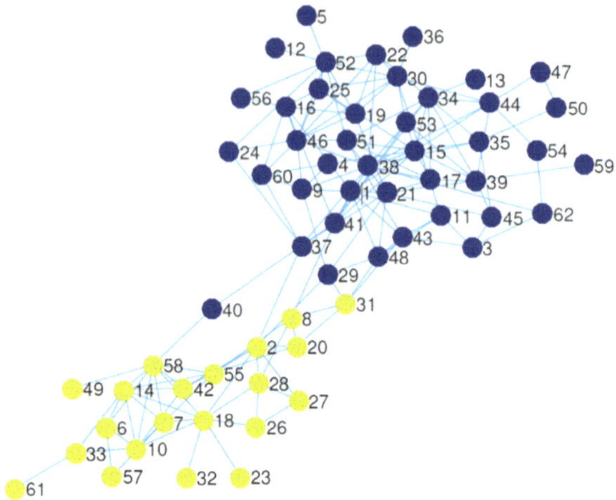

Fig. 5.21 Communities extracted from the dolphin social network using LADE

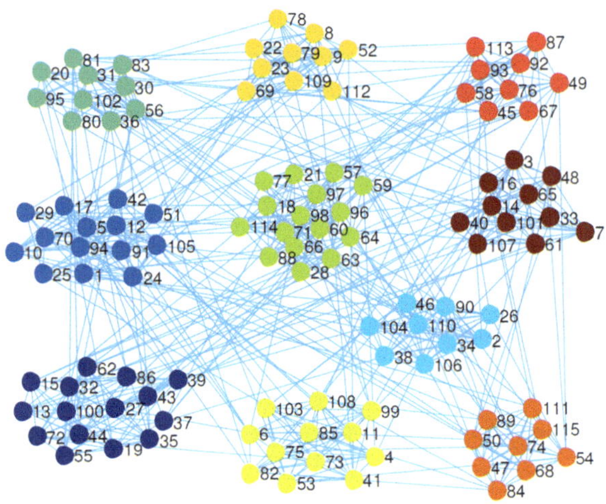

Fig. 5.22 Communities extracted from the US college football network using LADE

into big ones. From Table 5.9 we see that despite of those misunderstanding our algorithm still outperforms the others in both modularity and NMI indexes.

5.4.2.6 Political Books Network

The identification of our algorithm in Fig. 5.23 shows that people tend to have relatively stable opinion on politics which is indicated by the obvious segregation of nodes. Blue and green colors represent the conservative and liberal group respectively while the yellow color indicates the neutral part which occupies a small number of nodes. However, it is not very uncommon that some neutral books look like more liberal or conservative in eyes of some critical readers, which explains why

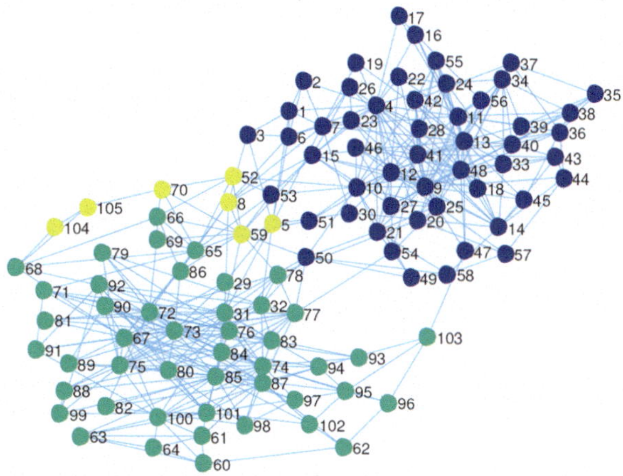

Fig. 5.23 Communities extracted from the political books network using LADE

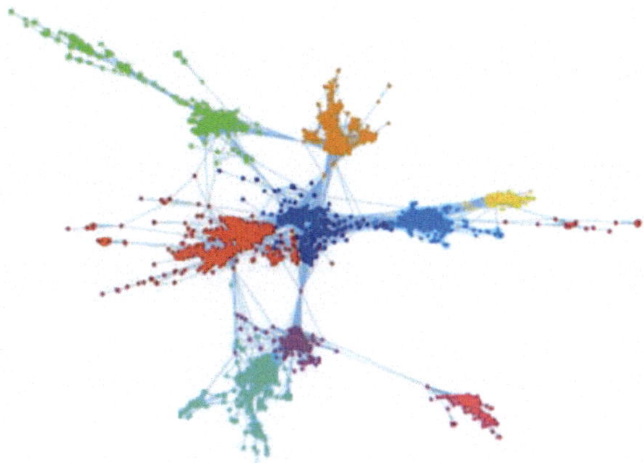

Fig. 5.24 Community structure extracted from the Facebook network using LADE

our algorithm makes most of mistakes on nodes belonging to the most small-scale community. The modularity and NMI results in Table 5.9 also demonstrate the accuracy of our algorithm which outweighs the other methods.

5.4.2.7 Facebook Networks

To validate the effectivity of this algorithm, we choose a so-called ego network composed of circles on Facebook, where people share common features like hometown, political affiliations, education background etc.

There are 4039 nodes representing anonymous user ids and 88,234 edges denoting interactions among those people. The clear number of communities in this network is not known, which gives us an opportunity to apply our algorithm on it for the scalability testament. Figure 5.24 shows the result of the community detection process using our algorithm on this dataset, getting rid of some isolated nodes having no connection to the main part of the network. Ten communities corresponding to the real circles have been extracted and marked by different colors, and the modularity reaches 0.7779 which illustrates a relative high density of community characteristics. Nevertheless, since the accurate community structure is unknown, there may still exist some small delicate groups inside the discovered communities, which leaves us an open space of interest for the future work.

5.5 Robustness and Fragility of Tree-Based Wireless Sensor Networks

The robustness and fragility of WSNs, whose underlying architecture patterns crucially affect the system functionality, is defined as the resilience to deletion of network nodes. It is equivalent to be a percolation model on a graph depicting the network. In this section, based on the analysis for robustness and fragility of network, we have derived an analytic solution of the site percolation model using generating function formalism on a Cayley-Tree based wireless sensor network structure with community features. Our solution provides predictions for quantities including the position of the percolation threshold and the average cluster size as functions of rewiring probabilities and occupation proportion respectively [10].

5.5.1 Cayley-Tree Based Wireless Sensor Networks and Model

Tree-based topology is a substantial and commonsense structure when constructing the wireless sensor network. Ever since the prevalence of LEACH (Low Energy Adaptive Clustering Hierarchy) protocol many of its famous descendants and competitors like SCT (Semantic/Spatial correlation-aware tree) and MST (minimal spanning tree) etc. have been coming out. Among these researches there has been one specifically attractive branch: Cayley-Tree based wireless sensor network structures.

We study the Cayley-Tree based network with community characteristics which means densely connected subsets of nodes with only sparser connections between them. Figure 5.25 shows a case in which there are two communities composed by two-layer Cayley-Trees. The underlying deployment of sensors within each community is a Cayley-Tree, whose definition is a simple connected undirected graphs $G = (V, E)$ (V is set of vertices and E is set of edges) with no cycles (a cycle is a closed path of different edges). A Cayley-Tree of order $K \geq 2$ with L levels means given a root vertex node 0, we link node 0 with K new vertices by means of K edges. This set of K vertices constitutes the level one of the Cayley-Tree (the origin node is on level 0). Then, to build up the $L \geq 2$ levels, each vertex on level one is linked to

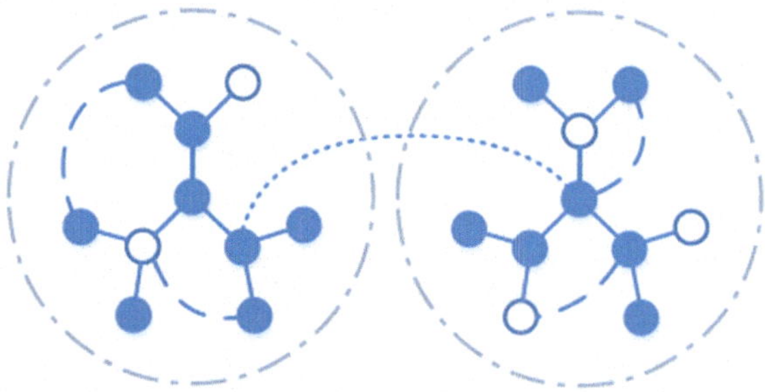

Fig. 5.25 An example of two communities built upon Cayley-Trees with order $K = 3$ and level $L = 2$.

$K - 1$ new vertices. Note that the vertices on the last level have degree $k = 1$, while all the other vertices have degree $k = K, L = 0,1,2, \ldots, n - 1$.

For any finite tree it holds

$$|V| = |E| + 1 \tag{5.46}$$

and for any graph the average degree is

$$\langle k \rangle = 2|E|/|V| \tag{5.47}$$

Therefore, for Cayley-Tree

$$\langle k \rangle = 2 - 2/|V| \tag{5.48}$$

in light of which the number of nodes in each community has been set to be equilibrium in our networks.

Given a Cayley-Tree of $L \geq 2$, the number of vertices on the level L is

$$N_L = K(K-1)^{L-1} \tag{5.49}$$

while the total number of vertices in the Cayley-Tree is

$$N_0 = \frac{K(K-1)^{L-1}}{K-2} + 1 \tag{5.50}$$

In our scenario each Cayley-Tree constructs a community abstracted underlying architecture. There are N_c communities each with N_0 nodes inside, so the total number of nodes in our network is

$$N = N_c N_0 \tag{5.51}$$

Random links are then added to the underlying network following the Newman-Watts style, which means we add with probability ϕ one shortcut for each bond on the original graph and explicitly allow there to be more than one bond between any two vertices, or a bond, which connects a vertex to itself. The shortcuts have two folds. The first one is the random links within the community, and the second one connects nodes from different groups.

Given two nodes i, j and a sub graph $V \subset G$, we define the rewiring probability as

$$\phi = \begin{cases} \phi_{\text{in}} & \forall i,j \in V \\ \phi_{\text{out}} & \forall i \in V, j \notin V \end{cases} \tag{5.52}$$

In Fig. 5.25, the dark nodes could be seen as susceptible sites, meaning it would be infected after being exposed to the worm virus, in perspective of network fragility to virus propagation. Under the circumstance of the network robustness, the dark ones may represent the working nodes while the white nodes are those dysfunctional sites. A node is occupied with probability p or not with probability $q = 1 - p$. Since the distribution of occupied sites is random, our problem of when the prevalence of cyber-attack takes place or at which moment the working network would be sabotaged to breakdown then becomes equivalent to a standard percolation problem: what fraction p must be to form a "giant component" of connected sites whose size scales extensively with the total number of sites. We use p_c to denote the percolation threshold. If $p > p_c$, the virus is capable of spreading throughout the network and infect all the susceptible nodes, or speaking of the robustness there still exists some tolerance for the mortality of nodes. Alternatively, for $p < p_c$, the virus dies out naturally after affecting a small portion of susceptible nodes, while on behalf of the robustness it means the network is no longer completely working well.

5.5.2 *Percolation and Generating Function Formalism*

We define a probability generating function $H(z)$ that

$$H(z) = \sum_{n=0}^{\infty} P(n) z^n \tag{5.53}$$

where $P(n)$ is the probability that a randomly chosen site belongs to a connected cluster of n sites other than the cluster inside communities. Note that if the probability distribution $P(n)$ is normalized properly, then $H(1) = 1$ below the transition and $H(1) = 1 - S$ above the transition, in which S is the fraction of the system occupied by the cluster inside communities.

We also define $P_0(n)$ to be the probability of a randomly chosen site belonging to a local cluster of size n. For our Cayley-Tree based underlying it has been proved that

$$P_0(n) = \begin{cases} 1-p & n=0 \\ np^n q^{2+(K-2)n} & 1 \le n \le \dfrac{N}{N_c} \end{cases} \tag{5.54}$$

when N is large enough.

The largest complete cluster throughout the whole network is composed of a set of such clusters inside communities, joined together by occupied shortcut bonds inner and inter communities. If we define $P(m|n)$ to be the probability that a cluster inside communities of n sites emitting exactly m shortcuts from it, then we can write the generating function $H(z)$ self-consistently as

$$H(z) = \sum_{n=0}^{\infty} P_0(n) z^n \sum_{m=0}^{\infty} P(m|n)(H(z))^m \tag{5.55}$$

Note that the derivation of Eq. (5.55) requires an assumption that all clusters excluding the percolating cluster, if there exists one, endure no closed loops except for those on the underlying structure, which means there are no loops encapsulating shortcut bonds. It is only strictly true in the limit of very large system size, and hence our results can be exact only in large scale WSNs.

In total, there are $N_s = 2(\phi_{\text{in}} + \phi_{\text{out}})(N - N_c)$ ends of occupied shortcuts in the model. Since all ends are uniformly distributed throughout the whole network, the probability that an end lands in a given cluster of size n is just n/N for very large N. Thus $P(m|n)$ could be given by the binomial distribution

$$P(m|n) = \binom{N_s}{m} \left(\frac{n}{N}\right)^m \left(1 - \frac{n}{N}\right)^{N_s - m} \tag{5.56}$$

Substituting Eq. (5.56) into Eq. (5.55) and sum over m, it gives

$$\begin{aligned} H(z) &= \sum_{n=0}^{\infty} P_0(n) z^n \left(1 + (H(z) - 1)\frac{n}{N}\right)^{N_s} \\ &= \sum_{n=0}^{\infty} P_0(n) z^{2n} e^{2n(\phi_{\text{in}} + \phi_{\text{out}})(H(z)-1)} \end{aligned} \tag{5.57}$$

where the last equality holds only for very large N.

Then we define an additional generating function

$$H_0(z) = \sum_{n=0}^{\infty} P_0(n) z^n \tag{5.58}$$

which is the generating function for the probability of cluster sizes for ordinary site percolation inside communities.

Accompanying with Eq. (5.54), we derive

$$H_0(z) = q + \frac{pq^K z}{(1 - pq^{K-2z})^2} \tag{5.59}$$

As a result, Eq. (5.57) could be written as

$$H(z) = H_0\left(z \cdot e^{2(\phi_{\text{in}} + \phi_{\text{out}})(H(z)-1)}\right) \tag{5.60}$$

which provides a self-consistency formation to evaluate $H(z)$. Consequently, it is available for us to calculate the probabilities $P(n)$ for cluster sizes in certain networks.

In substance, although it is possible to evaluate $H(z)$ by numerical iteration, solving Eq. (5.60) for $H(z)$ in closed form is often infeasible. However, we can derive other closed-form expressions for quantities of interest. Particularly, the average cluster size is given by

$$\begin{aligned} \langle n \rangle &= \sum_{n=0}^{\infty} nP(n) = H'(1) \\ &= H_0'(1)(1 + 2(\phi_{\text{in}} + \phi_{\text{out}})H'(1)) \end{aligned} \tag{5.61}$$

Rearranging Eq. (5.61) we derive

$$\begin{aligned} \langle n \rangle &= \frac{H_0'(1)}{1 - 2(\phi_{\text{in}} + \phi_{\text{out}})H_0'(1)} \\ &= \frac{pq^K(1 + pq^{K-1})}{(1 - pq^{K-2})^3 - 2(\phi_{\text{in}} + \phi_{\text{out}})pq^K(1 + pq^{K-2})} \end{aligned} \tag{5.62}$$

This quantity diverges when

$$2\ (\phi_{\text{in}} + \phi_{\text{out}})H_o'(1)H_o'(1) = 1 \tag{5.63}$$

or equivalently when

$$\phi_{\text{in}} + \phi_{\text{out}} = \frac{\left(1 - p_c(1 - p_c)^{K-2}\right)^3}{2p_c(1 - p_c)^K\left(1 + p_c(1 - p_c)^{K-2}\right)} \tag{5.64}$$

This is the phase transition moment at which a largest cluster first appears to scale as the size of the entire network.

5.5.3 Simulations and Results

We have undertaken extensive computer experiments of percolation and virus propagation on the Cayley-Tree based network with community characteristics, both to check our analytic results, and as a further investigation on the behavior of the model.

Firstly, we measured the percolation threshold positions and performed comparisons with our analytic results. The basic idea for doing this simulation is to fill in the sites one by one randomly and to calculate the size of either the average or the largest cluster of connected sites at each step. Positions of percolation thresholds could then be estimated from the points at which the derivatives of these sizes with respect to the numbers of occupied sites take their maximum values.

In Fig. 5.26, we show numerical simulation results for p_c from calculations of the largest cluster size with the variation of ϕ_{in} while ϕ_{out} is fixed to different probabilities for system of more than six thousand sites located in two communities with Cayley-Tree order $K = 3$. Figure 5.27 is the situation where ϕ_{in} is fixed while ϕ_{out} changes its values. As Fig. 5.27 shows, the results agree with our analytic expressions in Eq. (5.64) even though there are some relatively small fluctuations. The larger ϕ_{in} and ϕ_{out} are, meaning there are more fluent communications, the smaller proportion of nodes need to be occupied to percolate the network.

In accordance with Eq. (5.64), Figs. 5.26 and 5.27 also indicate the symmetrical character of the relationship between the percolation threshold p_c and the two different kinds of shortcut rewiring probabilities ϕ_{in} and ϕ_{out}. Even as $\phi_{\text{out}} = 0$ means there is no connection between two communities, stochastically speaking they still affect the percolation process symmetrically. As a consequence of such symmetry, it is not compulsory for us to launch all the same quantities comparison on different Cayley-Tree orders. Instead we pick out the function between p_c and ϕ_{in} with ϕ_{out} fixed to 0.5 as a demonstration in Fig. 5.28 being in consistent with our

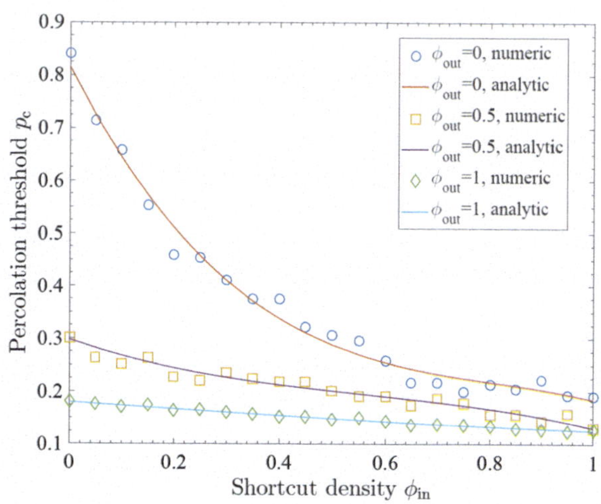

Fig. 5.26 The percolation threshold positions as a function of shortcut density ϕ_{in}

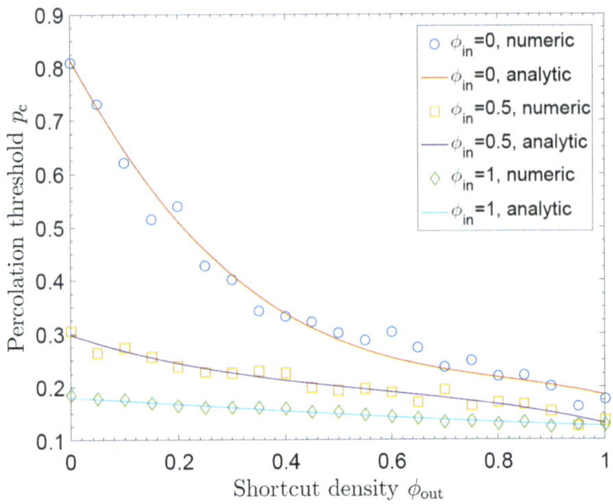

Fig. 5.27 The percolation threshold positions as a function of shortcut density ϕ_{out}

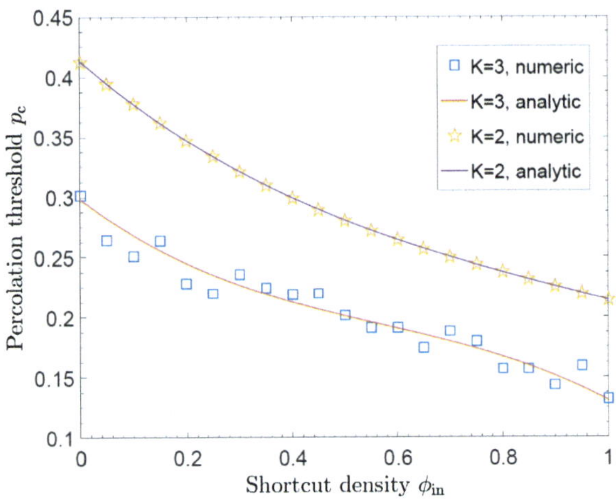

Fig. 5.28 The percolation threshold positions as a function of shortcut density ϕ_{in} with ϕ_{out} fixed to 0.5

instinct, larger Kcreates more opportunities for transmissions between nodes and leads to lower percolation threshold. Note that for $K = 2$ the topology is becoming a chain, in which each node is connected with two nearest neighbors except for the head and the tail. This case is easy to find in WSNs routing protocols and has been thoroughly studied in the past articles. On the other hand, for $K \geq 4$ the network scale grows into enormous that the overhead of the calculation has to be taken into account, and hence we leave it here as an open problem for readers who have interests (and the courage).

In Figs. 5.29 and 5.30, we show the fitness of average cluster size$\langle n \rangle$ as a function of occupation proportion p. It is analyzed by Eq. (5.62) against numerical results calculated again and each point is the average of 1000 runs. We deploy the

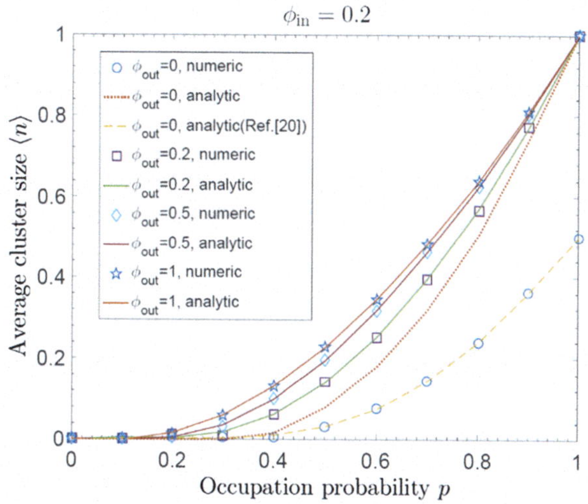

Fig. 5.29 The average cluster size as a function of the occupation probability for percolation process

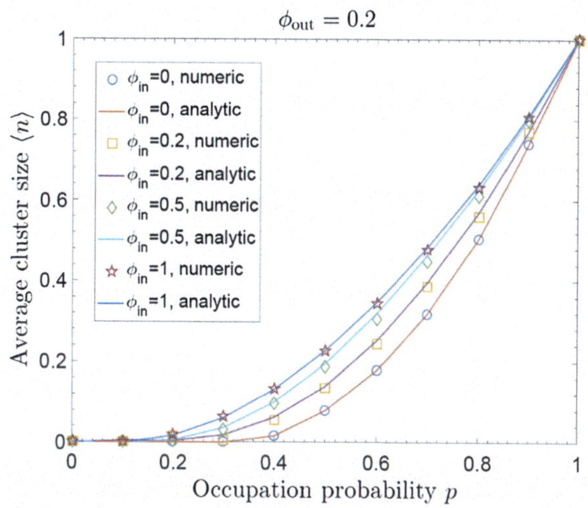

Fig. 5.30 The result for $<n>$ as a function of p

experiment with one of ϕ_{in} and ϕ_{out} fixed to 0.2 and the other one varies from 0 to 1 on the same underlying topology as in Fig. 5.26 ($K = 3$, $L = 10$ and $N_c = 2$).

As shown in Figs. 5.29 and 5.30, the agreement is excellent except for a singularity in which $\phi_{out} = 0$ leading to the disconnection between the two communities. When $\phi_{out} = 0$, our model transforms to be two independent Cayley-Trees with shortcut inside, which is equilibrium to the small-world Cayley-Tree model with ϕ_{in} being its rewiring probability. Obviously, without the assistance of inter communities' connections the largest cluster would never overwhelm the whole network, which means the largest cluster number is impossible to jump over N_0. In Fig. 5.29 we see the analyzed result derived according to small-world Cayley-Tree

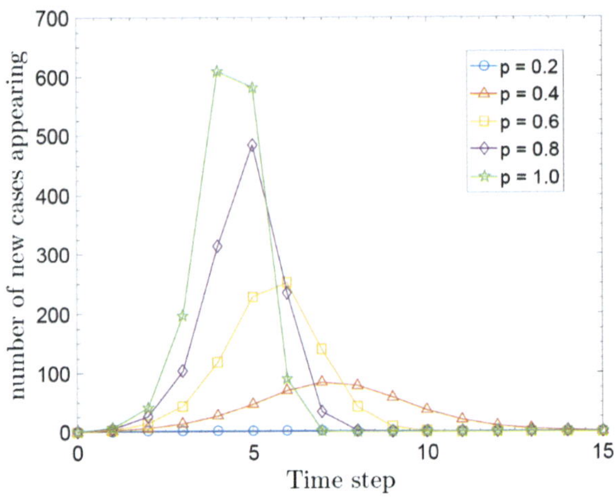

Fig. 5.31 The number of new appearing cases affected by the virus as a function of time step

model (dash line) corresponds to the numerical points when $\phi_{\text{out}} = 0$ as well as all the other correspondences where our formulation Eq. (5–62) fits with all the numerical results. Moreover, the symmetrical character exists between Figs. 5.29 and 5.30 again.

On behalf of network robustness, the percolation threshold p_c is the point which marks the network achieving its large-scale connectivity, and furthermore, functioning as an effective distribution network. In the language of network fragility to worm propagation p_c is the moment at which an epidemic of the virus first occurs and the sizes of clusters are equivalent to the sizes of outbreaks of a virus among susceptible nodes. In contrast, if we are approaching the transition place from values of p above p_c, it is the point where a sufficient number of nodes have been removed from a distribution network to prevent the large-scale distribution, or the point at which a large enough number of nodes are immune to the virus to prevent it from spreading throughout the whole network. Direct confirmation that the percolation places in our model are indeed in agreement with the threshold at which the prevalence of virus first occurs can also be illustrated by numerical simulations.

Figure 5.31 shows results for the number of new appearing victims of a cyber-attack as a function of time step in simulations of our site percolation model. In these simulations we configured the system as $K = 3$, $L = 8$, $N_c = 2$, $\phi_{\text{in}} = 0.6$ and $\phi_{\text{out}} = 0.2$, which means, according to Eq. (5–64), there should be epidemics once the proportion of susceptible nodes exceeds $p_c = 0.258$. At the very beginning we set only one node infected in the network and during each time step all the susceptible neighbors of the contaminated nodes are becoming infected.

The points in Fig. 5.31 are for p is 0.2, 0.4, 0.6, 0.8 and 1.0 respectively from the bottom upwards. As we can see, when $p = 0.2$ the number of new appearing cases of the virus almost shows no peak of activity before the virus fizzles out, which is barely visible along the horizontal axis in the figure. Once the occupation proportion is above the percolation threshold, as expected, a large number of new cases appear

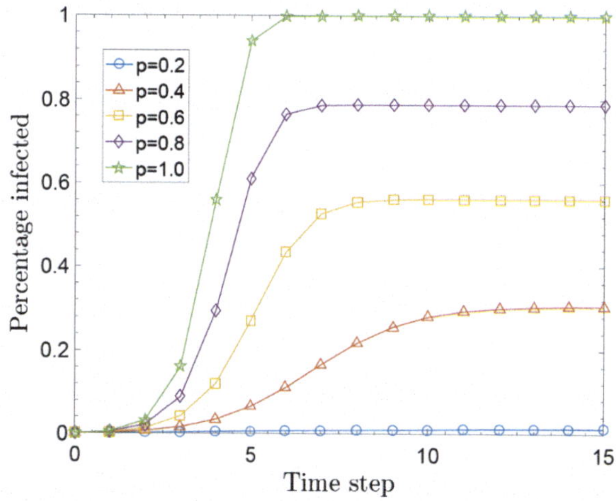

Fig. 5.32 The percentage of the infected nodes as a function of time step

which indicates the onset of an epidemic behavior. As the simulations depict, the proportion of susceptible nodes affected by the prevalence of virus is typically between 50% and 100% (the upper four curves), compared with only about 5% in the sub-critical case (the other curve). It also indicates a significant tendency that a higher susceptibility could speed up the spread of virus, which correspond with our intrinsic explanation that more susceptible nodes mean an infection has more chances to change an uninfected node into an infected one.

In Fig. 5.32 we show the total percentage of the nodes affected by the virus as a function of time step in the same simulations as Fig. 5.31. It is obviously to tell a sigmoidal form taken by this quantity, which shares a similarity as in random-graph models, small-world disease models, and in real-world data.

Generating function formalism has been used to derive exact solutions to the percolation model on Cayley-Tree based WSNs with community characteristics. Such percolation systems on graphs have been proposed as models for the robustness of networks to breakdown or sabotage, and for the fragility to cyber-attack. The exact analytic expressions for the percolation threshold and average cluster size we derived are in consistent with their numerical simulation counterparts. We have also performed extensive numerical simulations of virus prevalence to conform the percolation threshold positions as well as the fact that outbreaks only take place above these thresholds.

5.6 Conclusion

In this chapter, we present several community detection methods based on methods of support vector clustering and proximity graph modelling. Based on the traditional support vector clustering methods, we introduce four new algorithms, PSVC,

DA-ELM, DAC and LADE. Meanwhile, generating function formalism has been presented to derive exact solutions to the percolation model on Cayley-Tree based WSNs with community characteristics.

References

1. Pettie S, Ramachandran V. A randomized time-work optimal parallel algorithm for finding a minimum spanning forest[J]. Siam Journal on Computing, 2002, 31(6): 1879-1895.
2. Wang F, Zhang B, et al. Community detection in complex networks using proximate support vector clustering[J]. Modern Physics Letters, B. Condensed Matter Physics, Statistical Physics, Applied Physics, 2018: 1850101.
3. Wang F, Zhang B, Chai S, et al. Lightweight support vector clustering algorithm for community detection in complex networks[C]. The 37th Chinese Control Conference. 2018, 2317-2322.
4. Newman M E J. Community detection in networks: modularity optimization and maximum likelihood are equivalent[J]. Physical Review E Statistical Nonlinear & Soft Matter Physics, 2016, 94: 052315.
5. Zachary W W. An information flow model for conflict and fission in small groups[J]. Journal of Anthropological Research, 1977, 33(4): 452-473.
6. Wang F, Zhang B, et al. Community detection in complex networks using deep auto-encoded extreme learning machine[J]. Modern Physics Letters, B. Condensed Matter Physics, Statistical Physics, Applied Physics, 2018, 1850180.
7. Wang F, Zhang B, Chai S, Xia Y. An extreme learning machine-based community detection algorithm in complex networks[J]. Complexity, 2018: 1-10.
8. Wang F, Zhang B, Chai S. Deep auto-encoded clustering algorithm for community detection in complex networks[J]. Chinese Journal of Electronics, 2019, 28(03): 489-496.
9. Wang F, Zhang B, Chai S, et al. Local aggregated differential evolution algorithm for community detection in complex networks[C]. The 37th Chinese Control Conference. 2018, 2384-2389.
10. Wang F, Zhang B, et al. A study on the robustness and fragility of tree-based wireless sensor networks with community characteristics[C]. IEEE International Conference on Unmanned Systems, 2017, 307-312.

Chapter 6
Small World and Information Dissemination

The small world phenomena is initially investigated in the sociology that individuals are often linked by a short chain of acquaintances. S. Milgram and his group conduct a series of mail delivery experiments and find that an average of six degrees of separation exists between senders and receivers. Recent researchers have shown that the small worlds are also observed in wireless sensor networks. In this chapter, we discuss the epidemic and immunization process in small world of wireless sensor networks and mainly centre on the tree topologies. We introduce some basic characteristics of small world networks and analyze epidemiogical process in small-world networks. Meanwhile, we present immunization strategies during immunization process.

6.1 Tree Topologies of Small World

Tree topology is a kind of architecture that is ubiquitous in the deployment of wireless sensor nodes. In this section, we specifically introduce the small world phenomena in the tree topologies. The characteristics of time synchronization of tree topologies is discussed as well.

6.1.1 Topologies of Small World

Recently, many topologies of WSNs based on small world have been proposed. For different requirements in applications of WSNs, there are many methods to construct WSN topological models on small world phenomena. Construction methods based on Newman-Watts (NW) small-world models, by adding random links between nodes, greatly reduce the average path length between nodes. Construction methods

S. Chai et al., *Wireless Sensor Networks*, Wireless Networks,
https://doi.org/10.1007/978-981-15-5757-6_6

of small-world WSNs based on sink nodes, such as Directed Angulation Toward the Sink Node Model (DAS) and Sink Node as Source/Destination Model (SSD), make lower energy consumption and delay during the data transition between the nodes. Construction method based on super-link, super-link data can achieve the reliable transmission shortcut to the sink node. The method based on topology optimization can significantly improve the clustering coefficient, reduce interference conflicts between networks and improve the communication efficiency across networks. In addition, there are many other methods to construct WSNs model based on small world phenomena, such as dynamic construction method, etc.

The topologies of wireless sensor networks do not belong to the category of relational graphs, so the α-model, which falls into the category of sociology, can't be used directly to analyze the small world characteristics of wireless sensor networks. In fact, they belong to the category of spatial graphs, where the link between two nodes depends on the node's radio range and the distance between two nodes. With more and more efficient ways of deployment for sensor nodes rising into view, Tree topologies, which construct spatial graphs with larger characteristic path lengths than random graphs and small clustering coefficients, are ubiquitous in wireless sensor networks [4]. Suffering from the link rewiring or the link addition, the characteristic path length of the tree topology reduces rapidly and the clustering coefficient increases greatly, which leads to a strong connection to the small world phenomena. Obviously, the time synchronization characteristics of wireless sensor networks are greatly influenced by characteristic path length varying.

6.1.2 *Topological Characteristics of Tree Topologies*

Small world networks between regular lattice and completely random graph areoriginally proposed by Watts and Strogaze. Its most notable characteristics are the small path length and the large clustering coefficient. These features of small-world network are also introduced into WSNs. The small path length can reduce the information dissemination in network, while the large clustering coefficient reduces the number of isolated clusters. It significantly improves the cover result and the life of networks. Tree topology is one of the main topologies of WSNs, and small world phenomena occurs in WSNs based on tree topology universally.

The small world phenomena existing in the tree topologies during the network worktime suffers from the influence of some random factors. Obstacles, adjustments of the radio energy, joins of new members and errors of the location precision all incur the link rewiring or the link addition in the tree topologies. The characteristic path length and the clustering coefficient vary according to the variety of the link rewiring or the link addition probability.

The characteristic path length or average distance in entire graph, denoted by $L(G)$, is the median of the average path lengths between any two vertices of the graph G. Since any vertex i in the connected graph G of size n can reach any other

$n-1$ vertex, an average distance between i and any other vertex j, denoted by $L(i)$, is given by the following equation.

$$L(i) = \frac{\sum_{j \in V(G)} d(i,j)}{n - 1} \tag{6.1}$$

where $V(G)$ is the set of the vertices of the graph G, $d(i,j)$ is the distance between vertex i and vertex j. The characteristic path length $L(G)$ in the connected graph is computed from the median of $\{L(i),\ i = 1,2\ldots n\}$. The characteristic path length reflects the global structure of a network graph.

The clustering coefficient denoted by C indicates the proportion of the neighborhood of vertex $i(\Gamma(i))$ adjacent to each other. Let the number of edges in the neighborhood of vertex i be $|E(\Gamma(i))|$. Each vertex i has a number of edges or degree k_i. The total number of possible edges in $(\Gamma(i))$ is $\binom{k_i}{2}$. Only those vertices in the neighborhood that have at least two edges are included. The clustering coefficient of vertex i is as follows:

$$C(i) = \frac{|E(\Gamma(i))|}{\binom{k_i}{2}}, \quad \forall i \in V'(G) \tag{6.2}$$

where $i \in V(G) \equiv \{i \quad \in V(G), \quad k_i \geq 2\}$. The clustering coefficient C of the whole graph or network, is the proportion of total actual connections within all vertices' neighborhoods relative to all possible connections between vertices in the graph. The value of C should be equal to the averaged $C(i)$ for each vertex in the graph. $C = 0$ would imply an empty graph and mean that no neighbor of any vertex i is directly connected to any other neighbor of i. On the other hand, $C = 1$ would imply a complete $\frac{n}{k-1}$ subgraph, or clique, in which every vertex is linked to every other vertex directly. The average degree of all vertices in graph G, denoted by k, is defined as follows:

$$k = \sum_{i \in V(G)} \frac{|V_i(G)|}{n} \tag{6.3}$$

The degree of vertex i is $k_i = |V_i(G)|$.

We conduct several numerical experiments to explore the basic characteristics, like characteristic path length and clustering coefficient of the small world phenonmena in different tree topologies.

In the experiments, link additions are conducted in the topologies of LEACH and Tree Based Energy Efficient Protocol for Sensor Information (TREEPSI). Thousand experiments are conducted and the results are analyzed. In the figures, the logarithmic coordinate is used for x-coordinate.

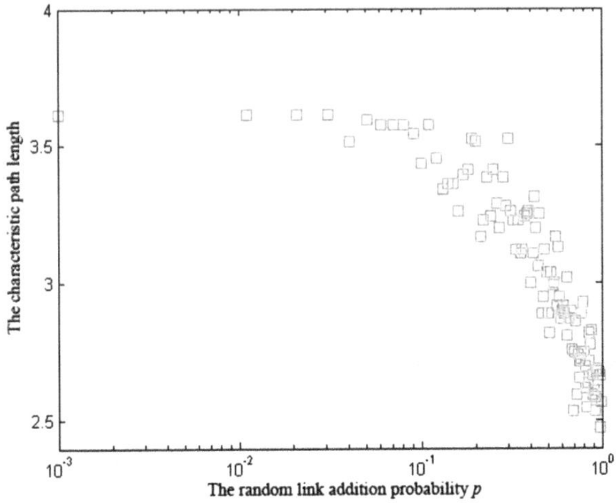

Fig. 6.1 The characteristic path length of the LEACH tree

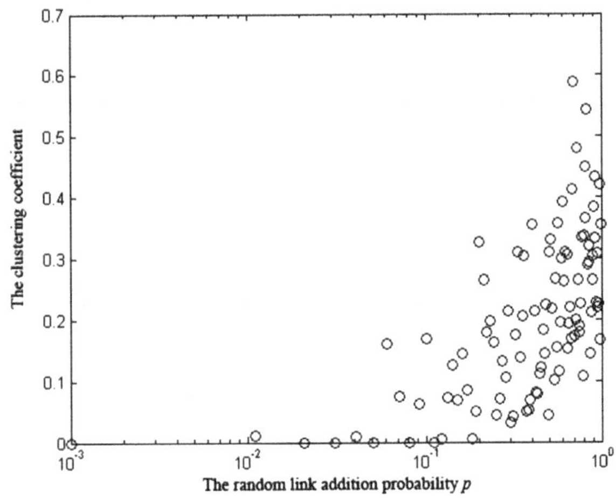

Fig. 6.2 The clustering coefficient of the LEACH tree

In the first experiment for the small world phenomena in the LEACH tree, 100 nodes are distributed randomly in a 100×100 rectangle field and no more than 6 cluster-heads are selected out. When the link addition probability p increases from 0 to 1, the variety of the characteristic path length or the clustering coefficient is shown in Fig. 6.1 or Fig. 6.2 respectively.

From Fig. 6.1 we can see that in this distribution situation the characteristic path length $L(G)$ of the graph is 3.65 at the initial state. Then it decreases with the increase of p. When $p = 1$, $L(G)$ is 2.45, which decreases by about 33% compared with the initial value when $p = 0$. When $p = 0$, the graph is a two-dimensional regular graph and it has large characteristic path length. When $p = 1$, the graph transforms into a random graph. At this time, the tree understratum can be neglected and it has a small characteristic path length. When $0 < p < 1$, the state of the graph is between the regular graph and the random graph, and the small world phenomena take place now. When p increases from 0 to 0.1 the characteristic path length of the graph remains unchanged. When p increases from 0.1 to 1, it decreases greatly. From the figure we can see that the characteristic path length of the graph varies mainly when p is between 0.1 and 1.

If we regard the graph with a large characteristic path length as a large graph and the graph with a small characteristic path length is a small graph, the topology experiences a transformation from a large graph to a small graph with the increase of the link addition probability p. At the same time, it transforms from a regular graph to a random graph. This disorder influences the length characteristic greatly.

From Fig. 6.2 we can see that the variety of the clustering coefficient of the LEACH tree. The clustering coefficient of the graph is 0 at the initial state for the special topology structure. It increases in the main trend when p increases. When p increases from 0 to 0.1 the clustering coefficient of the graph remains unchanged. When p increases from 0.1 to 1 it increases greatly. The clustering coefficient of the graph varies mainly when p is between 0.1 and 1 as well.

Different from the ring understratum, the clustering coefficient keeps on increasing when p is between 0.1 and 1. The graph transforms from a regular graph to a random graph in this process. Between the two extremes, the topology is in disorder. Compared with Fig. 6.1, we see that the graph has large characteristic path length like a regular graph and large clustering coefficient like a random graph simultaneously in the middle state. The graph transforms into a small world now.

The topology of the LEACH tree is a regular graph, which constructs a two-hop tree. For the member node, or the leaf node, its degree is 1 in the regular graph. The degree of a cluster-head node is $m + 1$ (m is the number of member nodes in its cluster). The degree of the base station is w (w is the number of cluster-heads). The clustering coefficient of every node is 0 because no neighbor of any node i is directly connected to any other neighbor of i. Then the clustering coefficient C of the regular graph is 0. The base station is a key point in the topology, without which the regular graph transforms into a forest.

The experiment results show that the small world phenomena exist in the LEACH tree, and the characteristic path length or the clustering coefficient suffers from the influence of the random link addition. But the TREEPSI tree, with multi-hop, multi-cluster and plenty of branches, suffers from random factors. Different from the LEACH tree, it falls into disorder sometimes.

The experiments below explore characteristics of TREEPSI, we can compare the differences between LEACH and TREEPSI through these two experiments. In this experiment, 1000 nodes are distributed in a 1000×1000 rectangle field. TREEPSI is conducted and a multi-hop tree topology is constructed. Suffering from the influence of some random factors, the small world phenomena exist during the work of the TREEPSI like LEACH. The link rewiring or the link addition takes place. They come from the communication between the parent node and its children nodes or two children nodes.

Figure 6.3 shows the variety of the characteristic path length of the TREEPSI tree when p increases from 0 to 1. When $p = 0$, the characteristic path length of the graph

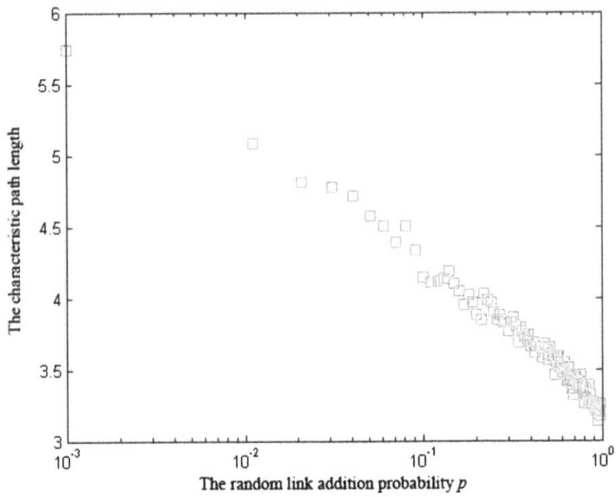

Fig. 6.3 The characteristic path length of the TREEPSI tree. (The topology experiences a transformation from a 'large' graph to a 'small' graph with the increase of the link addition probability p. It transforms from a regular graph to a random graph)

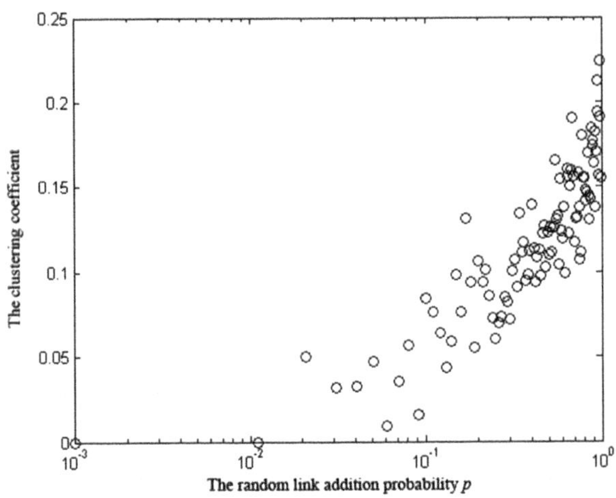

Fig. 6.4 The clustering coefficient of the TREEPSI tree. (The clustering coefficient increases clearly with the increase of the link addition probability p. For a multi-hop and multi-cluster topology, the increase is obviously. Compared with Fig. 6.3, we see that the graph has large characteristic path length like a regular graph and large clustering coefficient like a random graph simultaneously when p is between 0.1 and 1. The graph transforms into a small world now)

is 5.75, and it is a regular graph now. When $p = 1$, the characteristic path length of the graph is 3.15, which decreases by about 45% compared with the value when $p = 0$, and now it becomes a random graph. The tree understratum can be neglected when $p = 1$. The characteristic path length keeps decreasing when p increases.

When $p = 0$, the clustering coefficient of the TREEPSI tree is 0. When p increases from 0 to 1, the clustering coefficient increases greatly. Fig 6.4 shows the increasing process during the increase of p.

The above experiment results show that the small world phenomena exist in the TREEPSI tree as well. The tree topologies are common in the deployment of wireless sensor networks. Under the influence of random interferences, the small world phenomena is ubiquitous in the tree topologies.

6.1.3 *Time Synchronization of Tree Topologies*

The length characteristic of the network plays an important role in time synchronization characteristics. For the existence of the small world phenomena in the tree topologies, the time synchronization error suffers from the influence of the random link rewiring or the random link addition. We conduct some experiments to show the influence in this section and simulate the synchronization process by these two algorithms, References Broadcast Synchronization (RBS) and Tinming-sync Protocol for Sensor Network (TPSN). Before implementing the experiments, we introduce these two time synchronization algorithms in our simulation at first.

6.1.3.1 Time Synchronization Algorithms

Random events, which lead to asymmetric round-trip message delivery delay, contribute directly to synchronization error. The sources of a message's latency include send time, access time, transmission time, propagation time, reception time and receive time. The fundamental property of RBS is that it synchronizes a set of receivers with one another. The nodes periodically send messages to their neighbors using the network's physical-layer broadcasts. Recipients use the message's arrival time as a point of reference for comparing their own clocks. The synchronized time error is analyzed in [1]. If two nodes A and B receive the common packet at $T2$ and $T3$ from node C at time $T1$, the following equations can be obtained:

$$T2 = T1 + S_c + P_{C->A} + R_A + D_{t1}^{C->A} \tag{6.4}$$

$$T3 = T1 + S_c + P_{C->B} + R_B + D_{t1}^{C->B} \tag{6.5}$$

S_C is the time for node C to send the packet, which includes send time, access time and transmission time. R_A is the time for node A to receive the packet, which includes reception time and receive time, and R_B has the similar meaning. $P_{C \to A}$ is propagation time between node C and node A, and $P_{C \to B}$ has the similar meaning. $D_{t1}^{C \to A}$ is the drift between node C and A at time $t1$, and $D_{t1}^{C \to B}$ has the similar meaning. Node B sends its timestamp information in $T3$ to node A, and A receives it in $T4$. Then node A calculates the drift $D_{t4}^{A \to B}$ between A and B as follows:

$$\Delta = T3 - T2 \tag{6.6}$$

$$\Delta = (P_{C\to B} - P_{C->A}) + (R_B - R_A) + \left(D_{t1}^{C\to B} - D_{t1}^{C-A}\right) \tag{6.7}$$

$$D_{t1}^{C\to B} - D_{t1}^{C->A} = D_{t1}^{A\to B} = D_{t4}^{A\to B} + RD_{t1\to t4}^{A\to B} \tag{6.8}$$

where $RD_{t1\to t4}^{A\to B}$ from $t1$ to $t4$. The synchronized time error can be calculated as follows:

$$Error = \Delta - D_{t4}^{A->sB} = P^{UC} + R^{UC} + RD_{t1-t4}^{A-B} \tag{6.9}$$

where $P_D^{UC} = P_{C\to B} - P_{C\to A}$ and $R^{UC} = R_B - R_A$.

In a multi-hop network structure, each conversion step may independently introduce the synchronization error in RBS. The average path error for an n-hop path is $\sigma\sqrt{n}$ if the average per-hop error along the path is σ.

TPSN works in two phases, the level discovery phase and the synchronization phase. In the former, a tree hierarchical structure is established in the network. In the latter, a pair wise synchronization is performed along the edges of this structure to establish a global timescale throughout the network. TPSN is suitable especially for the tree topologies of wireless sensor networks, the construction of the tree topology and the time synchronization can be conducted at the same time. If node A sends a packetto node B at time $T1$, and B receives it at time $T2$. At time $T3$, B sends back a reply, and node A receives it at $T4$. The following equations are derived:

$$T2 = T1 + S_A + P_{A\to B} + R_B + D_{t1}^{A\to B} \tag{6.10}$$

$$T4 = T3 + S_B + P_{A\to B} + R_A - D_{t4}^{A\to B} \tag{6.11}$$

where $D_{t1}^{A->B} = D_{t4}^{A\to B} + RD_{t1->t4}^{A->B}$

$$2\Delta = (T2 - T1) - (T4 - T3) \tag{6.12}$$

$$= S^{UC} + P^{UC} + R^{UC} + RD_{t1-3+4}^{A-B} + 2D_{t4}^{A->B} \tag{6.13}$$

$S^{UC} = S_A - S_B$, $R^{UC} = R_B - R_A$ and $P^{UC} = P_{A\to B} - P_{B-A}$ in above equations. The synchronized time error can be calculated as follows:

$$Error = \Delta - D_{t4}^{A\to B} \tag{6.14}$$

$$= \frac{S^{UC}}{2} + \frac{P^{UC}}{2} + \frac{R^{UC}}{2} + \frac{RD_{t1-t4}^{A-B}}{2} \tag{6.15}$$

The error over multi-hop is independent and the worst-case mean error for an n-hop network is $2 \times n$ times more in RBS compared to TPSN.

6.1.3.2 Time Synchronization Characteristics

After the time synchronization algorithms are introduced, we can evaluate their performance in small world phenonmena and find out the time synchronization characteristics by experiments.

The length characteristic of the network plays an important role in time synchronization characteristics. For the existence of the small world phenomena in the tree topologies, the time synchronization error suffers from the influence of the random link rewiring or the random link addition. The following experiments show this influence. 1000 times of experiments are conducted and the results are analyzed.

If the network is implemented with RBS and the average per-hop error along the path is σ, Figure 6.5a shows the variety of the ratio of average time synchronization error to s in the LEACH tree under the influence of the random link addition. From the figure we can see that the initial ratio is 1.9 when $p = 0$. When $p = 1$, the ratio

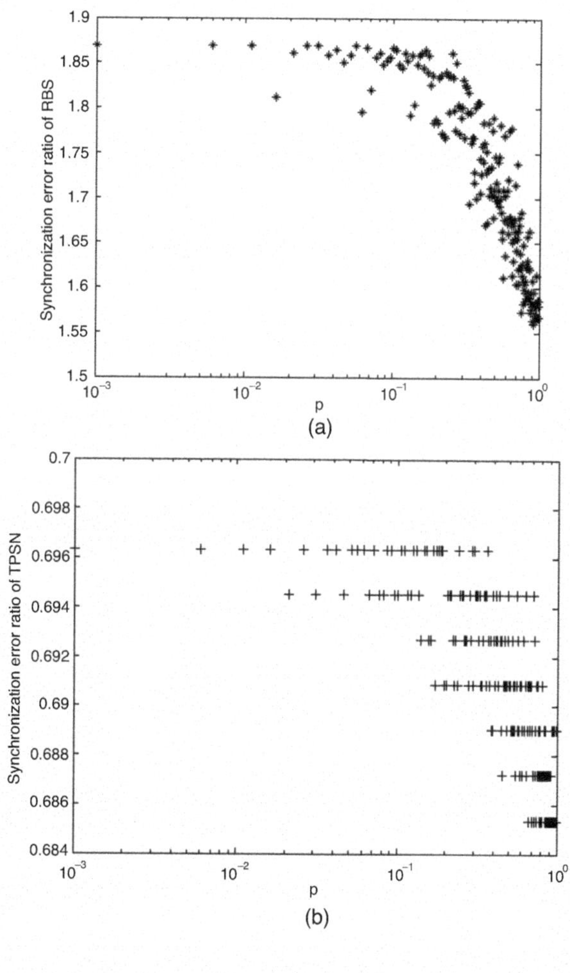

Fig. 6.5 (**a**) The time synchronization error ratio of RBS in the LEACH tree. (It relies on the length characteristic of the network. With the decrease of characteristic path length, the time synchronization error ratio decrease simultaneously). (**b**) The time synchronization error ratio of TPSN in the LEACH tree. (The synchronization error is measured referring to the root node)

reduces to be 1.55, which decreases by about 18.4% compared with the value when $p = 0$. The ratio decreases drastically when p increases from 0 to 1. It implies that there is a large time synchronization error in a regular network while there is a small time synchronization error in a random network, where the understratum can be neglected. In the small world, the time synchronization error decreases in the main trend when the random link addition probability increases.

If the network is implemented with TPSN and the average per-hop error along the path is σ, Figure 6.5b shows the variety of the ratio of average time synchronization error to σ in the LEACH tree. For the tree hierarchical structure, the time synchronization error of TPSN is measured referring to the root node (the base station node) according to its special synchronization procedure. The initial ratio is 0.696 when $p = 0$. When p increases to 1, the ratio decreases to be 0.685, which decreases by about 1.58% compared with the value when $p = 0$. The big trend indicates the decrease of the time synchronization error when p increases.

The time synchronization error of the TREEPSI tree reflects the similar characteristics with that of the LEACH tree. When $p = 0$, the ratio of average time synchronization error to s is large. When p increases, the ratio keeps decreasing. When $p = 1$, the graph becomes a random graph and it has a small time synchronization error.

Figure 6.6a, b show the varieties of the ratio of average time synchronization error to s in the TREEPSI tree under the influence of the random link addition. Figure 6.6a shows the variety of the ratio of average time synchronization error to s when the network is implemented with RBS. The ratio is 2.4 when $p = 0$, and it is 1.75 when $p = 1$, which decreases by about 27% compared with the value when $p = 0$. Figure 6.6b shows the variety of the ratio of average time synchronization error to s when the network is implemented with TPSN. The ratio is 0.865 when $p = 0$, and it is 0.705 when $p = 1$, which decreases by about 18.5% compared with the value when $p = 0$.

From analysis above, we can see that the small world characteristics in the tree topologies of wireless sensor networks affect the time synchronization results greatly. The existence of some random factors brings impacts to the time synchronization characteristics of the network drastically. The time synchronization error decreases with the increase of the random link rewiring or the random link addition probability in the main trend. Random factors exist in the wireless sensor network ubiquitously and their influences on the time synchronization play an important role in the work of network. We use the analysis results to improve the time synchronization algorithms or the network protocols.

In summary, we can see that tree topology is frequently used architecture, which is ubiquitous in the deployment of wireless sensor nodes. Because of the existence of some random factors, the link rewiring or the link addition takes place universally. Moreover, the characteristic path length and the clustering coefficient of the topology vary drastically with the variety of the link rewiring or the link addition probability. Obviously, the variety of the characteristic path length affects the time synchronization characteristics of the network greatly.

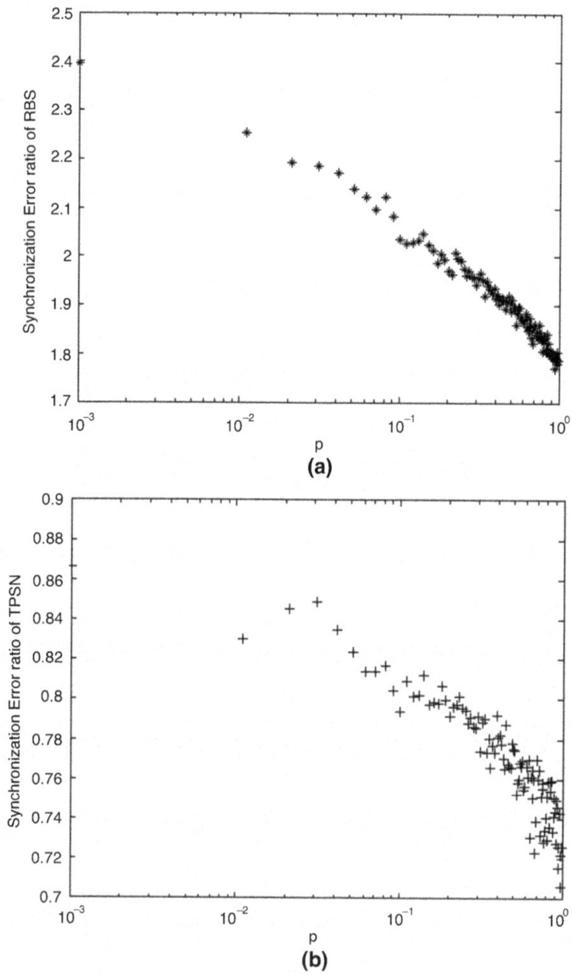

Fig. 6.6 (**a**) The time synchronization error ratio of RBS in the TREEPSI tree. (With the decrease of characteristic path length, the time synchronization error ratio decreases simultaneously). (**b**) The time synchronization error ratio of TPSN in the TREEPSI tree. (All the nodes can synchronize their local clocks to the root node in the multi-hop and multi-cluster network. The error is measured referring to the root node and it relies greatly on the length characteristic of the network. In this disorder network, the time synchronization error varies with the probability p. A global timescale throughout the network is established)

6.2 Epidemic Analysis of Small World

Since sensor does not have complicated hardware architecture or operating system to protect its safety due to cost and resource constraints, sensor nodes in the same network are homogeneous in hardware and software. The characteristic makes it easier for sensors to be compromised by virus attacks compared with regular computer systems. The improvement of WSNs based on small world phenomena enhances the transmission performance but aggravates the dissemination of the virus in the same time. Therefore, analyzing the epidemiological processes is of great significant helping us defend the virus attacks.

In this section, we consider the viruses as sensor worms that stand as perfect example of most kind of viruses. If the sensor worm attacks the network, the epidemic propagates rapidly from one side to another. Due to shortcuts in small worlds, the epidemic propagation becomes much drastic in the network. To analyze the epidemiogical process, we describe the network model and calculate the epidemic threshold. The immunization procedure is also conducted at the end of this section.

6.2.1 Network Model

We consider a network composed of N nodes distributed in the network. Its properties of the wireless sensor network are supposed as

1. Nodes are not moveable after they are distributed randomly in the network.
2. All nodes are symmetric. They have similar properties (e.g., range of radio coverage, energy of batteries, etc.).
3. There is only one root, from which only one tree is constructed in the sensor network.
4. Link additions occur on the tree-based network to construct the small-world network.

Cayley tree is considered in dynamics, which is such an abstract with one root node and K children nodes for each parent node with $K \geq 2$. It neglects the actual distance information between any pair of nodes. Random link additions are conducted on the tree-based model to construct a small-world network. It simplifies extremely the problem and represents a spatial graph model. This abstract reflects homogeneous properties of small worlds, in which each node has approximately the same number of links. If p is defined as the average number of shortcuts per bond (link) on the underlying tree, there are $p(N - 1)$ link additions and altogether $(p + 1)(N - 1)$ links in the small world. Figure 6.7 shows small worlds of tree-based networks. The small world of an semantic/spatial correlation-aware tree (SCT) is shown in Fig. 6.7a. SCT is a simple, scalable and distributed tree structure that address the pratical challenges in the context of aggregation in WSNs [2]. Circles denote nodes and green ones denote the nodes linked by link additions. The Black line denotes the link between a pair of nodes. Dashed lines denote link additions of the small world. The actual location and distance information are neglected, and it only describes logical links of the SCT. Cayley tree with link additions is shown in Fig. 6.7b. In this figure, each node has three nearest neighbors on the underlying topology with $K = 3$, and dashed lines denote link additions.

The epidemic propagation depends on the underlying structure and the immunization strategies. Plentiful researches focus on the propagations on the random network or the small-world network with $p \to 1$ in a susceptible-infected-recovered (SIR) process. In a hypothetical scene, all nodes are randomly movable or they flood messages under no rule, and then the hypothesis is suitable. The mean-field theory is taken into consideration of the analysis of virus propagations. The proposed small-world abstract considers the hierarchical tree-based structure with random

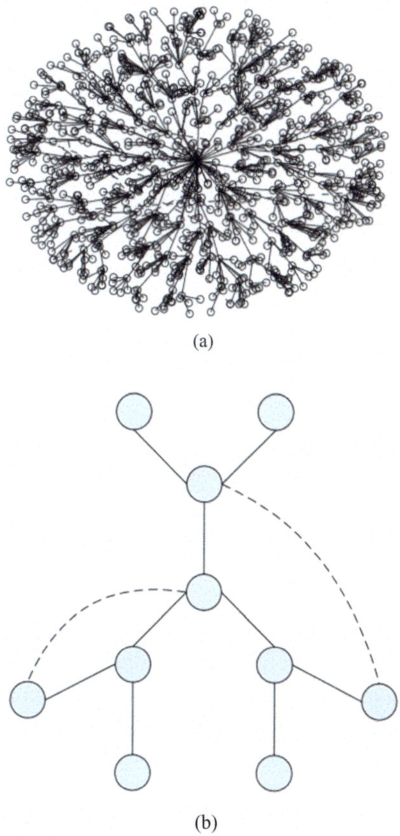

Fig. 6.7 Small worlds of tree-based networks. (**a**) The small world of an SCT. Bonds denote logical links, which neglect the actual distance information. (**b**) Cayley tree with link additions

connections occurring among nodes, which describes complex dynamics of wireless sensor networks. If parts of nodes are movable among most unmovable nodes distributed in the network, random connections occur in the hierarchical topology. As we have analyzed, obstacles, adjustments of radio energy, joins of new members, and errors of location precision all incur small-world phenomena. If there is no immunization in the network, the epidemic experiences a susceptible-infected (SI) process. With the uniform immunization, the epidemic extends in the susceptible nodes. The immune nodes cannot become infected or infect their neighbors.

6.2.2 *Epidemic Thresholds*

The wireless sensor network is vulnerable to computer viruses. The high-density deployment of wireless sensors implies that any virus can be highly contagious. Sensor nodes are severely resource constrained. They lack sophisticated defense mechanisms to fight virus attacks. Due to shortcuts in small worlds, the epidemic propagation becomes much drastic in the sensor networks. In our small-world abstract, we start with Cayley tree and add shortcuts among nodes chosen uniformly at random. More than one bond between any two nodes, or a bond which connects a node to itself, is allowed.

A certain fraction h of nodes in the tree-based network is assumed to be susceptible to the disease, and the bonds represent the physical contacts by which a disease can spread. We assume that each node could be connected with any other node in the network. The epidemic propagation begins with a single infected node. The nodes could be occupied (infected) or not depending on whether they are susceptible to the disease. If the distribution of occupied nodes is random, the problem is equivalent to a standard percolation problem when an epidemic occurs. The node is denoted by a site. The percolation probability h_c can be calculated when the outbreak of the epidemic occurs. In this section, the site percolation is only considered.

In Cayley tree, each node has K nearest neighbors, and some properties of Cayley tree have been analyzed in. The percolation probability h_c is given by:

$$h_c = \frac{1}{K-1} \tag{6.16}$$

The tree-based abstract reflects different characteristics with the ring understratum. The parameter p is defined as the average number of shortcuts per bond on the underlying topology. The probability that two sites chosen randomly have a shortcut between them is calculated as:

$$\begin{aligned} \theta &= 1 - \left(1 - \frac{2}{N^2}\right)^{p(N-1)} \\ &\approx \frac{2p(N-1)}{N^2} \\ &\approx \frac{2p}{N} \end{aligned} \tag{6.17}$$

Occupied sites' connecting constructs the local clusters in the small world, and they can be connected by shortcuts. The average number of local clusters of size i in the network can be obtained by:

$$X_i = h^i(1-h)^{2+(K-2)i}N \tag{6.18}$$

To construct a so-called "giant component" as in the random graph, we start with one particular local cluster and add all other local clusters to it, which can be reached by traveling along a single shortcut. Then all other local clusters, which can be reached by traveling along a single shortcut, are added to the new ones. This process continues until the connected cluster, the giant component, is constructed.

To calculate the percolation probability h_c, a vector V is defined at each step in this process, whose component v_i is the probability that a local cluster of size i is added to the overall connected cluster. We define another vector $V^{'}$, whose component v_i can be gotten in terms of the value of V at the previous step. At or below the percolation threshold the component $v_i^{'}$ is small and we can calculate the vector $V^{'}$ using a transition matrix M. The following formula reflects the relationship between V and $V^{'}$

$$v_i^{'} = \sum_{j=1}^{N} M_{ij} v_j \tag{6.19}$$

where

$$M_{ij} = X_i\left[1 - (1-\theta)^{ij}\right] \tag{6.20}$$

X_i is still the number of local clusters of size i. $[1-(1-\theta)^{ij}]$ is the probability of a shortcut, which connects a local cluster of size i with one of size j, and there are ij possible pairs of sites by which these can be connected.

The largest eigenvalue λ of the transition matrix M is considered. For $\lambda < 1$, the vector V tends to 0 according to Eq. (6.18). The rate at which new local clusters are added falls off exponentially, and the connected clusters are finite with an exponential size distribution. Conversely, for $\lambda > 1$, V keeps growing until the size of the connected cluster becomes limited by the size of the whole system. The percolation threshold occurs at the point $\lambda = 1$.

It is difficult to find the largest eigenvalue of the transition matrix M for finite N. If p is a constant, θ tends to 0 with $N \to \infty$. Eq. (6.19) can be simplified through the relation

$$M_{ij} = ij\theta X_i \tag{6.21}$$

If we set $v_i = \lambda v_i$, (6.18) is rewritten as

$$\lambda v_i = i\theta X_i \sum_{j=1}^{\infty} j v_j \tag{6.22}$$

Then,

$$v_i = C\lambda^{-1} i\theta X_i \tag{6.23}$$

where $C = \sum_{j=1}^{\infty} j v_j$ is a constant.

And

$$\sum_{i=1}^{\infty} v_i = C\lambda^{-1}\theta \sum_{i=1}^{\infty} iX_i \tag{6.24}$$

$$\sum_{i=1}^{\infty} iv_i = C\lambda^{-1}\theta \sum_{i=1}^{\infty} i^2X_i \tag{6.25}$$

$$C = C\lambda^{-1}\theta \sum_{i=1}^{\infty} i^2X_i \tag{6.26}$$

$$\lambda = \theta \sum_{i=1}^{\infty} i^2X_i \tag{6.27}$$

For $K = 2$,

$$X_i = h^i(1-h)^2N \tag{6.28}$$

This formula is the same as that of the ring understratum proposed by Duncan. In this state, λ can be derived by

$$\lambda = \theta Nh\frac{1+h}{1-h} = 2ph\frac{1+h}{1-h} \tag{6.29}$$

We set $\lambda = 1$ to get the value of p at the percolation threshold hc,

$$p = \frac{1-h_c}{2h_c(1+h_c)} \tag{6.30}$$

This special topology with $K = 2$ is frequently seen in wireless sensor networks. Power efficient gathering in sensor information systems (PEGASIS) is a well-known routing protocol, which constructs a chain in the network. In this chain, each node has two nearest neighbors. If the end and the head of the chain are connected, it becomes a ring. Tang, et al. proposed a routing algorithm named chain-cluster based mixed routing (CCM), which divided a wireless sensor network into several chains and caused a longer delay for the data transmission. Pal, et al. introduced a chain structure among sensor nodes in every cluster. Cluster heads got accumulated data from this chain and sent them to the nearest base station. Figure 6.8 shows a chain-based network. The nodes only communicate with their nearest neighbors. The black node denotes the leader of the chain, which collects and conducts data from other nodes before it transmits them to the sink.

For $K = 3$, Eq. (6.17) can be rewritten as

$$X_i = h^i(1-h)^{2+i}N = (1-h)^2[h(1-h)]^iN \tag{6.31}$$

λ can be calculated through the relation

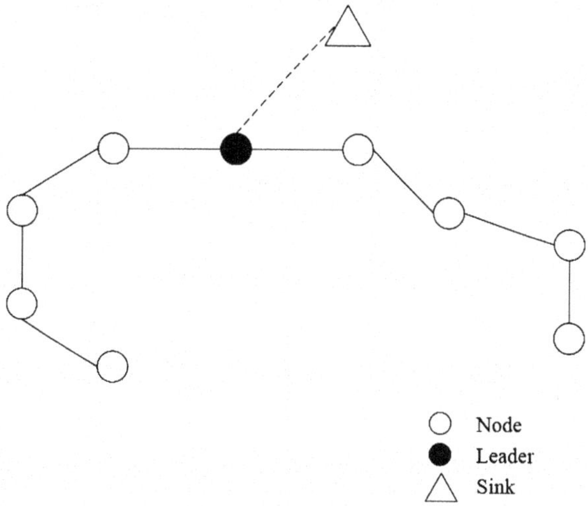

Fig. 6.8 A chain-based network

$$\lambda = \theta N(1-h)^2 \frac{g(1+g)}{(1-g)^3} \tag{6.32}$$

where $g = h(1-h)$. Then,

$$\begin{aligned}\lambda &= \theta N(1-h)^2 \frac{h(1-h)[1+h(1-h)]}{[1-h(1-h)]^3} \\ &= \theta N \frac{h(1-h)^3[1+h(1-h)]}{[1-h(1-h)]^3}\end{aligned} \tag{6.33}$$

We set $\lambda = 1$ to get the value of p at the percolation threshold h_c,

$$p = \frac{[1-h_c(1-h_c)]^3}{2h_c(1+h_c)^3[1+h_c(1-h_c)]} \tag{6.34}$$

For general K, the average number of local clusters of size i in the network can be given by

$$X_i = h^i(1-h)^{2+(K-2)i}N = (1-h)^2\left[h(1-h)^{K-2}\right]^i N \tag{6.35}$$

Thus,

$$\lambda = \theta N(1-h)^2 \frac{g'(1+g')}{(1-g')^3} \tag{6.36}$$

where $g' = h(1-h)^{K-2}$. Then,

$$\lambda = \theta N(1-h)^2 \frac{h(1-h)^{K-2}\left[1+h(1-h)^{K-2}\right]}{\left[1-h(1-h)^{K-2}\right]^3} \tag{6.37}$$

We set $\lambda = 1$ to get the value of p at the percolation threshold h_c,

$$p = \frac{\left[1-h_c(1-h_c)^{K-2}\right]^3}{2h_c(1-h_c)^K\left[1+h_c(1-h_c)^{K-2}\right]} \tag{6.38}$$

p is a constant here, and the percolation threshold h_c for general K can be calculated by Eq. (6.38).

With the vectors and the transition matrix, the percolation threshold h_c is obtained. The largest eigenvalue λ of the transition matrix is considered, and h_c occurs at the point $\lambda = 1$. We can get the same result if we focus on the distribution of local clusters. The quantity $P(n)$ is defined as the probability that a randomly chosen site belongs to a connected cluster of n sites. Then we define

$$H(z) = \sum_{n=0}^{\infty} P(n) z^n \tag{6.39}$$

For $h < h_c$, the distribution of clusters falls off exponentially with the cluster size. In this state, the probability of two shortcuts connecting the same pair of clusters can be neglected. $H(z)$ satisfies the Dyson equation-like iterative condition. Eq. (6.39) is calculated by

$$H(z) = \sum_{n=0}^{\infty} P_0(n) z^n \sum_{n=0}^{\infty} P(m/n)[H(z)]^m \tag{6.40}$$

where m is the number of shortcuts, which connect to other clusters. $P_0(n)$ is the probability that a randomly chosen site belongs to a local cluster of size n. We know

$$P_0(n) = \begin{cases} 1-h, & n=0 \\ nh^n(1-h)^{2+(K-2)n}, & n \geq 1 \end{cases} \tag{6.41}$$

$P(m/n)$ is the probability that there is exactly m shortcuts emerging from a local cluster of size n. There are $2p(N-1)$ ends of shortcuts in the network, and

$$P(m/n) = \binom{2p(N-1)}{m}\left[\frac{n}{N}\right]^m\left[1-\frac{n}{N}\right]^{2p(N-1)-m} \tag{6.42}$$

When N is large enough, (6.39) is rewritten as follows:

$$\begin{aligned} H(z) &= \sum_{n=0}^{\infty} P_0(n)z^n\left[1+(H(z)-1)\frac{n}{N}\right]^{2p(N-1)} \\ &= \sum_{n=0}^{\infty} P_0(n)\left[ze^{\frac{2(N-1)p[H(z)-1]}{N}}\right]^n \\ &= \sum_{n=0}^{\infty} P_0(n)\left[ze^{2p[H(z)-1]}\right]^n \end{aligned} \tag{6.43}$$

We define $H_0(z)$ as

$$H_0(z) = \sum_{n=0}^{\infty} P_0(n)z^n \tag{6.44}$$

Compared Eq. (6.42) with Eq. (6.43), $H(z)$ can be calculated as

$$H(z) = H_0\left(ze^{2p[H(z)-1]}\right) \tag{6.45}$$

From Eq. (6.38), the mean outbreak size can be gotten by the first derivative of $H(z)$,

$$\langle n\rangle = H'(1) \tag{6.46}$$

From Eq. (6.44), $H(z)$ is calculated through the relation

$$\begin{aligned} H'(z) &= H_0'\left(ze^{2p[H(z)-1]}\right)\left[ze^{2p[H(z)-1]}\right]' \\ &= e^{2p[H(z)-1]}H_0'\left(ze^{2p[H(z)-1]}\right)[1+2pzH'(z)] \end{aligned} \tag{6.47}$$

We know $H(1) = 1$ from Eq. (6.38), and

$$\langle n\rangle = H'(1) = \frac{H_0'(1)}{1-2pH_0'(1)} \tag{6.48}$$

From Eq. (6.40) and Eq. (6.43), $H_0(z)$ can be derived by

$$H_0(z) = 1 - h + \frac{h(1-h)^K z}{\left[1 - h(1-h)^{K-2} z\right]^2} \tag{6.49}$$

The first derivative of $H0(z)$ is calculated by

$$H_0'(z) = \frac{h(1-h)^K \left[1 + h(1-h)^{K-2} z\right]}{\left[1 - h(1-h)^{K-2} z\right]^3} \tag{6.50}$$

For $z = 1$, $H_0(1)$ is derived as follows:

$$H_0'(1) = \frac{h(1-h)^K \left[1 + h(1-h)^{K-2}\right]}{\left[1 - h(1-h)^{K-2}\right]^3} \tag{6.51}$$

Then,

$$\langle n \rangle = \frac{H_0'(1)}{1 - 2pH_0'(1)}$$
$$= \frac{h(1-h)^K \left[1 + h(1-h)^{K-2}\right]}{\left[1 - h(1-h)^{K-2}\right]^3 - 2ph(1-h)^K \left[1 + h(1-h)^{K-2}\right]} \tag{6.52}$$

The mean outbreak size diverges at the percolation threshold hc. This threshold marks the onset of the epidemic and occurs at the zero of the denominator of Eq. (6.51). Thus,

$$p = \frac{\left[1 - h_c(1-h_c)^{K-2}\right]^3}{2h_c(1-h_c)^K \left[1 + h_c(1-h_c)^{K-2}\right]} \tag{6.53}$$

which agrees with Eq. (6.37).

Now, we consider $h > h_c$, and there is a giant component of connected nodes with a large number of smaller clusters whose distribution falls off exponentially with the cluster. $P(n)$ is redefined to be the probability that a site chosen randomly belongs to a cluster of size n which is not part of the giant component. The volume of the giant component is $x = 1 - H(1)$, and $P(n)$ sum not to 1 now. From (6.44), x is derived by

$$x = 1 - H_0\left(e^{-2px}\right) \tag{6.54}$$

The derivatives of both sides are calculated as

$$1 = 2pH_0'\left(\mathrm{e}^{-2px}\right)\mathrm{e}^{-2px} \tag{6.55}$$

From Eq. (6.49), Eq. (6.54) can be calculated as follows:

$$1 = 2p\mathrm{e}^{-2px}\frac{h(1-h)^K\left[1+h(1-h)^{K-2}\mathrm{e}^{-2px}\right]}{\left[1-h(1-h)^{K-2}\mathrm{e}^{-2px}\right]^3} \tag{6.56}$$

All values of h are suitable for $x = 0$. For $x < 0$, it is unphysical. The threshold is gotten at $x = 0$. Then,

$$p = \frac{\left[1-h_c(1-h_c)^{K-2}\right]^3}{2h_c(1-h_c)^K\left[1+h_c(1-h_c)^{K-2}\right]} \tag{6.57}$$

The result agrees with Eq. (6.37) and (6.52).

With the transition matrix of vectors and the distribution of local clusters, we calculate the percolation threshold of Cayley tree with link additions. The results are used to approximatively analyze percolation thresholds of tree-based networks with multi-branches and homogeneous structures. In practical context, the larger the scale of the network is, the more the real percolation threshold approaches the ideal value. When the sensor network is attacked by viruses, the infection dies out fast with the infection probability $h < h_c$. The infection spreads with $h \geq h_c$, and viruses attack the network from one side to another.

6.2.3 *Epidemiological Processes*

We are interested in the epidemiological process with $h \geq h_c$. Supposing that there is only one infected node in the initial stage, all other nodes are susceptible (healthy), and the infected nodes only attack all their nearest neighbors in each time unit. The number of infected nodes, $I(t)$, has two contributions: The first comes from the neighbor nodes on the underlying topology; The second comes from the nodes which can be reached via shortcuts. Figure 6.9 describes this epidemiological process on the tree-based sensor network. At the initial stage, all nodes are susceptible ones. The infection spreads on the underlying topology and shortcuts denoted by dashed lines. The infected nodes are denoted by black ones.

Following the description above, $I(t)$ can be calculated by

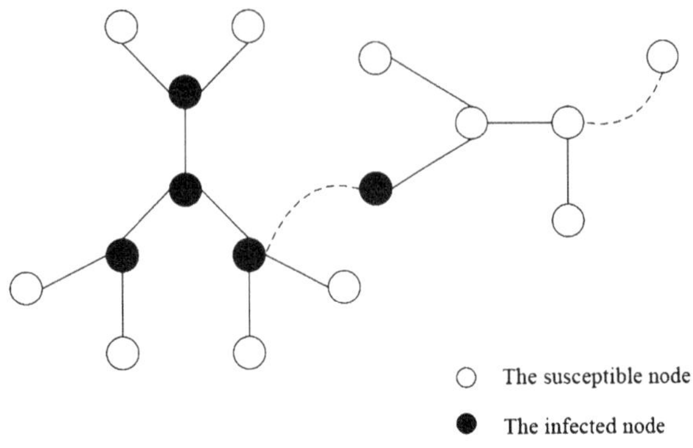

Fig. 6.9 The epidemiological process on the tree-based network

$$I(t) = \sum_{t'=0}^{t} ha(t')\left[1 + 2\xi^{-1}I(t-t')\right] \quad (6.58)$$

where $a(t)$ is the number of infected neighbors on the underlying topology, and ξ is the average distance between the ends of the shortcut, which is calculated as $\xi = \frac{N}{(N-1)p}$. If $a(t)$ is a constant C_K, the epidemiological process is considered, and

$$I(t) = \sum_{t'=0}^{t} hC_K\left[1 + 2p\frac{N-1}{N}I(t-t')\right] \quad (6.59)$$

The sum is approximated with an integral,

$$\begin{aligned} I(t) &= \int_0^t hC_K\left[1 + 2p\frac{N-1}{N}I(t-t')\right]dt' \\ &= hC_K t + \frac{2phC_K(N-1)}{N}\int_0^t I(t')dt' \end{aligned} \quad (6.60)$$

The first-order derivative with respect to t can be obtained by

$$\frac{dI(t)}{dt} = hC_K + \frac{2phC_K(N-1)}{N}I(t) \quad (6.61)$$

and the solution is gotten as

$$I(t) = \left[1 + \frac{N}{2p(N-1)}\right]\mathrm{e}^{\frac{2phC_K(N-1)t}{N}} - \frac{N}{2p(N-1)} \quad (6.62)$$

With $N \to \infty$,

$$I(t) = \left(1 + \frac{1}{2p}\right) e^{2phC_K t} - \frac{1}{2p} \tag{6.63}$$

We can see that the number of infected nodes increases exponentially with t, and the result coincides with that of the ring topology. With $I(t) = N$, all nodes are infected and we can get a threshold $t_c \approx \frac{N}{2phC_K(N-1)} \ln \frac{2pN(N-1)+N}{2p(N-1)+N}$.

In fact, $a(t)$ has more complex mathematical forms when viruses spread on the underlying topology. For $a(t) = C_K t$, $I(t)$ is calculated as

$$\begin{aligned} I(t) &= \sum_{t'=0}^{t} hC_K t' \left[1 + 2p\frac{N-1}{N} I(t-t')\right] \\ &= \int_0^t hC_K t' \left[1 + 2p\frac{N-1}{N} I(t-t')\right] dt' \\ &= hC_K \int_0^t t' dt' + \frac{2phC_K(N-1)}{N} t \int_0^t I(t')dt' \\ &- \frac{2phC_K(N-1)}{N} \int_0^t t' I(t')dt' \end{aligned} \tag{6.64}$$

Both sides are differentiated with respect to t, and

$$\frac{dI(t)}{dt} = hC_K t + \frac{2phC_K(N-1)}{N} \int_0^t I(t')dt' \tag{6.65}$$

The second-order derivative with respect to t can be obtained by

$$\frac{d^2 I(t)}{dt^2} = hC_K + \frac{2phC_K(N-1)}{N} I(t) \tag{6.66}$$

It is a second-order linear differential equation is solved as

$$I(t) = C_1 e^{\sqrt{\frac{2phC_N(N-1)}{N}}t} + C_2 e^{-\sqrt{\frac{2phC_K(N-1)}{N}}t} - \frac{N}{2p(N-1)} \tag{6.67}$$

where C_1 and C_2 are two constants. There is only one infected node in the network in the initial stage at $t = 0$, i.e., $C_1 + C_2 = 1 + \frac{N}{2p(N-1)}$. With $N \to \infty$,

$$I(t) = C_1 e^{\sqrt{2phC_K}t} + C_2 e^{-\sqrt{2phC_K}t} - \frac{1}{2p} \tag{6.68}$$

and $C_1 + C_2 = 1 + \frac{1}{2p}$.

The infected parent node may attack all its children nodes in each time unit with $a(t') = C_K(K-1)^{t'}$ and

$$
\begin{aligned}
I(t) &= \sum_{t'=0}^{t} hC_K(K-1)^{t'}\left[1 + 2p\frac{N-1}{N}I(t-t')\right] \\
&= \int_0^t hC_K(K-1)^{t'}\left[1 + 2p\frac{N-1}{N}I(t-t')\right]dt' \qquad (6.69)
\end{aligned}
$$

$$
= hC_K\int_0^t (K-1)^{t'}dt' + \frac{2phC_K(N-1)}{N}(K-1)^t\int_0^t (K-1)^{-t'}I(t')dt'
$$

Let $F(t) = (K-1)^t\int_0^t (K-1)^{-t'}I(t')dt'$. both sides are differentiated with respect to t,

$$
F'(t) = \ln(K-1)\cdot F(t) + I(t) \qquad (6.70)
$$

From Eq. (6.68), $I(t)$ can be rewritten as

$$
I(t) = hC_K\int_0^t (K-1)^{t'}dt' + \frac{2phC_K(N-1)}{N}F(t) \qquad (6.71)
$$

Then,

$$
F(t) = \frac{N}{2phC_K(N-1)}I(t) - \frac{N}{2p(N-1)}\int_0^t (K-1)^{t'}dt' \qquad (6.72)
$$

Differentiate both sides with respect to t,

$$
F'(t) = \frac{N}{2phC_K(N-1)}I'(t) - \frac{N}{2p(N-1)}(K-1)^t \qquad (6.73)
$$

From Eqs. (6.69), (6.71), and (6.72), we obtain

$$
\begin{aligned}
I'(t) - hC_K(K-1)^t = \ln(K-1)&\left[I(t) - hC_K\int_0^t (K-1)^{t'}dt'\right] \\
&+ \frac{2phC_K(N-1)}{N}I(t) \qquad (6.74)
\end{aligned}
$$

The second-order derivative with respect to t can be obtained by

$$I''(t) = \left[\ln(K-1) + \frac{2phC_K(N-1)}{N} \right] I'(t) \tag{6.75}$$

It can be solved as

$$I(t) = C_3 e^{\left[\ln(K-1) + \frac{2phC_K(N-1)}{N} \right] t} + C_4 \tag{6.76}$$

where C_3 and C_4 are two constants. At $t = 0$, there is C_K infected nodes in the network, i.e., $C_3 + C_4 = C_K$. With $N \rightarrow \infty$,

$$I(t) = C_3 e^{[\ln(K-1) + 2phC_K]t} + C_4 \tag{6.77}$$

and $C_3 + C_4 = C_K$.

From the analysis above, we can see that although different epidemiological processes occur on the underlying topology, the total number of infected nodes increases exponentially as the infection spreads. It is an SI process if there is no any immune strategy.

The mean-field theory is taken into consideration of the analysis. Viruses propagate on small worlds traditionally in the hypothetical scene that all nodes are randomly movable or they flood messages under no rule. Traditional models ignore the spatial proceeding of the infection propagation. The analysis in this section focuses on the spatial-temporal dynamics of the epidemic prevalence on the hierarchical tree-based small world. As it has depicted, the virus rapidly spreads till all susceptible nodes are infected. The analysis can strengthen the understanding of the epidemic proceeding on small worlds of tree-based wireless sensor networks.

6.2.4 *Simulation and Analysis*

Experiments are designed to test the characteristic path length and the clustering coefficient of the tree-based wireless sensor network. In one experiment, 100 nodes are distributed randomly in a 100 × 100 surveillance area and LEACH is constructed. Supposing the small world phenomenon occurs with the increase of the link addition probability p from 0 to 1, the characteristic path length decreases by about 33%, and in the meanwhile the clustering coefficient increases from 0 to 0.5. In another experiment, 1000 nodes are distributed randomly in a 1000 × 1000 surveillance area and TREEPSI is constructed. Supposing the small world phenomenon occurs with the increase of the link addition probability p from 0 to 1, the characteristic path length decreases by about 45%, and the clustering coefficient increases from 0 to 0.2 simultaneously. Both the characteristic path length and the clustering coefficient suffer from the small world phenomenon, which greatly influence the information propagation and the epidemic in some way.

Simulations of percolation thresholds and epidemic propagations on small worlds are presented in this section. We observe the time evolution of the infected number in the small world of SCT by plentiful experiments, supposing that there is a portion of infected nodes in the initial stage. Say that the node i is susceptible, and it has k_i neighbors, of which k_{inf} are infected. Then, i becomes infected with probability k_{inf}/k_i. Propagations based on the mathematical analysis are also presented in this section. Simulations show that the mathematical analysis reveals dynamic characteristics of epidemic propagations with or without the uniform immunization strategy.

6.2.4.1 Percolation Threshold

The percolation probability at which the outbreak of the epidemic occurs is observed. The percolation threshold of Cayley tree is 1 when there is no shortcut in the network. The tree topology transforms into a chain structure with $K = 2$. If the random link addition probability p increases from 0 to 1, the percolation threshold can be calculated as$h_c = \frac{\sqrt{(2p+1)^2+8p-(2p+1)}}{4p}$. As shown in Fig. 6.10, the percolation threshold h_c keeps decreasing with the increase of the link addition probability p. It shows $h_c = 0.67$ with $p = 0.15$, $h_c = 0.61$ with $p = 0.2$, $h_c = 0.52$ with $p = 0.3$, and $h_c = 0.41$ with $p = 0.5$. Table 6.1 shows the variety of the percolation threshold h_c.

The simulation result coincides with the reality. In a regular tree topology, the epidemic propagates along fixed bonds with a large percolation threshold. Due to

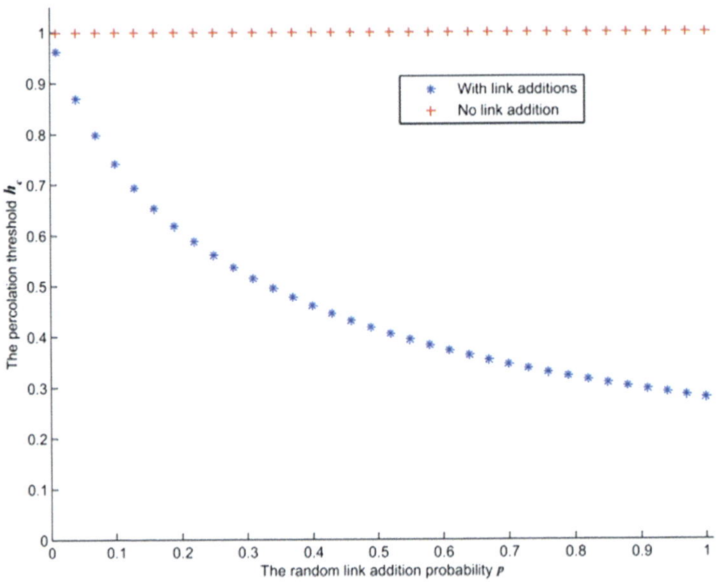

Fig. 6.10 The variety of the percolation threshold h_c hc with $K = 2$

Table 6.1 The percolation threshold of small world

p	0.1	0.15	0.2	0.25	0.3	0.35	0.4	0.45	0.5
h_c	0.74	0.67	0.61	0.56	0.52	0.49	0.46	0.44	0.41

shortcuts in small worlds, the epidemic propagation becomes much drastic in tree-based networks when the random link addition probability p increases from 0. The percolation threshold decreases at the same time. In an entire random network, the percolation threshold is small. The viruses easily attack the network from one side to another.

For $K > 2$, the variety is similar with $K = 2$. When the random link addition probability p increases, the percolation threshold h_c keeps decreasing. At $p = 1$, there exists the smallest percolation threshold in the tree-based sensor network.

6.2.4.2 Epidemiological Processes

Figure 6.11 shows epidemiological processes on a small world of the SCT. There is one infected node in the initial stage. Assuming that the node i is susceptible, and it has k_i neighbors, of which $k$inf are infected. Then, i will become infected with probability k_{inf}/k_i. In the simulation, certain parameters are set with $p = 0.3$ and $N = 2000$. For the theoretical parameters based on [3] and in Fig. 6.11a, the infection extends exponentially in most time on the small world of the SCT. In the late stage of the epidemic, the exponential evolution process experiences a decline for the reduction of remaining susceptible nodes in the network. The number of infected nodes keeps increasing until all nodes are infected. Figure 6.11b shows that the total number of infected nodes still increases exponentially in most time although the uniform immunization reduces the prevalence speed. The larger the immunization factor ρ is, the slower the speed of the prevalence is. The uniform immunization is equivalent to the removal of individuals from the relevant population and reduces the effective average number of neighbors per node.

The epidemiological processes and prevalence behaviors with the uniform immunization on the proposed small-world abstract are observed in Figs. 6.12 and 6.13. The simulations show that our theoretical analysis reflects dynamic characteristics of the virus propagation in small-world networks.

For $a(t') = C_K$, a certain number of nodes are infected on the underlying tree topology in each time unit. Due to shortcuts, the infection extends exponentially with t on the basis of (6.61). In the simulation, certain parameters are set with $p = 0.3$, $h = 0.8$, and $C_K = 3$. Each time unit includes 10 s. Figure 6.12a shows the time evolution of the infected number in the epidemiological process with $N = 500, 2000$, and ∞. Figure 6.13a shows the time evolution of the infected number with the uniform immunization factor $\rho = 0.05, 0.15$, and 0.3 in the epidemiological process for $N = 1000$. From this figure we see that the speed of the infection decreases as ρ increases. At $t = 43$ s, the infected number in the network is 950 with $\rho = 0.05$, 511 with $\rho = 0.15$, and 201 with $\rho = 0.3$. Figure 6.13b shows the prevalence process

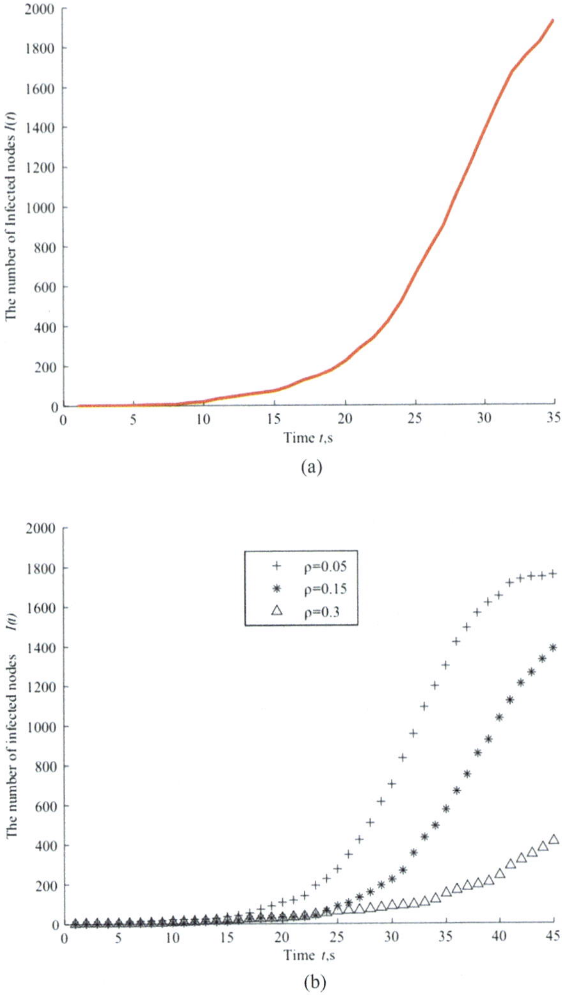

Fig. 6.11 The epidemic on the small world of the SCT. (**a**) The number of infected nodes without the immunization. (**b**) The number of infected nodes without the immunization

with $\rho = 0.05, 0.15$, and 0.3 when N is infinity. At $t = 30$s, the infected number in the network is 160 with $\rho = 0.05$, 103 with $\rho = 0.15$, and 53 with $\rho = 0.3$.

For $a(t') = C_K(K-1)^{t'}$, certain parameters are set with $p = 0.3$, $h = 0.8$, and $C_K = 3$ in the simulation. Each time unit includes 10 s. $C1$ and $C2$ can be approximately calculated at $t = 0$ and $t = 10$s. Figure 6.12b shows the time evolution of the infected number in the epidemiological process with $N = 50$, 500, and ∞ on the basis of (49). Figure 6.13c shows the time evolution of the infected number with the uniform immunization factor $\rho = 0.05$, 0.15, and 0.3 in the epidemiological process for $N = 1000$. At $t = 53$ s, the infected number in the network is 924 with $\rho = 0.05$, 657 with $\rho = 0.15$, and 377 with $\rho = 0.3$. Figure 6.13d shows the prevalence process with $\rho = 0.05, 0.15$, and 0.3 when N is infinity. At $t = 30$s, the

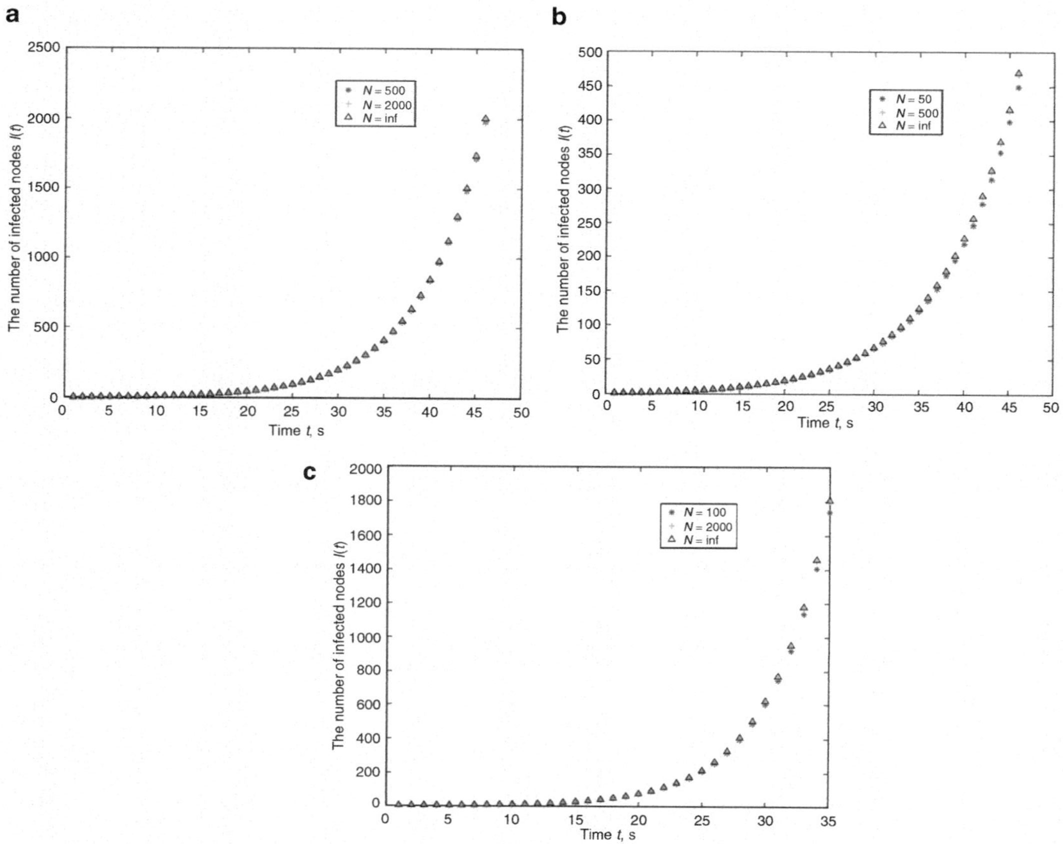

Fig. 6.12 Epidemiological processes on the proposed small-world abstract. **(a)** The number of infected nodes with $a(t') = C_K$. **(b)** The number of infected nodes with $a(t') = C_K t'$. **(c)** The number of infected nodes with $a(t') = C_K(K-1)^{t'}$

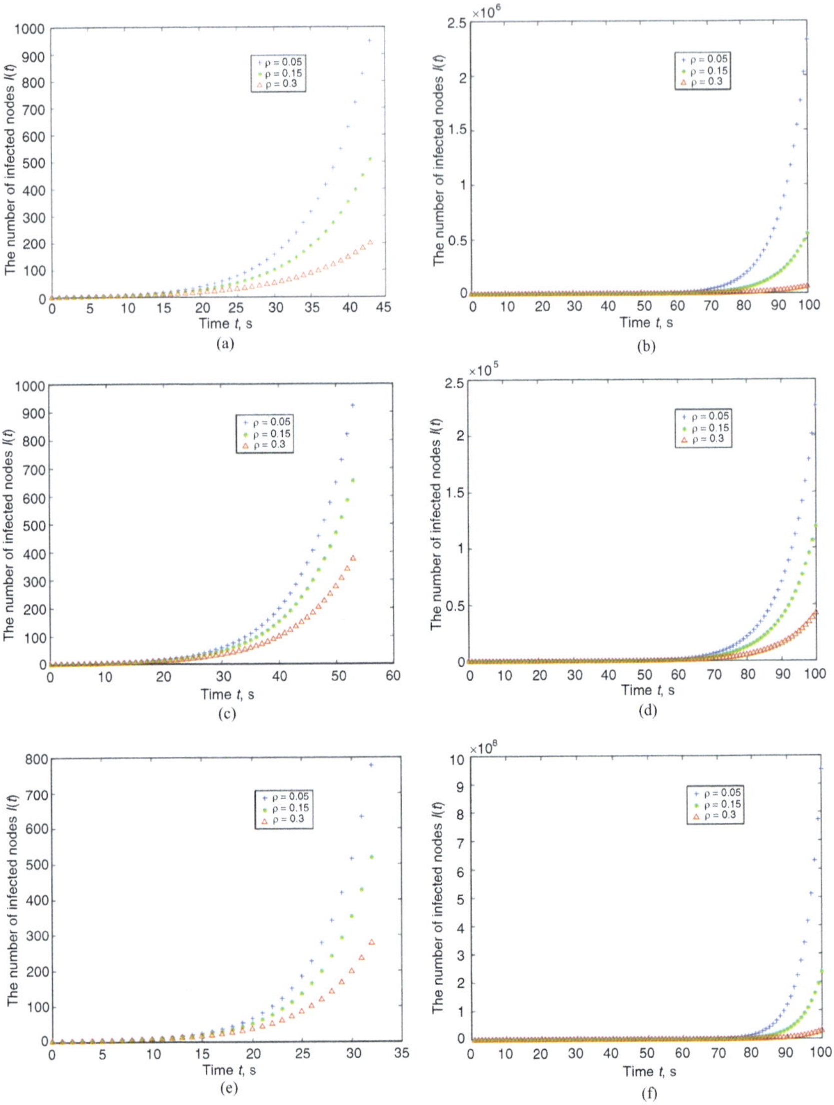

Fig. 6.13 Epidemiological processes with the uniform immunization on the proposed small-world abstract. (**a**) The number of infected nodes with $a(t') = C_K$ and $N = 1000$. (**b**) The number of infected nodes with $a(t') = C_K$ and $N = \infty$. (**c**) The number of infected nodes with $a(t') = C_K t'$ and $N = 1000$. (**d**) The number of infected nodes with $a(t') = C_K t'$ and $N \to \infty$. (**e**) The number of infected nodes with $a(t') = C_K (K-1)^{t'}$ and $N = 1000$. (**f**) The number of infected nodes with $a(t') = C_K (K-1)^{t'}$ and $N = \infty$

Table 6.2 Infected numbers with the uniform immunization

ρ	0.05	0.1	0.15	0.2	0.25	0.3	0.35	0.4	0.45
$a(t') = C_K$	160	129	103	83	66	53	43	34	27
$a(t') = C_K t'$	61	56	50	45	40	36	32	28	25
$a(t') = C_K(K-1)^{t'}$	518	430	356	294	242	199	163	133	108

infected number in the network is 61 with $\rho = 0.05$, 50 with $\rho = 0.15$, and 36 with $\rho = 0.3$.

For $a(t') = C_K(K-1)^{t'}$, certain parameters are set with $p = 0.3$, $h = 0.8$, and $C_K = 3$, and $K = 3$ in the simulation. Each time unit includes 10 s. C_3 and C_4 can be approximately calculated at $t = 0$ and $t = 10$ s. Figure 6.12c shows the time evolution of the infected number in the epidemiological process with $N = 100$, 2000, and ∞ on the basis of Eq. (6.75). Figure 6.13e shows the time evolution of the infected number with the uniform immunization factor $\rho = 0.05$, 0.15, and 0.3 in the epidemiological process for $N = 1000$. At $t = 32$ s, the infected number in the network is 778 with $\rho = 0.05$, 520 with $\rho = 0.15$, and 279 with $\rho = 0.3$. Figure 6.13f shows the prevalence process with $\rho = 0.05$, 0.15, and 0.3 when N is infinity. At $t =$ 30s, the infected number in the network is 518 with $\rho = 0.05$, 356 with $\rho = 0.15$, and 199 with $\rho = 0.3$.

Table 6.2 shows the number of infected nodes with $p = 0.3$, $h = 0.8$, $C_K = 3$, $K = 3$, and $N = \infty$ at $t = 30$ s when the uniform immunization factor ρ increases. As we have analyzed, the total number of infected nodes increases exponentially when different epidemiological processes occur on the underlying topology. The larger N is, the closer the result approaches that when N is infinity. Although the uniform immunization reduces the prevalence speed, the total number of infected nodes still increases exponentially. The larger the immunization factor ρ is, the slower the speed of the prevalence is.

We can see the epidemic propagation becomes much drastic when the sensor worm attacks the network due to the shortcuts in small worlds. The infection dies out fast when the infection probability is smaller than the percolation threshold. On the contrary, the virus easily attacks the network from one side to another while the infection probability is larger than the percolation threshold. The number of infected nodes increases exponentially when different epidemiological processes occur on the underlying tree topology.

6.3 Immunizations of Small World

As attacker can simply send a single packet to compromise the entire sensor network, the sensor virus is a serious threat. With the former strategy, epidemics become drastic with link additions among sensors when the small world phenomena occur. The immunization is one of the most common strategies for combating the outbreak of sensor viruses. In this section, we focus on the immunization strategies for combating epidemic propagations on small worlds of tree-based wireless sensor networks. These two immunization strategies, uniform immunization and temporary immunization, are conducted on small worlds of tree-based wireless sensor networks to combat the sensor viruses.

6.3.1 Immunization Strategies

In network G, $S(t)$, $I(t)$, and $R(t)$ are the nodes in the states of susceptible (S), infectious (I), and immune/recovered (R) at time t, respectively. We know that $S(t) + I(t) + R(t) = N$. Let S_t, I_t, and R_t be the proportions of susceptible, infective, and recovered nodes in the network at time t, respectively, we have $S_t + I_t + R_t = 1$. We can partition the infected nodes in the network into $I(t_0)$, $I(t_1)$,$I(t_2)$... at timet_0;t_1; t_2..., respectively. With the uniform immunization, $I(t + 1)$ contains more nodes than $I(t)$ if the infected nodes attack their neighbors in each time unit. The $I(t)$ increases rapidly with time t as the epidemic spreads on the network, although the immunization reduces the propagation speed. With the temporary immunization, $I(t)$ increases or decreases periodically with time t if the ratio of the infectious period τ_I to the immune period τ_R is suitable. Shortcuts are uniformly added besides the inherent edges in the original tree topology with the average degree $\langle k \rangle = (1 + p)K$. If node i is susceptible, and it has k_i neighbors, of which k_{inf} are infected. Then, node i becomes infected with probability $k_{\text{inf}} = k_i$. Note that i become infected with probability 1 if all its neighbors are infected. Besides this parameter-free mechanism, there may be other reasonable choices. For example, if the susceptible has a probability h of contagion with each infected neighbor, its probability of infection becomes $\left[1 - (1 - h)^{k_{\text{inf}}}\right]$.

6.3.2 Uniform Immunization

We now analyze the epidemiological process with the uniform immunization, which is used frequently in homogeneous small-world networks. We assume that there is only one infected node in the initial stage, ρ is defined as the fraction of immune nodes present in the network, and the remaining nodes are susceptible (healthy). The immune nodes cannot become infected or transmit the infection to their neighbors. The infected nodes only attack their nearest susceptible neighbors in each time unit. During the propagation, the understratum plays a key role and is an essential factor considered in the immunization strategies. Two sources contribute to the number of infected nodes, one is the susceptible neighbor nodes on the underlying topology, and the other is the susceptible nodes reachable via the shortcuts. Following the above description, $I(t)$ can be calculated by

$$I(t) = \sum_{t'=0}^{t} (1-\rho)ha(t')\left[1 + 2\xi^{-1}I(t-t')\right] \tag{6.78}$$

where $a(t')$ is the number of attacked neighbors on the underlying topology, h is the infection probability, and ξ is the average distance between the ends of shortcuts, $\xi = \frac{N}{(N-1)p}$. For mathematical analysis of epidemics on the small world, three cases are considered. If $a(t')$ is a constant C_K, the epidemiological process is considered, and

$$I(t) = \sum_{t'=0}^{t} (1-\rho)hC_K\left[1 + 2p\frac{N-1}{N}I(t-t')\right] \tag{6.79}$$

The sum can be approximated by an integral

$$\begin{aligned} I(t) &= \int_0^t (1-\rho)hC_K\left[1 + 2p\frac{N-1}{N}I(t-t')\right]\mathrm{d}t' \\ &= (1-\rho)hC_K t + \frac{2p(1-\rho)hC_K(N-1)}{N} \\ &\quad \times \int_0^t I(t')\mathrm{d}t' \end{aligned} \tag{6.80}$$

The first-order derivative with respect to t can be obtained as

$$\frac{\mathrm{d}I(t)}{\mathrm{d}t} = (1-\rho)hC_K + \frac{2p(1-\rho)hC_K(N-1)}{N}I(t) \tag{6.81}$$

and the solution is obtained as

$$I(t) = \left[1 + \frac{N}{2p(N-1)}\right]\mathrm{e}^{\frac{2p(1-\rho)hC_K(N-1)t}{N}} - \frac{N}{2p(N-1)} \tag{6.82}$$

With $N \to \infty$

$$I(t) = \left(1 + \frac{1}{2p}\right)\mathrm{e}^{2p(1-\rho)hC_K t} - \frac{1}{2p} \tag{6.83}$$

We can see that the number of infected nodes increases exponentially with t, and the result coincides with that on the ring topology. When $I(t) = N$, all nodes are infected, we obtain the threshold

$$t_c \approx \frac{N}{2p(1-\rho)hC_K(N-1)} \ln \frac{2pN(N-1)+N}{2p(N-1)+N} \tag{6.84}$$

In fact, $a(t')$ has more complex mathematical forms when the viruses spread on the underlying topology. If $a(t') = C_K t'$, $I(t)$ can be calculated as

$$\begin{aligned} I(t) &= \sum_{t'=0}^{t} (1-\rho)hC_K t' \left[1 + 2p\frac{N-1}{N} I(t-t')\right] \\ &= \int_0^t (1-\rho)hC_K t' \left[1 + 2p\frac{N-1}{N} I(t-t')\right] \mathrm{d}t' \\ &= (1-\rho)hC_K \int_0^t t'\mathrm{d}t' + \frac{2p(1-\rho)hC_K(N-1)}{N} t \int_0^t I(t')\mathrm{d}t' \\ &- \frac{2p(1-\rho)hC_K(N-1)}{N} \int_0^t t' I(t')\mathrm{d}t' \end{aligned} \tag{6.85}$$

Both sides are differentiated with respect to t, we obtain

$$\frac{\mathrm{d}I(t)}{\mathrm{d}t} = (1-\rho)hC_K t + \frac{2p(1-\rho)hC_K(N-1)}{N} \times \int_0^t I(t')\mathrm{d}t' \tag{6.86}$$

The second-order derivative with respect to t can be obtained as

$$\frac{\mathrm{d}^2 I(t)}{\mathrm{d}t^2} = (1-\rho)hC_K + \frac{2p(1-\rho)hC_K(N-1)}{N} I(t) \tag{6.87}$$

It is a second-order linear differential equation, and can be solved as

$$I(t) = C_1 \mathrm{e}\sqrt{\frac{2p(1-\rho)hC_K(N-1)}{N}} t + C_2 \mathrm{e}^{-\sqrt{\frac{2p(1-\rho)hC_K(N-1)}{N}} t} - \frac{N}{2p(N-1)} \tag{6.88}$$

where C_1 and C_2 are two constants. There is only one infected node in the network at $t = 0$, i.e., $C_1 + C_2 = 1 + \frac{N}{2p(N-1)}$.

With $N \to \infty$,

$$I(t) = C_1 \mathrm{e}\sqrt{2p(1-\rho)hC_K} t + C_2 \mathrm{e}^{-\sqrt{2p(1-\rho)hC_K t}} - \frac{1}{2p} \tag{6.89}$$

And $C_1 + C_2 = 1 + \frac{1}{2p}$.

The infected parent node may attack all its children nodes in each time unit, $a(t') = C_K(K-1)^{t'}$, then

$$
\begin{aligned}
I(t) &= \sum_{t'=0}^{t} (1-\rho)hC_K(K-1)^{t'} \times \left[1 + 2p\frac{N-1}{N}I(t-t')\right] \\
&= \int_0^t (1-\rho)hC_K(K-1)^{t} \times \left[1 + 2p\frac{N-1}{N}I(t-t')\right]\mathrm{d}t' \\
&= (1-\rho)hC_K \int_0^t (K-1)^{t'}\mathrm{d}t' + \frac{2p(1-\rho)hC_K(N-1)}{N}(K-1)^t \\
&\times \int_0^t (K-1)^{-t'} I(t')\mathrm{d}t'
\end{aligned}
\tag{6.90}
$$

Let $F(t) = (K-1)^t \int_0^t (K-1)^{-t'} I(t')\mathrm{d}t'$, and differentiate both sides with respect to t, we obtain

$$
F'(t) = \ln(K-1) \cdot F(t) + I(t) \tag{6.91}
$$

From Eq. (6.90), we can rewrite $I(t)$ as

$$
I(t) = (1-\rho)hC_K \int_0^t (K-1)^{t'}\mathrm{d}t' + \frac{2p(1-\rho)hC_K(N-1)}{N}F(t) \tag{6.92}
$$

Then,

$$
F(t) = \frac{N}{2p(1-\rho)hC_K(N-1)}I(t) - \frac{N}{2p(N-1)}\int_0^t (K-1)^{t'}\mathrm{d}t' \tag{6.93}
$$

Differentiate both sides with respect to t, we have

$$
F'(t) = \frac{N}{2p(1-\rho)hC_K(N-1)}I'(t) - \frac{N}{2p(N-1)}(K-1)^t \tag{6.94}
$$

From Eqs. (6.91), (6.93), and (6.94), we obtain

$$
\begin{aligned}
I'(t) - (1-\rho)hC_K(K-1)^t &= \ln(K-1)\left[I(t) - (1-\rho)hC_K \int_0^t (K-1)^{t'}\mathrm{d}t'\right] \\
&\quad + \frac{2p(1-\rho)hC_K(N-1)}{N}I(t)
\end{aligned}
\tag{6.95}
$$

The second-order derivative with respect to t can be obtained as

$$
I''(t) = \left[\ln(K-1) + \frac{2p(1-\rho)hC_K(N-1)}{N}\right]I'(t) \tag{6.96}
$$

It can be solved as

$$I(t) = C_3 \exp\left(\left[\ln(K-1) + \frac{2p(1-\rho)hC_K(N-1)}{N}\right]t\right) + C_4 \tag{6.97}$$

where C_3 and C_4 are two constants. At $t = 0$, there are C_K infected nodes in the network, i.e., $C_3 + C_4 = C_K$.

With $N \to \infty$,

$$I(t) = C_3 e^{[\ln(K-1)+2p(1-\rho)hC_K]t} + C_4 \tag{6.98}$$

and $C_3 + C_4 = C_K$

The uniform immunization is equivalent to the removal of individuals from the relevant population. It reduces the effective average number of neighbors per node and slows down the propagation. From the above analysis, we can see that although different epidemiological processes occur on the underlying tree topology, the total number of infected nodes increases exponentially with the uniform immunization. The presence of the immunization can effectively reduce the prevalence speed by a factor $(1 - \rho)$.

6.3.3 *Temporary Immunization*

The success of the epidemic to invade a network is strongly dependent on the network's specific connectivity structure. We are interested in a further theoretical understanding of the spatial–temporal dynamics of the epidemic on the small world of tree based networks. A period of temporary immunization is conducted on the node after it is infected, which is then removed after an intrinsic time delay. As a result, recurrent propagation oscillations of infection occur in the network. The epidemic cycles arise because of the delayed susceptible–infectious–recovered–susceptible (SIRS) process. In the SIRS process, the susceptible nodes become infected, recover with the temporary immunization, but return to the susceptible state when the immunization wears off.

The state of the node at the $(i, j) - th$ location of the lattice, $x_{ij}(t)$, can be S, I, or R. The following rules guide the state transitions:

$$x_{i,j}(t) \in \mathrm{S} \to \begin{cases} x_{i,j}(t+1) \in \mathrm{I}, & \text{with probability } 1-(1-h)^{k_{\inf}^{i,j}(t)}(t) \\ x_{i,j}(t+1) \in \mathrm{S}, & \text{otherwise} \end{cases} \tag{6.99}$$

$$x_{i,j}(t_0) \in \mathrm{I} \to x_{i,j}(t_0+1) \in \mathrm{I} \to \cdots \to x_{i,j}(t_0+\tau_1) \in \mathrm{R} \tag{6.100}$$

$$\to x_{i,j}(t_0+\tau_1+1) \in \mathrm{R} \to \cdots \to x_{i,j}(t_0+\tau_0) \in \mathrm{S} \tag{6.101}$$

where h is the infection probability, $k_{\inf}^{i,j}(t)$ is the total number of infectious contacts of the node at time t, t_0 is the initial time, τ_I is the time period that a node remains infectious, and τ_R is the period that a node remains immune. We assume that $\tau_I = 1$, the proportion of recovered nodes can be described by the sum $\sum_{i=1}^{\tau_0 - 1} I_{t-i}$, and S_t can be calculated as

$$S_t = 1 - I_t - \sum_{i=1}^{\tau_0 - 1} I_{t-i} \tag{6.102}$$

where $\tau_I = \tau_I + \tau_R$. For the homogeneous small-world abstract, the number of infected neighbors of each node is the sum of the number of neighbors on the underlying topology and that of the nodes connected with shortcuts, which at time t are I_tK and I_tpK, respectively.

Let h_K be the probability that such a node is infected by a neighbor in a given time step. Let h_p be the probability that such a node is infected by a node connected with the shortcut. The spread of the epidemic can be described with the following model of the SIRS dynamics:

$$\begin{aligned} I_{t+1} &= \left[1 - (1 - h_K)^{I_tK}\left(1 - h_p\right)^{I_tpK}\right] S_t \\ &= \left[1 - (1 - h_{\text{eff}})^{I_tK}\right] S_t \end{aligned} \tag{6.103}$$

where heff is defined as

$$1 - h_{\text{eff}} = (1 - h_K)\left(1 - h_p\right)^p \tag{6.104}$$

Equation (6.103) captures both the time delay dynamics resulting from the temporary immunization of the infection and the effects of the shortcuts. The recurrent propagation oscillations relay on the infection, the recovery, and the reinfection of the nodes in the small world. The shortcuts of the small world and the temporary immunization each play an important role in this procedure.

6.3.4 Simulations and Analysis

Simulations of epidemic propagations with immunizations on small worlds are presented in this section. We study the time evolutions of the infected numbers in small worlds of SCT and minimal spanning tree (MST) by large number of experiments assuming a portion of nodes infected in the initial stage. We assume that node i is susceptible, and it has k_i neighbors, of which kinf are infected. Then, node i becomes infected with probability k_{inf}/k_i. The network is conducted with the uniform immunization or the temporary immunization. The epidemiological behaviors with the immunizations are studied when the small-world phenomena occur. The propagations based on mathematical analyses are also presented in this section. The simulations show that the mathematical analyses reveal the dynamic characteristics of the epidemic propagations with the two immunization strategies.

6.3.4.1 Simulations of the Uniform Immunization

Epidemiological behaviors with the uniform immunization are studied when the small-world phenomena occur in the tree-based networks. Fig 6.14a, b show the time evolutions of the infected numbers with the uniform immunization on small worlds of the SCT and the MST, respectively.

There is only one infected node in the initial stage. The parameters from Ref. [5] are used in Fig. 6.14. The epidemics spread exponentially with the uniform immunization in most time on the two small worlds. In the late stage, the exponential evolution process experiences a decline due to the reduction of the remaining susceptible nodes in the network. The infected number keeps increasing until all nodes are infected in the network. The simulations show that the larger the immunization factor ρ is, the slower the speed of prevalence is. The uniform immunization reduces the effective average number of neighbors for each node and slows down the propagation.

The following simulations depict the epidemiological propagations with the uniform immunization on the small world of the Cayley tree based on the mathematical analyses. For $a(t') = C_K$, a certain number of nodes are attacked on the tree-based topology in each time unit. Due to the shortcuts, the infection extends exponentially with t, as shown in Eq. (6.82). The duration of each time unit is 10 s. Figure 6.15a shows the time evolution of the infected number in the epidemiological process. From the figure, we can see that the speed of the infection decreases as ρ increases. At $t = 43s$, the infected number in the network is 950 for ρ= 0:05, 511 for ρ= 0:15, and 201 for $\rho = 0.3$.

For $a(t') = C_K t'$, C_1 and C_2 can be approximately calculated at $t = 0s$ and $t = 10s$. The duration of each time unit is 10 s. Figure 6.15b shows the time evolution of the infected number in the epidemiological process. At $t = 53s$, the infected number in the network is 924 for $\rho = 0.05$, 657 for $\rho = 0.15$, and 377 for $\rho = 0.3$.

For $a(t') = C_K(K-1)^t$, C_3 and C_4 can be approximately calculated at $t = 0$ s and $t = 10$ s. The duration of each time unit is 10 s. Figure 6.15c shows the time evolution of the infected number in the epidemiological process. At t= 32 s, the infected number in the network is 778 for ρ= 0.05, 520 for ρ= 0.15, and 279 for ρ= 0.3.

Table 6.3 shows the numbers of infected nodes at $t = 30$ s with $p = 0.3$, $h = 0.8$, $C_K = 3$, $K = 3$, and $N = \infty$ for different ρ. Although the uniform immunization reduces the prevalence speed, the total number of infected nodes increases exponentially. The larger the immunization factor ρ is, the slower the speed of the prevalence is.

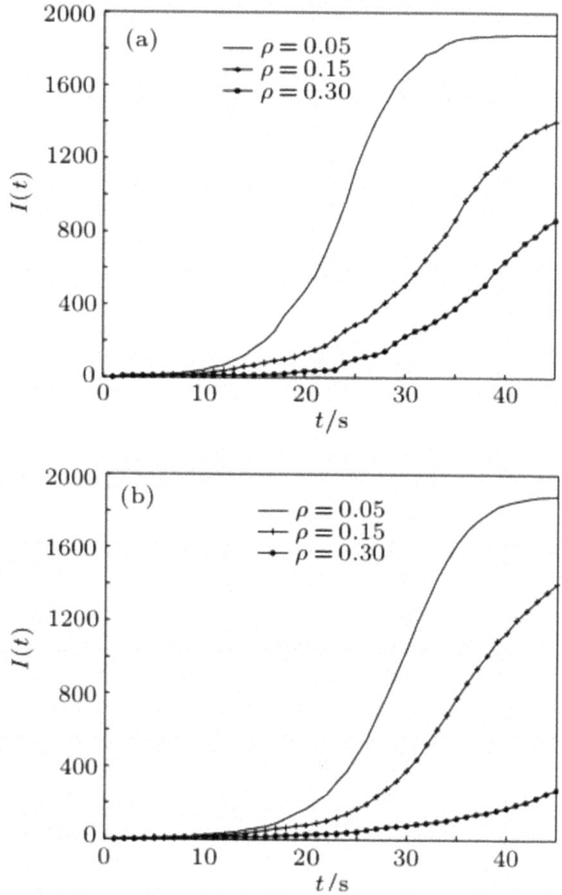

Fig. 6.14 Time evolutions of the infected numbers on the small worlds of (**a**) the SCT and (**b**) the MST with $p = 0.3$ and $N = 2000$

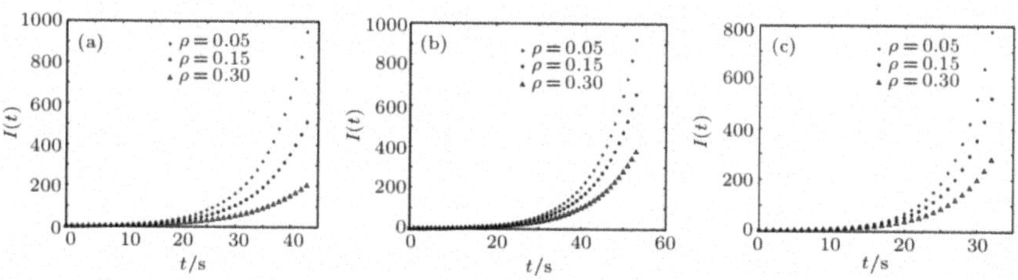

Fig. 6.15 Epidemiological propagations on the small world of the Cayley tree with $p = 0.3$, $h = 0.8$, $C_K = 3$, $K = 3$, and $N = \infty$. The time evolutions of the infected number are shown for (**a**) $a(t') = C_K$, (**b**) $a(t') = C_K t'$, and (**c**) $a(t') = C_K (K-1)^{t'}$

Table 6.3 Infected numbers with the uniform immunization

p	0.05	0.1	0.15	0.2	0.25	0.3	0.35	0.4	0.45
$a(t') = C_K$	160	129	103	83	66	53	43	34	27
$a(t') = C_K t'$	61	56	50	45	40	36	32	28	25
$a(t') = C_K(K-1)^{t'}$	518	430	356	294	242	199	163	133	108

6.3.4.2 Simulations of the Temporary Immunization

The epidemiological behaviors of sensor worms with the temporary immunization are investigated in the following simulations. Figure 6.16a, b show the time evolutions of the infected numbers with the temporary immunization on small worlds of the SCT and the MST, respectively. The parameters from Ref. [5] are used in Fig. 6.16. The recurrent waves are observed both spatially and temporally. The periodicity of the wave comes from the small-world structure and the temporary immunization, the reinfection of the recovered nodes after a certain time delay introduced by the temporary immunization in the SIRS procedure.

Figure 6.17 depicts the recurrent oscillations of the spatio–temporal epidemiological behaviors on the small world of the Cayley tree based on Eq. (6.103). In the simulations, 2000 nodes are distributed randomly in the network, and 10% nodes are infected in the initial stage. A certain number of neighbors ($K = 4$) are connected directly with each node on the tree-based topology, and pK shortcuts link it with the other ones with $p = 0.4$. The spread of the prevalence is affected by the infection probabilities $h_K = 0.35$ and $h_p = 0.12$. In Fig. 6.17a, $\tau_I = 1$ and $\tau_R = 8$, while in Fig. 6.17b, $\tau_I = 1$ and $\tau_R = 6$. The simulations lead to a better understanding of the epidemic spread with the temporary immunization. The infection probabilities h_K, h_p and the average number p of the shortcuts per bond have great influences on generating the long-term cycles. The periodicity of the waves is the outcome of both the small-world structure and the temporary immunization. The reinfection at the center of the wave after a certain amount of time is due to the shortcuts and the intrinsic time delay introduced by the temporary immunization.

On the other hand, the ratio of time period τ_I to time period τ_R has a great influence on generating the long-term cycles. In Fig. 6.17b, the number of infected nodes remains a constant after $t = 260$ s. It is the outcome of both the small-world structure and the temporary immunization. An equilibrium is reached between the reinfection and the infection due to the intrinsic time delay introduced by the temporary immunization. The periodic waves based on the mathematical analyses coincide with the recurrent propagation oscillations of epidemics in the small worlds of the SCT and the MST.

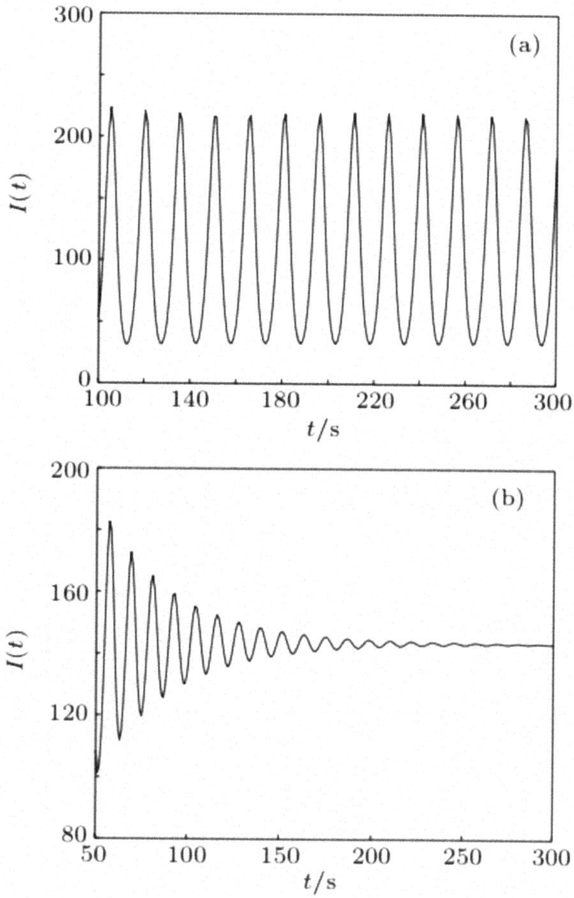

Fig. 6.16 Time evolutions of the infected numbers on the small worlds of (**a**) the SCT and (**b**) the MST with $p = 0.9$ and $N = 2000$

6.4 Conclusion

Small world phenomena not only increases the connectivity of WSNs, but also aggravates the transmission of the epidemic viruses in WSNs. In this chapter, the epidemic threshold at which the outbreak of the epidemic occurs is figured out. Epidemiological processes are analyzed when the infection probability is larger than the percolation threshold. Two immunization strategies, uniform immunization and temporary immunization, are conducted on small worlds of tree-based wireless sensor networks to combat the sensor viruses. Although the immunization effectively reduces the contagion speed, the infection still extends exponentially for the former strategy. For the latter strategy, recurrent contagion oscillations occur in the small world when the spatial–temporal dynamics of the epidemic are considered.

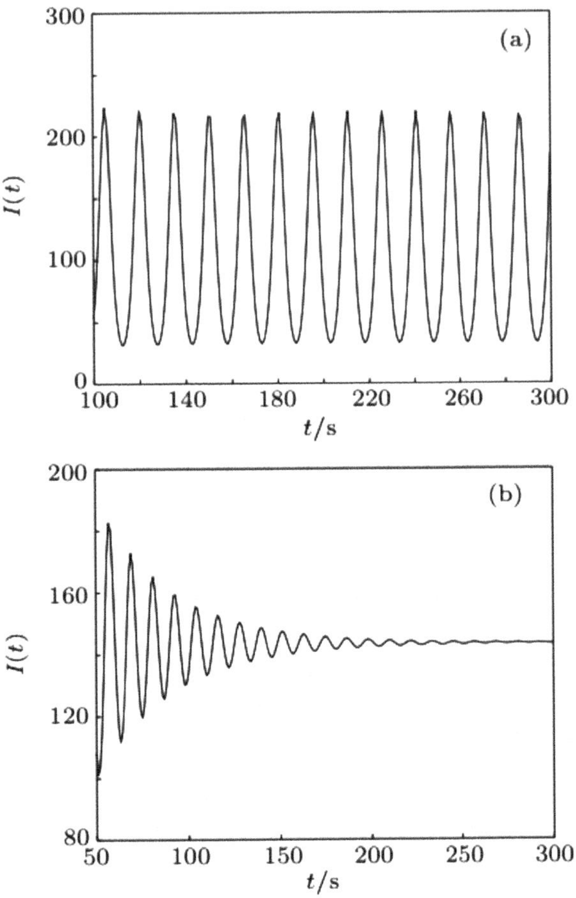

Fig. 6.17 Epidemiological propagations with the temporary immunization on the small world of the Cayley tree. The time evolutions of the infected number are shown on the small world of the Cayley tree for (**a**) $\tau_I = 1$ and $\tau_R = 8$ and (**b**) $\tau_I = 1$ and $\tau_R = 6$

References

1. S. Ganeriwal, R. Kumar, M. B. Srivastava. Timing-sync protocol for sensor networks[C]. Proceedings of the 1st international conference on Embedded networked sensor systems, 2003: 138-149.
2. Zhu Y J, Vedantham R, Park S J, et al., A scalable correlation aware aggregation strategy for wireless sensor networks[C]. First International Conference on Wireless Internet, 2005, pp. 122-129.
3. LI Q, Zhang B, Cui L, et al. Epedemics on Small Worlds of Tree-based Wireless Sensor Networks[J]. Journal of Systems Science and Complexity, 2014,27(6): 1095-1120.
4. Li Q, Cui L, Zhang B, et al. Small Worlds in the Tree Topologies of Wireless Sensor Networks [C]. The 29th Chinese Control Conference, 2010: 1-7.
5. Li Q, Zhang B, Cui L, et al. Immunizations on small worlds of tree-based wireless sensor networks[J]. Chinese Physics B, 2012,21(5): 25-33.

The manufacturer's authorised representative in the EU is Springer Nature Customer Service Centre GmbH, Europaplatz 3, 69115 Heidelberg, Germany. If you have any concerns regarding our products, please contact ProductSafety@springernature.com

Printed and bound by CPI Group (UK) Ltd, Croydon, CR0 4YY
30/09/2026
02225247-0002